AGRICULTURE AND ENVIRONMENT

THE PHYSICAL GEOGRAPHY OF TEMPERATE AGRICULTURAL SYSTEMS

AGRICULTURE AND ENVIRONMENT

THE PHYSICAL GEOGRAPHY OF TEMPERATE AGRICULTURAL SYSTEMS

David J. Briggs   Frank M. Courtney

Longman   London and New York

Longman Group Limited
Longman House, Burnt Mill, Harlow
Essex CM20 2JE, England
Associated companies throughout the world

Published in the United States of America
by Longman Inc., New York

First published 1985

British Library Cataloguing in Publication Data
Briggs, David J.
   Agriculture and environment.
   1. Agricultural ecology
   I. Title   II. Courtney, Frank M.
   630′.912   S589.7

ISBN 0-582-30000-2

Library of Congress Cataloging in Publication Data
Briggs, David J. (David John), 1948-
   Agriculture and environment.

   Bibliography: p.
   Includes index.
   1. Agricultural ecology. 2. Agricultural geography.
3. Agriculture.   I. Courtney, Frank M., 1945-
II. Title.
S589.7.B75   1985   630′.912   84-10094
ISBN 0-582-30000-2

Set in 10/11pt Times Roman Linotron 202
Produced by Longman Singapore Publishers (Pte) Ltd.
Printed in Singapore

For Matthew Briggs and Claire and Rebecca Courtney

CONTENTS

PREFACE

Modern temperate agriculture is a complex scientific activity, but the developments of recent years have been grafted onto generations of human experience since men ceased their nomadic pastoral economy. The scientific developments of this century have also wrought a substantial change in the way in which agricultural activity affects the pre-existing environment. The purpose of this book is to evaluate and, as far as possible, to explain the interrelationships between agriculture and the physical environment. Thus this book draws together research from a variety of sources on the environmental relations of temperate agriculture. Over the past decade there has been a considerable development in the literature: nevertheless, in the case of the environmental effects of agriculture some of this has been somewhat emotive or, at best, somewhat selective in its scientific basis.

Anyone with even the slightest knowledge of modern farming will realise that the task we have set ourselves is vast. Most people experienced in these areas will have different ideas of what should be included in a book of this kind and what should be left out. (It is perhaps instructive to admit that we ourselves sometimes initially disagreed with each other on the relative significance of particular topics.) Although the context of this book is temperate agriculture, many of the examples cited are from Britain. Lowland Britain is one of the most intensively farmed areas in the temperate zone and the agricultural practice is among the most technologically advanced. Hence pressures on natural ecosystems are certainly as great as anywhere. It is likely that many of the lessons being learnt in Britain will be of increasing significance elsewhere in the temperate world in years to come.

The book is divided into four parts. In Part 1, following an introduction to the conceptual background, there is an examination of the historical context of modern agriculture. Part 2 describes the nature and basis of modern agricultural operations whilst Part 3 examines the environmental relationships of the major types of agriculture. In Part 4 the environmental impact of agriculture is described from what may broadly be called an ecological perspective.

When writing this book we had in mind groups of second- and third-year geography and environmental studies undergraduates we were teaching. It is clear that the issues are important to various others, such as students of agriculture, soil science, conservation and planning. In addition we hope that the book may be of interest to farmers, agricultural advisers, professional and amateur conservationists and to teachers.

This book, like almost any other, is derivative to the extent that we have drawn on a wide variety of sources. To the writers of these papers, articles and books we are very grateful: without their earlier efforts we would not have been able to develop this volume.

Particular thanks are due to: Carl Bendelow (Soil Survey of England and Wales, Penrith), Jim France (Grassland Research Institute, Hurley) and Michel Cornaert (Environmental Service, EC, Brussels) for supplying much useful information; and to Rick Cryer, David Grigg, Peter Smithson, Steve Trudgill and Paul White (all of the Geography Department at Sheffield University) for putting up with perpetual pestering by DJB in search of books, references, data, encouragement and sympathy. FMC thanks his former colleagues at Manchester Polytechnic, in particular Don Bayliss, for encouragement in the early stages, together with colleagues in Wiltshire for their continuing interest. We are indebted to Bev Arrand for her help with the index. Above all, however, we are indebted to Ann and Catherine and to our respective families for their continued tolerance, patience and support.

David Briggs

Frank Courtney

ACKNOWLEDGEMENTS

We are grateful to the following for permission to reproduce copyright material:

American Geophysical Union for fig 9.7 from fig 1 p 772 (Gifford & Hawkins 1978) copyright by the American Geophysical Union; Applied Science Publishers Ltd for fig 11.6 from figs 1–5 pp 285–7 (Webb & Haskins 1980); Association of Applied Biologists for fig 6.3 from figs 1,2,3,5,9,10 pp 59–60 (Edwards et al 1967); the author, David Ball for tables 12.2, 12.3 from Glascwm Study (Ball et al 1981a,b, 1982); the author, Keith Beven for fig 9.14 from fig 19 p 22 (Beven 1980); Blackwell Scientific Publications Ltd for figs 13.1, 13.2, 13.3 from figs 1,2–4, 5, pp 243, 247–9, 250–1 (Boorman & Fuller 1981), fig 7.19 from fig 5 p 262 (Clement & Williams 1958), fig 7.3 from fig 2.5 p 32 (Williams 1980), fig 7.5 from fig 4.4 p 152 (Holmes 1980), table 7.5 from table 5.2a p 177 (Murdoch 1980), fig 9.9 from fig 7 p 236 (McGowan & Williams 1980a), figs 9.10, 9.11 from figs 1, 2 p 235 (McGowan & Williams 1980b), fig 12.4 from fig 3a, b pp 666–7 (Rawes 1981), table 7.2 from table 12 p 362 (Allen 1964); British Birds Ltd for fig 10.9 from fig 6 p 204 (Davies & Davis 1973); the author, S Broekhuizem for fig 11.8 from fig 2 p 95 (Broekhuizem 1982); Butterworth & Co Ltd for fig 8.14 from fig 2.6 p 55 (Briggs 1977), fig 8.18 from fig 5 p 32 (Moss & Strath Davis 1982); Cambridge University Press for fig 1.2 from fig 8.2 p 103 (Bayliss-Smith 1982), fig 3.1 from fig 3 p 293 (Thow 1963), table 8.9 from fig 5 p 464 (Webster et al 1977); Centre for Agricultural Strategy for fig 7.27 from fig 2 p 404 (Wilkins et al 1981); Chapman & Hall for fig 11.4 from fig 1.8 p 27 (Duffey et al 1974); Wm Collins & Sons Co Ltd for data adapted from D A Ratcliffe, 1970 (Brown 1976), fig 11.10 from fig 40 (redrawn from Marshall 1967), fig 11.11 from fig 42 p 177 (Pollard et al 1974); Commission of the European Communities for fig 13.5 & table 13.1 (ICBP 1981); Commonwealth Agricultural Bureaux SL2 3BN UK for fig 2.4 from fig 2 p 70 (Arnold 1980), fig 7.15 from fig 9.4 p 62 (Jewiss 1972), fig 2.5 & 2.7 from fig 1 p 345 (Hawkins 1980); the author, G W Cooke for fig 5.3 from fig 13 p 218, table 23 p 85, table from p 72 (Cooke 1975); Entomological Society of America for fig 6.4 from fig 3 p 1224 (Harris 1966); Elsevier Science Publishers B V for fig 9.5 from fig 1 p 168 (Baier 1969); Faber & Faber Ltd for fig 8.1 from figs 51, 54, 69, 74 pp 81, 91, 96 (Coppock 1976); W H Freeman & Co for fig 6.2 from p 92 Copyright (c) 1969 by Scientific American Inc (Edwards 1969b); the author, M J Frissel & Elsevier Science Publishers B V for fig 1.5 from fig 3 p 28 (Frissel 1978); the authors,

G F Gifford & R H Hawkins for fig 9.8 from fig 2 p 310 (Gifford & Hawkins 1978); Gleadthorpe Experimental Husbandry Farm for fig 4.6 from
fig 7 p 19 (MAFF 1977); Grassland Research Institute for fig 7.16 from
fig 5a,c,d,e, p 59 (Garwood et al 1977); Grassland Research Institute &
The Agricultural Development & Advisory Service for fig 7.12 from fig 1
p 2 (Patto et al 1978), table 7.14 from tables 13, 14 pp 28/29 (Forbes et al
1977); the Controller of Her Majesty's Stationery Office for fig 2.6 from
figs 1A, 1B p 79 HMSO (Austin 1978), fig 3.2 from fig 2 p 441 (Cannell
1976), fig 3.3 from figs 3,4 pp 131–2, fig 3.4 from fig 5 p 133, table 3.2
from table 2 p 130 (Spoor 1975), fig 3.5 from fig 3.4 p 170–2, fig 3.7 from
fig 10 p 179, fig 3.8 from fig 9 p 178 (Soane 1975), fig 4.2 from fig 1 p 425
(Trafford 1975), fig 8.11 from fig 8 p 457 (Welbank 1975), fig 9.15 (Burford 1976); Hodder & Stoughton Ltd for 2.2 from fig 15 pp 156–7 (Rowley 1972); the editor, G E Hollis for fig 9.12 from fig 4A p 24 (Reid
1979); ICI for figs 3.12, 3.13 from figs 8, 10 pp 382–3 (Cannell 1981), figs
5.6, 5.7 from figs 1,3 pp 145–6 (Burns 1977); Journal of Environmental
Management for fig 4.7 from fig 1 pp 378 (Green 1973), fig 4.8 from fig 3
p 152 (Green 1980); the author, Dr W K Lauenroth & Springer Verlag
NY for fig 7.25 from fig 1.4 p 7 (Lauenroth 1979); the author, Dr J
Ledieu for fig 7.26 from fig 2 p 210 (Harrath et al 1977); the author, J Lomas & Elsevier Science Publishers B V for fig 2 p 243 (Lomas et al 1974);
Longman Group Ltd for fig 11.13 from fig 13.1 p 237 (Curtis et al 1976);
Macmillan Publishing Co for fig 5.8 from figs 6–10 p 228 (Tisdale & Nelson 1975), table 6.6 from table 21.3 p 558 (Brady 1974); McGraw-Hill
Book Co (UK) Ltd for fig 9.4 from fig 3.3 p 62 Vol 2 (Ward 1975); Nature Conservancy Council for fig 10.1 from p 19, fig 14.2 (Nature Conservancy Council 1977); Oliver & Boyd for fig 12.5 from figs 27, 28 pp 104–5
(Tivy 1973); The Open University for fig 13.4 from fig 6 p 14 (Open University 1972); Oxford University Press for fig 8.10 from fig 6.1 p 68, table
4.8 from table 5.3 p 55 (c) Oxford University press (Eddowes 1976),
fig 12.3 from p 393 ed by Stanley Cramp Maps (c) Oxford University
Press 1980 (Cramp & Simmons 1980), table 7.8 from table 7.8 p 115
(Spedding 1971); the author, M L Parry for table 12.1 from table 7.2 p 49
(Parry et al 1982a); T & A D Poyser Ltd for figs 10.2, 10.3, 10.5, 10.6,
10.7, 10.8 from figs 49, 39, 45, 47, 37, 36 pp 256, 233, 242, 249, 221, 220
(Newton 1979), fig 12.1 from fig 13.1 p 164 (Fuller 1982), fig 12.2 from
p 133 (Sharrock 1976); D A Ratcliffe & Nature Conservancy Council for
tables 11.2, 11.3 from pp 182, 142 (Ratcliffe 1977); Rothamsted Experimental Station for fig 3.15, table 3.5 from fig 3, table 1 pp 20, 4–5
(Cannel et al 1979), table 3.3 from table 7 p 28 (Jones 1979), fig 44 from
fig 14 p 61 (Trafford & Walpole 1975), fig 5.2 from fig 1 p 63 (Mattingley
et al 1975), fig 3.9 from fig 2 p 138 (Johnston 1973), table 8.2 (Widdowson et al 1980); the author, K A Smith, Macmillan Journals Ltd & Agricultural Research Council for fig 4.1 from fig 1 p 148 (Smith & Robertson
1969) Copyright (c) 1969 Macmillan Journals Ltd; Soil Science Society of
America for fig 8.8 from fig 1 p 597 (Rasmussen et al 1980); the author,
D Scholefield for fig 7.8; the authors, F Sturrock and J Cathie for fig 11.9
from fig 2 p 12 (Sturrock & Cathie 1980); University of Chicago Press for
fig 2.3 from fig 3 p 486 (Jones 1965), fig 14.1 from fig 147 p 726 (Curtis
1956); University of Texas Press for fig 9.13 from fig 11 p 46 (Glymph &
Holtan 1969); the author, A Walker for fig 6.6 from fig 1 p 62 (Walker
1976b); Water Research Centre for fig 9.17 from figs C.1.24, C.2.4,
C.4.1, C.5.4, from pp 51, 54, 58, 59 (Young & Gray 1978); The Williams
& Wilkins Co for fig 8.12 from fig 1 p 299 (Jenkinson & Raynor 1977).

PART 1

CONCEPTUAL AND HISTORICAL BACKGROUND

1 INTRODUCTION

1.1 SCOPE AND OBJECTIVES

Recent decades have seen great changes in the character of farming in many parts of the temperate world. Developments in crops and livestock, increased mechanisation, increased use of agricultural chemicals and improvements in drainage and irrigation have all led to considerable increases in yield. At the same time, social, economic and political changes have occurred. The agricultural labour force has declined, farms have been consolidated and amalgamated, the character of tenure has altered as agricultural consortia have grown up and large commercial institutions have taken over farms for investment purposes. Nowhere have these changes been more apparent than in Britain; indeed it can be argued that Britain is today almost unique in its agricultural structure (Table 1.1). But similar developments are occurring throughout much of the temperate world, albeit at a slower rate than in Britain, and there is reason to believe that these changes will continue.

The implications of these changes are considerable. Agriculture is, and always has been, an activity involving a close interaction with the environment. Soil, climatic, topographic, hydrological and biological conditions together exert a major control upon farming operations and upon the profitability of agriculture. Many of the developments which have taken place have been intended to overcome the limitations imposed by such environmental factors, but the deterministic influence of these effects nevertheless remains. In addition, however, agriculture modifies the environment; the history of man's use of the land for agriculture has been a history of environmental modification. Forests have been cleared, wetlands drained, heathlands and wasteland enclosed and taken into cultivation. Recent changes in agriculture have thus been grafted on to generations of farming practice, and the effects upon the environment have merely continued or accelerated effects which have been felt for many centuries. Nonetheless, concern about the speed and scope of changes which are currently taking place and about their impact on the environment has been growing (e.g. Shoard 1980).

The importance of these issues cannot easily be overstated. The direction of agricultural change may be much as it has been for many centuries, but the magnitude of change and the potential implications for the environment have undoubtedly increased significantly in recent times. At the same time, these changes have to be seen in a wider context of

Table 1.1 Characteristics of European agriculture

	UK	France	West Germany	Netherlands	Denmark	Ireland
% of labour force in agriculture[a]	2.2	8.9	5.2	5.5	7.3	20.4
% of agricultural labour force over 60 years[a]	3.6	22.7	17.3	13.0	18.5	52.3
% agricultural labour force family members[b]	29	48	55	43	37	43
% farmland owner-occupied[b]	57.6	52.3	69.7	57.5	85.4	96.3
average farm size (ha)[b]	65.6	25.5	14.3	15.0	23.5	22.5
% holdings with tractors[b]	74	89	91	94	87	48
land/man ratio (ha/man unit)[c]	26.3	15.1	10.0	8.2	16.8	15.6
% farmland under woods[b]	1.3	8.0	11.0	2.0	5.0	0.7

Notes: [a] – 1980; [b] – 1977; [c] – 1975.

Source: Data from CEC 1982

conflicting human priorities. On the one hand there are (globally) increasing populations calling for more and cheaper food. On the other hand many individuals in the developed countries are becoming more directly concerned with questions of environmental quality, and hence with the effects of modern agriculture on the cultural and historical landscape and on natural and semi-natural ecosystems. Thus there exist simultaneously pressure to modernise and intensify agricultural systems, and an eagerness to minimise or nullify the effects of such systems on the environment. If these apparently contradictory aspirations are to be reconciled, then it is essential to understand more clearly the way in which agriculture operates, the factors that control farming decisions and practices, the nature of farming methods, and the ways in which the various components of the environment respond to these methods. Ultimately, such an understanding must encompass not only physical aspects of the system, but also the cultural, social, economic and political aspects. No single book can do justice to all these topics, and there are, in any case, many texts which already consider the sociological and economic implications and relations of modern agriculture (e.g. Newby 1979; Bowers and Cheshire 1983), or which describe the 'human geography' of modern farming systems (e.g. Grigg 1974). The purpose of this volume is to concentrate instead upon what might be referred to as the 'physical geography' of temperate agricultural systems: their relationships with the soil, the climate, the biota and the hydrological cycle. As such, the aim is to examine the changes in farming methods which have given rise to modern farming systems (Ch. 2), the main farming practices applied in these systems (Ch. 3 to 6), the character of selected farming systems (Ch. 7 and 8), and the impacts of modern agriculture upon the environment (Ch. 9 to 14).

1.2 AGRICULTURAL SYSTEMS

Like any natural system, agricultural enterprises may be described as open systems, receiving inputs from outside and losing energy and matter as outputs. Internally, the system comprises a number of interrelated components through which this energy and matter flow (Fig. 1.1), whilst the matter involved in these flows includes water, nutrient elements (solutes) and solids (e.g. soil particles).

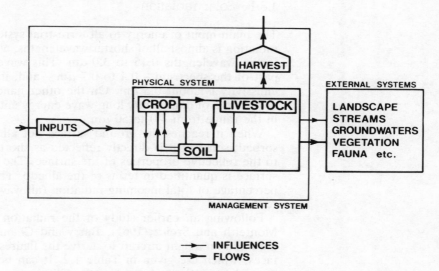

Fig. 1.1 The general structure of an agricultural system and its relationship with external systems.

Inputs to the system occur in a number of ways: by weathering of bedrocks (which produces solids and nutrient elements), by solar radiation (energy), precipitation (water and nutrient elements), by transfers from adjacent land surfaces (e.g. by erosion or runoff which together bring water, solids and solutes), and, above all, by deliberate inputs by the farmer. These last inputs take the form of seeds, livestock, manures, fertilisers, pesticides, animal feeds and fuel energy. Indeed it is partly through these inputs that man controls the agricultural system, and maintains its capacity to produce the high levels of outputs which characterise many modern farming enterprises.

Man also controls many of the outputs from the agricultural system. Through land drainage he affects the magnitude and pathways of water loss from the system. Through practices such as tillage, crop rotation and soil conservation he controls the rate of erosion and thus the losses of solids. Most important of all, through harvesting and the management of livestock and crop residues, he influences the losses of nutrients from the system. In fact, as we will see later in this chapter and more fully in Chapters 7 and 8, losses in harvesting often represent the major output of nutrients from agricultural systems, and it is to replenish these losses that large inputs of fertilisers are required.

As this discussion implies, one of the main concerns in this book is the nature of cycling processes (of energy, water and nutrients) within agricultural systems, and the way in which farming practices affect these cycles. To provide background to discussion of these aspects it is useful here to outline the main features of these cycles.

1.3   ENERGY CYCLES

1.3.1   Solar radiation

The main input of energy to all terrestrial systems is from the sun. Solar radiation is almost all of shorter wavelengths, and 98 per cent is contained in the wavelengths 0.15 to 3.0 μm. This waveband includes the visible part of the spectrum – 0.4 to 0.7 μm – and, in fact, the maximum solar emissivity is about 0.5 μm. On the other hand, the earth's surface acts as a source of dominantly long-wave energy and most radiation is emitted in the range from 4.0 to 50 μm.

When it reaches the ground surface not all incoming radiation is absorbed; some of it is directly reflected as short-wave radiation according to the reflective properties of the surface. The reflective character of the surface is quantified in terms of the albedo. The albedo is defined as the percentage of total incoming radiation (all wavelengths) which is directly reflected.

Following an earlier study of the radiation balance of vegetation by Monteith and Szeicz (1961), Barry and Chambers (1966) used a meter mounted on a light aircraft to derive the figures for albedo of various surfaces, which are given in Table 1.2. It can be seen that there are substantial variations between the different surfaces. Stewart (1971) examined the seasonal variation of albedo in the course of an intensive study of a pine forest at Thetford, Norfolk. He noted a variation from 8.9 per cent in June to 11.7 per cent in December, with an average figure of about 10 per cent. This variation appeared related to solar zenith angle. Other factors which may influence albedo include time of day and, possibly, topography. (For a general discussion see Jones 1976).

Radiation which is not reflected is absorbed and heats the surface of the ground. Sensible heat transport is then responsible for heating the lower layers of the soil and there is further sensible heat loss from the

Table 1.2   Average albedo values for England and Wales

Cover type	Average albedo (%)
Festuca downland	24
Agricultural land (i.e. mixed farmland) with:	
(a) more than 84% grassland	24
(b) 40–80% grassland	23
(c) less than 40% grassland	22
Molinia grassland	20
Deciduous woodland	18
Coniferous plantation	16
Heather moor	15
Peat and moss	12

Source: After Barry and Chambers 1966

surface in the long-wave emission. Some of the available energy is absorbed in the form of latent heat so that water may evaporate from the surface. Later, in the atmosphere, this latent heat will be given up as water condenses. The latent heat is then used in heating the atmosphere.

A second heat source which is, in many situations, of greater magnitude than radiant heat is that derived from other parts of the earth's surface or lower atmosphere by advection (transportation by atmospheric motion).

1.3.2   The biological significance of radiation

Plants use light to fix energy into organic molecules in the process known as photosynthesis. Within ecosystems photosynthesis is the source of all energy and is thus of critical importance. This photosynthetic transformation of energy by green plants is called primary production and the plants themselves are referred to as autotrophs. The photosynthesis equation summarises a very complex series of operations:

$$2816 \text{ kJ}$$

$$6CO_2 \quad + \quad 12H_2O \quad \rightarrow \quad C_6H_{12}O_6 + \quad 6O_2 + \quad 6H_2 \qquad [1.1]$$
carbon dioxide   water   energy   carbohydrate oxygen water

Light is only involved in the first part of the process in which chlorophyll uses light energy to split a water molecule thereby releasing oxygen. The second part of the process uses the energy thus provided in a number of steps to reduce carbon dioxide to carbohydrates.

The various factors which can influence the rate of photosynthesis have been summarised by, among others, Simmons (1979) and Watts (1971). The most important of these factors is light intensity and, generally speaking, increasing light intensity gives a greater photosynthetic rate. However, beyond a certain critical level the plant effectively becomes light saturated, reaching the so-called photosynthetic capacity. The photosynthetic rate does not then increase if the light intensity increases further.

The rate of photosynthesis is also controlled by temperature and it has been shown that the rate of photosynthesis is higher at higher temperatures, with an optimal temperature level for many plants of about 25°C. The effect of temperature changes on a specific crop has been studied by Peacock (1975a, b). In a study of perennial ryegrass (*Lolium perenne*), Peacock used heating cables in plots to investigate the different effects of soil and air temperature and also the temperatures of specific parts of the plant. The conclusion was that temperature of the stem apex was much more important than soil temperature. Notwithstanding detailed research of this kind, it is generally assumed that a mean daily air temperature of 6.1°C (43°F) is necessary before plant growth can commence. Although in a technical sense this figure is only a 'rule of thumb' and, in reality, the precise figure varies from species to species, it nevertheless provides a useful average value. Using this figure, maps of the onset of spring have been produced (MAFF 1964a).

A further factor in controlling photosynthetic rate is carbon dioxide concentration, but this only appears to assume importance at middle and higher light intensities. Plants not only absorb carbon dioxide, but also emit it during respiration. (Work on carbon dioxide balances has been summarised by Baumgartner (1965) and Monteith (1973).)

Respiration involves the use of carbohydrates as an energy source for the plants themselves in order that they may grow and maintain tissue. It may be summarised as follows:

$$C_6H_{12}O_6 \quad + \quad 6O_2 \quad \rightarrow 6CO_2 \quad + \quad 6H_2O + 2830 \text{ kJ} \qquad [1.2]$$

carbohydrate   oxygen   carbon   water   energy
                          dioxide

The balance between the energy taken in during photosynthesis and that emitted in respiration is that which is subsequently available within the ecosystem. This balance is the net energy production within the ecosystem.

The crop efficiency relationship between energy input (in terms of solar radiation) and energy output (carbohydrate productivity) has been discussed by Monteith (1977). Using data from Collingbourne (1976), Monteith pointed out that solar radiation was relatively uniform over Britain (except in coastal areas where it is sometimes higher). Over most agricultural parts of Britain daily mean insolation (on an annual basis) is within ±10 per cent of 9 MJm^{-2}. Variation is even less in summer and, for example, in June it ranges between about 17 and 20 MJm^{-2}. Winter insolation is of course much less and is also much less uniform. Because net solar radiation is closely correlated with potential evaporation it follows that the main control on available moisture is actually rainfall amount. The problems of ensuring an adequate but not excessive moisture supply for crops will be examined in Chapter 4.

1.3.3 Energy balance modelling

In recent years a number of investigators have attempted calculations of the energy budgets of agricultural systems. The objectives of this work are partly to assess the relative efficiency of various types of agriculture and also to identify significant (and often thereby expensive) components of the system. Slesser (1975) has summarised work in this field, and an example of an energy flow model for an intensive arable farm in southern England is given in Fig. 1.2.

In addition to the major input of solar energy, modern farming has substantial inputs deriving from the use of artificial fertilisers and pesticides. The energy input of these can be assessed, firstly, in terms of the energy necessary to produce the chemical and, secondly, in terms of the energy necessary to apply it to the soil or crop. For example, Leach and Slesser (1973) have estimated the energy input figures given in Table 1.3. These data show the large relative quantities of energy which are used in some mechanical farming operations. Wilkins and Bather (1981) provide a breakdown of energy inputs to grassland and grazing systems in Britain (Table 1.4).

It should also be noted that energy cycles in many agricultural systems are apparently inefficient, in that a relatively small proportion of the total energy inputs is ultimately consumed by man. Duckham and Masefield (1970) show that only 0.25 per cent of energy inputs in a potato cropping system are available as human food, and in a cattle-ranching system the figure falls to 0.002 per cent (Table 1.5).

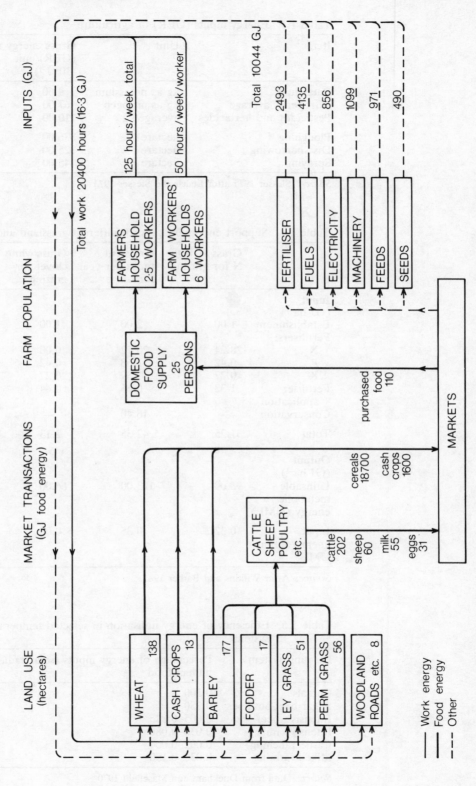

Fig. 1.2 Energy flows in a 460 ha intensive arable farming system in southern England (from Bayliss-Smith 1982).

Table 1.3   Energy inputs (GER) of agriculture

Item	Unit	Gross energy requirement (GER)/unit (10^6 J^{-1})
Potassium	kg as potassium	9.60
Nitrogen – average	kg as nitrogen	67.00
Pesticides and herbicides	average/kg	110.00
Ploughing	hectare	836.00
Disc harrowing	hectare	325.00
Spraying	hectare	45.00

Source: Slesser 1975 after Leach and Slesser 1973

Table 1.4   Support energy budgets for different grassland and grazing systems

	Grass + high N for grazing	Grass cut 3 times per year for silage	Grass/white clover for grazing	Lucerne for silage
Input (GJ ha^{-1})				
Establishment	1.00	2.00	1.00	2.67
Fertilisers:				
N	28.24	20.72	0.00	0.00
P	0.42	1.68	0.42	3.36
K	0.27	1.89	0.27	2.70
Fertiliser application	1.32	0.66	0.44	0.44
Conservation	—	16.40	—	23.70
Total	31.25	42.35	2.13	32.87
Output (GJ ha^{-1})				
Utilisable metabolic energy (UME)	98.00	123.00	66.00	122.00
Input energy/output UME	0.32	0.35	0.03	0.27

Source: After Wilkins and Bather 1981

Table 1.5   Efficiency of energy utilisation in selected temperate farming systems

Farming system	Percentage of energy inputs (solar radiation) available as human food
Cereals	0.200
Sugar beet/potatoes	0.250
Intensive beef	0.005–0.025
Intensive milk	0.030–0.080
Cattle ranching	0.002–0.004
Mixed farming	0.030–0.150

Source: Data from Duckham and Masefield 1970

1.4   THE HYDROLOGICAL CYCLE

As has been mentioned, water plays a fundamental part in all agricultural systems. Water is essential to maintain plant turgor, and it also transports nutrient elements to and through the plants. In addition, water has a vital role in weathering, leaching and erosion, and thus controls to a great extent inputs of nutrients to and losses from the system. At the same time, agricultural practices have a major impact upon these hydrological processes.

The main input of water to agricultural systems is by rainfall. Where these inputs are insufficient to meet the requirements of the crop, however, they are often supplemented by irrigation. The main losses are by evaporation, transpiration (the two generally referred to together as evapotranspiration), drainage to groundwater, and lateral flow (runoff and throughflow) to streams. Storages of water occur in the soil, in plant tissues and in the bodies of livestock.

The hydrological system is highly dynamic, such that inputs and outputs of water vary markedly over short periods, causing responding variations in soil water storage. Because the soil moisture store represents the main source of water for crops, these fluctuations in water availability have considerable significance for crop growth. Indeed, Austin (1978) and Cooke (1979) have argued that moisture deficiencies are one of the main limitations upon yields of agricultural crops in Britain. Short-term variations in soil water availability depend upon a large number of factors, including the amount and distribution of rainfall, the evaporative potential of the atmosphere (and thus on air temperature, humidity and wind speed), the water-holding capacity and drainage status of the soil, and the local topography (which influences rates of runoff and throughflow). In general terms, however, an indication of the annual water balance at any location can be provided by the so-called water budget diagrams initially constructed by Thornthwaite (1948). These show the monthly inputs of precipitation and losses by evapotranspiration. During periods when precipitation exceeds evapotranspiration a moisture surplus is considered to exist, and water accumulates in the soil until field capacity (maximum soil storage capacity) is reached. Any further excess is assumed to be lost by drainage or runoff. During periods when evapotranspiration exceeds rainfall, a moisture deficit is said to occur. Plants then deplete the soil moisture store until, under extreme circumstances, the soil reaches wilting point, at which time evapotranspiration ceases and plants begin to wilt. Figure 1.3 shows examples of moisture budget diagrams for selected locations in the temperate world, illustrating the difference between more arid regions where considerable soil moisture deficits build up during the growing season (e.g. Pola, Jugoslavia), and moister areas where moisture deficits rarely, if ever, occur (e.g. Arastook, Maine).

We will examine the significance of both moisture deficits and excesses for agriculture, and the ways in which man tries to control soil moisture budgets, in Chapter 4. It is clear, however, that the detailed character of the hydrological cycle in agricultural systems is fundamentally affected by farming practices. The character and extent of crop cover, tillage, land drainage and irrigation practices, and – more indirectly – even the use of fertilisers and pesticides all influence the amount of water stored in the system and the quantities lost by drainage, runoff and evapotranspiration. Moreover, as we will see in Chapter 9, these losses have wider

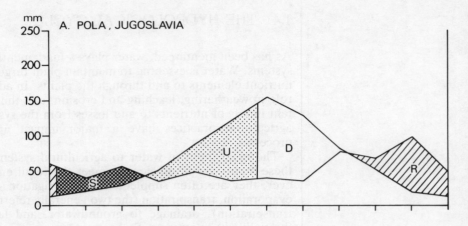

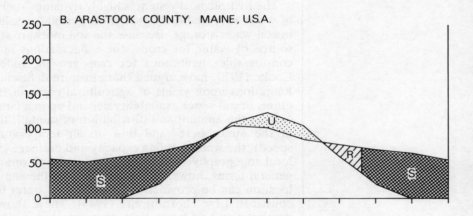

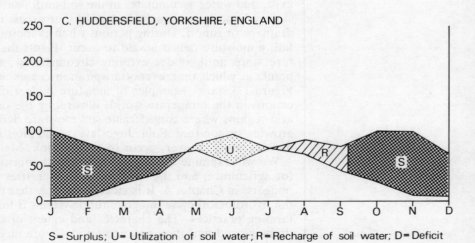

S= Surplus; U= Utilization of soil water; R=Recharge of soil water; D=Deficit

Fig. 1.3 Water budget diagrams for selected temperate locations.

implications, and agricultural impacts on the hydrological cycle are a major concern in temperate areas.

1.5 NUTRIENT CYCLES

The main features of nutrient cycles in agricultural systems are summarised in Fig. 1.4. Three main pools or reservoirs of nutrients can be defined: crop plants, livestock and soil.

The tissues of plants and animals are made up of carbohydrates, fats, proteins and nucleoproteins. In order to grow and develop tissue, plants need considerable amounts of carbon, hydrogen, oxygen, nitrogen, phosphorus and sulphur together with much smaller quantities of iron, manganese, zinc, copper, boron and molybdenum. To assist in tissue development, and for other purposes, varying quantities of potassium, magnesium and calcium are needed. These major and minor elements are

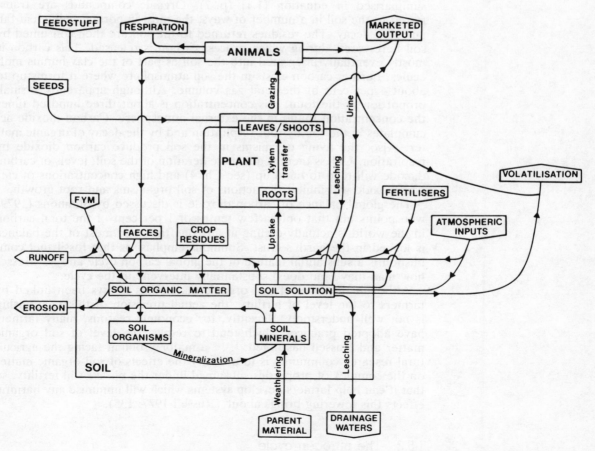

Fig. 1.4 Nutrient cycles in agricultural systems (FYM = Farmyard manure).

usually referred to as nutrients and the movement or cycling of these nutrients around the ecosystem is of crucial importance.

The driving force for nutrient cycling is energy and nutrients are usually moved in solution. Hence nutrient cycling generally follows the same paths as energy and water flows.

There are a number of different cycles that may be described, each of which determines the movement of various of the nutrients. In the present chapter the more important cycles will only be described in broad terms: in Part three of the book more detailed case-studies of the relation of agricultural activity to the cycles will be examined.

1.5.1   The carbon or organic cycle

The carbon cycle is of fundamental significance in ecosystems, for carbon represents a major state of energy in chemical form. The atmosphere acts as a reservoir of carbon dioxide (CO_2). (The balance of carbon dioxide in the atmosphere is maintained by the oceans which absorb excesses and also replenish deficits.) As we have seen in section 1.3.2, this gas is used by plants in photosynthesis to produce carbohydrates, the process being summarised in equation [1.1] (p. 7). Organic compounds are transferred to the soil in a number of ways, the most important being leaf fall and root decay. The residues returned to the soil are then consumed by soil herbivores, during which process carbon is released. This carbon is mostly eventually integrated into the soil as part of the clay-humus molecules. Further carbon exists in the soil atmosphere where it forms up to about 4 per cent of the total gas volume. Although apparently a small proportion of the total, this concentration is about three hundred times the concentration found in the external atmosphere. Carbon dioxide accumulates in the soil by root respiration and by the decay of organic matter. Also, the living organisms in the soil produce carbon dioxide by respiration. Unless there is adequate aeration of the soil, levels of carbon dioxide will tend to build up (see Ch. 4) and high concentrations of carbon dioxide will inhibit the actions of soil organisms and root growth.

The global balance of carbon dioxide is discussed by Simmons (1979) who points out that only a few tenths of 1 per cent of the total carbon in the world is actually cycling at any one time and most of the balance is locked in the earth's crust. Simmons emphasises that fossil fuel combustion is a substantial feature of the global carbon cycle and he stresses how man may (and does) substantially intervene in the cycle.

Although the level of soil organic matter has always been linked by farmers to the level of fertility, the actual precision of this relationship is not well understood. Recently, for economic reasons, many farmers have adopted practices which tend to reduce the level of soil organic matter and Russell has noted: '. . . a major problem facing the agricultural research community is to quantify the effects of soil organic matter on the complex of properties subsumed under the phrase soil fertility, so that it can help farmers develop systems which will minimise any harmful effects this lowering brings about' (Russell 1977: 135).

1.5.2   The nitrogen cycle

The nitrogen cycle is outlined in Fig. 1.5 and Table 1.6. Plants require

Table 1.6 The nitrogen cycle in selected farming systems (see Fig. 1.5 for pathways referred to)

	Extensive cattle (France)	Mixed (France)	Intensive sheep (UK)	Intensive arable (Netherlands)
Inputs				
1 Feed to livestock	0	0	16	0
2 Litter used indoors	0	0	—	0
10 Manure	—	—	—	0
11 Fertilisers	—	0	120	305
12 N-fixation	50	60	150	0
13 Litter used in field	—	—	—	—
14 Irrigation	—	—	—	—
15 Dry and wet deposition	t	t	17	14
29 Seeds and seedlings	0	0	—	2
31 Uptake from the atmosphere	0	0	—	—
Transfers				
3 Consumption of harvested crops by livestock	—	160	56	—
4 Grazing of forage	90	155	98	—
8 Animal manure returned to field	—	126	52	—
9 Faeces/urine left on field	70	126	89	—
16 Weathering of soil minerals	—	—	—	—
17 Mineralisation of organic matter	95	0	69	45
24 Fixation of mineral nutrients	—	—	—	—
25 Immobilisation from organic matter	t	0	?	45
26 Decomposition of plant residues	80	115	30	71
27 Return of seeds from crops in field	—	—	—	—
30 Uptake of nutrients from soil	170	430	184	195
Outputs				
5 Animal products	20	63	17	—
6 Ammonia from manure	t	t	12	—
7 Manure removed	—	—	—	—
18 Harvested crops	—	—	—	126
19 Denitrification	t	t	?	71
20 Volatilisation of soil ammonia	t	t	?	0
21 Leaching	50	50	30	58
22 Mineral nutrients in runoff	0	0	—	—
23 Dust	0	0	—	—
28 Organic matter in runoff	0	0	—	—

Source: Data from Frissell 1978

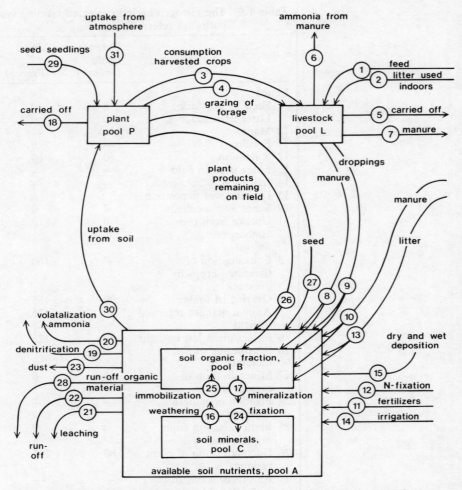

Fig. 1.5   The nitrogen cycle of an agricultural system (from Frissel 1978); for definition of numbered transfers see Table 1.6.

a substantial amount of nitrogen which is generally taken up through the roots in soluble nitrates. Nitrates are used in the synthesis of amino acids, which combine together to form proteins. Animals obtain their protein in a secondary way by eating plants or other animals (or animal products). The diagram demonstrates that, within the soil, there is a chain of processes leading from organic nitrogen in amino acides, through ammonia, nitrites and back to nitrates. Each of these transitions involves the soil microflora. A relatively small number of plants have attached bacterial organisms (of the genus *Rhizobium*) which are able to fix gaseous nitrogen into nitrates. These bacteria occur in nodules on the roots of leguminous plants (e.g. peas, alfalfa and clover).

In modern agricultural systems much of the nitrate is supplied as fertiliser. Simmons (1979) summarises the global budget for nitrogen and cites Söderlund and Svensson (1976) who estimated that industrial fixation of nitrogen from the atmosphere would exceed biological fixation from 1989.

Soluble nitrates in the soil are, of course, vulnerable to removal by being leached. This process and its effects will be examined more fully in Chapter 5 but, in summary, it is clear that significant quantities of artificial nitrate fertiliser can be wasted by leaching. A further danger is that this surplus nitrate may cause unwanted biological effects at its eventual destination (see Ch. 9). Hence eutrophication or nutrient enrichment of inland waters causing 'blooming' of algae has become a problem in recent years. In addition, considerable quantities of nitrogen are lost from the soil by denitrification and volatilisation, though the environmental effects of this may be small.

1.5.3 Nutrient cycling: some example studies

Because agricultural ecosystems by definition involve artificial outputs (the removal of the crop directly or by animals) then, in order to avoid nutrient depletion, some level of artificial input is inevitably necessary. The necessity for fertiliser application of nitrates has already been mentioned and fertiliser applications will be more fully discussed in Chapter 5. (Nutrient cycling in the individual systems will be discussed in Chapters 7 and 8.) The purpose of the present section is to make some more general points about agricultural nutrient cycling, and the difficulties of quantification, by reference to a particular set of data.

At a symposium held in the Netherlands (see Frissel 1978), Newbould and Floate summarised a considerable amount of data on nutrient cycling within agricultural ecosystems in the United Kingdom, drawn from a variety of sources. Although not entirely representative of British farming, six different ecosystems were described and commented on:

1. Moorland;
2. Traditional hill sheep farming;
3. Improved hill sheep farming, comprising:
   (a) hill grassland:
   (b) improved paddocks;
4. Deciduous woodland;
5. Intensive grassland husbandry, comprising:
   (a) intensive sheep system using mixed pastures of grass and clover with little use of nitrogen fertiliser;
   (b) intensive sheep system using pure grass swards with heavy applications of nitrogen fertiliser;
6. Intensive arable farming with winter wheat continuously.

Transfers of the nutrients nitrogen (N), phosphorus (P) and potassium (K) were estimated throughout each ecosystem. By assuming that the plant and livestock pools (see Fig. 1.1, p. 5) were in a steady state, Newbould and Floate were able to estimate possible changes in the nutrient status of the soil pool. The soil pool was further subdivided into an available soil nutrient pool, a soil organic pool and a soil mineral pool (Fig. 1.5).

A summary of some of their results is given in Table 1.7. In the case of the high elevation moorland (1) Newbould and Floate noted that the total soil appeared to be near to steady state in N and P and suggested that the apparent loss in K might be compensated for by release of K by weathering. They commented that the soil organic pool appeared to be

Table 1.7   Summary of estimated net nutrient gains (+) or losses (−) of the
total soil pool (kg ha^{-1} y^{-1}) under various land use types

	N	P	K
1 Moorland	−2	−0.7	−9
2 Traditional sheep farming	+2	−0.3	−6
3 Improved hill farming, paddock			
+ hill grassland	+2	+4.6	−7
(a) hill grassland	−1	−0.3	−6.3
(b) paddock	0	+14.4	−11.0
4 Deciduous woodland	+78	−1.3	−15.0
5 Intensive sheep farming			
(a) grass + clover	+244	+29.8	+46
(b) grass only	+308	+50.1	−89
6 Intensive arable	+15	+8.0	−35

Source: Newbould and Floate in Frissel 1978

gaining N, P and K from the available nutrient pool and pointed out that
there is a steady accumulation of soil organic matter by litter and in peat
formation. In the long term, however, a steady state may be maintained
by the compensating effect of peat erosion.

The soil pool in the traditional hill sheep farming (2) appears to be
more or less in a steady state. However, in the improved hill sheep farming (3) the total soil pool seems to be gaining N and P but losing K. Newbould and Floate considered that the net N gain may be anomolous but
that the net P gain is caused by fertiliser addition, not wholly balanced
by losses in animal product and leaching, whilst the K loss may be being
made good by weathering.

In the deciduous woodland (4) there is an apparent increase in N in the
soil pool which is supported by experimental evidence from the wilderness
area on the Broadbalk at Rothamsted Experimental Station, Herts.

The intensive sheep farming (5) shows gains in N, P and K. Newbould
and Floate consider that the gains for P and K are probably genuine although the loss of N by denitrification and volatilisation is not known, and
hence the figure given may be suspect. The intensive arable (6) appears
to gain N but again no estimate for denitrification is possible. The gain
in P is probably genuine but the loss in K despite the addition of potassium fertilisers was unexpected.

1.6   MODERN FARMING SYSTEMS AND THE ENVIRONMENT

1.6.1   Environmental influences on agriculture

The farming systems which have emerged throughout the temperate world
are clearly different in detail – in the specific crops they grow, for example, and in their degree of intensity. They are, however, based upon
similar agricultural principles, they use similar farming methods, and they
are constrained by the same factors. In particular, they are linked by a

common, close interrelationship with the environment. It might be anticipated that the developments in farming techniques which have occurred in recent decades have released agriculture from its dependence on the environment, and to some extent this is true. Yields are no longer so severely limited by soil and climatic conditions; these limitations have been relieved by the use of fertilisers, by drainage and by irrigation. Nor do pests and diseases pose the threat they once did now that chemical pesticides are widely available. Nevertheless, environmental factors still influence agricultural activity. Practices such as tillage, drainage and irrigation, fertiliser and pesticide application, are responses to specific environmental conditions or problems. They are an attempt to modify the environment in order to make it more favourable for plant growth. These practices, however, must be cost-effective, in that benefits from increased output or improved crop quality must outweigh the costs of applying them. This exerts a major constraint on farming activity; in many cases environmental factors cannot be ameliorated economically, with the result that they continue to limit productivity and they inhibit the range of crops that can be grown. The limitation often acts indirectly. As L. P. Smith (1972) has shown, barley yields in England and Wales are correlated with sowing dates, later sowing giving rise to lower yields. The reason for late sowing in most cases is that the soil is too wet, due to excess rainfall and inadequate drainage, to allow preparation of the land. Thus, the influence of the environment on productivity is felt through its control of management procedures. In this way, by controlling the yield and the flexibility of cropping, environmental factors limit the agricultural capability of the land (e.g. Bibby and Mackney 1969; Wilkinson 1974).

The trend of increasing agricultural output in recent decades has already been noted. To some extent this increase has been derived from expansion of agriculture on to previously virgin land; for example, ploughing up of moorlands, heath and forest, and the drainage and reclamation of wetlands (see, for example, Lloyd and Wibberley 1979; Curtis *et al*. 1976). The opportunities for such expansion are limited today, but it is clear that if such extension and reclamation is to be profitable, it is essential that it be undertaken with due regard to the environmental factors, so that the most suitable land is taken into cultivation, using appropriate methods and for the appropriate crops. The main contribution to increasing yields in recent decades, however, has come not from expansion of the cultivated area, but from intensification of the existing agricultural land. Until the early 1970s this produced a progressive rise in average yields, but since then evidence has accrued to suggest that the gains from intensification were diminishing. Several studies have indicated that for the last ten years, in Britain and the USA, yields of crops and livestock have been on a plateau (Centre for Agricultural Strategy 1976; Jensen 1978; Wittwer 1978). According to E. W. Cooke (1979) further increases in yield are likely to come from a better understanding of the factors limiting productivity, including environmental factors such as moisture and nutrient availability. A similar point is made by Sibma (1977), who states that '. . . high yields are due rather to applied experience on the correct cultivation measures, fertilizer application, disease control, choice of varieties, etc. . . . than to the application of growing quantities of energy and labour . . .' Thus, the need is to develop a yet clearer understanding of the environmental factors which influence crop growth and of their interaction with farming practices.

1.6.2   Effects of Agriculture on the Environment

Just as environmental factors influence farming activity, so that activity affects the environment. Agricultural systems are to that extent artificial systems; they can be seen as attempts to maintain the environment in an artificial, and more usefully productive, state by control of soil fertility, vegetation, fauna and microclimate. The success of these attempts to date is indicated by the recent history of agricultural activity, and by the much greater biomass production of agricultural systems than of natural systems in similar environments (e.g. Moss and Strath Davis 1982). Nevertheless, not all the effects of agriculture on the environment can be regarded as beneficial, and in the long run it is apparent that agricultural activities may damage the environment and possibly undermine the basis of agriculture itself.

In many ways, the major impact of agriculture is felt on the soil, and in a number of cases the effects have been to reduce soil fertility and inhibit yields. The Strutt Report (MAFF 1970a), for example, noted that inappropriate timing of tillage was causing structural damage on clay soils in England and Wales, with possible consequences of reduced yields. Loss of organic matter from soils under continuous arable cultivation has been seen as a process leading to reduced aggregate stability and increased risk of erosion (e.g. Russell 1977). Richardson and Smith (1977) have described the effect of modern agriculture on the fertile peatlands of the English Fens; between 1965 and 1984, at current rates of wastage, overdrainage and erosion will have reduced the area of peat soils from 142,000 ha to only 6000 ha (see also Ch. 12). In Australia, Conacher *et al*. (1983) have demonstrated the way in which farming may increase the salinity of rangeland soils. In all these cases, damage caused by agriculture may reduce the fertility of the soil, increase the problems of soil management, and ultimately inhibit yields and the flexibility of cropping. In the long run, therefore, agricultural systems are only sustainable if they are adapted to take account of the impacts of the systems upon the environmental resources – especially the soil – on which they depend.

The effects of agriculture on the environment extend more widely, however; they influence other ecological and socio-economic systems. Many of the inputs to agriculture, such as fertilisers and pesticides, pass through the system and contaminate surrounding environments. Nitrates, for example, have been shown to contribute to pollution of groundwaters (MAFF 1976; Young and Gray 1978) and surface-waters (Green 1973; Walling and Foster 1978). Organic wastes from livestock systems are occasionally a major pollutant of water resources (Loehr 1977; Terry *et al*. 1981). Pesticides have been found to accumulate in natural food chains, until they ultimately threaten higher organisms (e.g. Newman 1976; Walker 1976). Similarly, agricultural practices physically modify the environment, often destroying the operation of other systems. Drainage, for example, destroys wetland habitats; cultivation of old pasture, rangeland or heath may lead to the loss of rare plant and animal species; removal of hedgerows and coppices reduces the faunal and floral diversity of the environment (Hooper 1979; Newbould 1979). Changes in the vegetation and the landscape caused by intensification or extension of agriculture may reduce the scenic quality of the environment (Leonard and Stokes 1979). Agriculture competes, therefore, with wildlife conservation and recreation, and increasingly in recent years these concerns have found themselves in conflict. To some extent, this conflict is inevitable; as Viets

(1977) states, food-producing systems cannot avoid affecting the environment, but, as Browning (1977) points out, there is a duty upon agriculturalists to develop farming methods and systems which are less environmentally damaging, without being significantly less productive. In other words, it should be possible to minimise the adverse environmental effects of agriculture without undermining its economic basis. The need is to identify more exactly the functions fulfilled by the various farming techniques, to define more closely the structure and dynamics of individual farming systems, and to consider in greater detail the ways in which these methods and farming systems affect the environment. From such an understanding it may then be possible to predict more accurately the environmental effects of modern farming and, where necessary, take action to protect susceptible environments.

2 THE DEVELOPMENT OF MODERN AGRICULTURE

2.1 THE HISTORICAL CONTEXT

Modern farming throughout much of the temperate world is highly capital-intensive. It is characterised by high levels of mechanisation, by large inputs of energy, fertilisers and pesticides, by a relatively small and declining labour force, and by an output, measured in terms of either yield per unit area or yield per man, far in excess of anything achieved in its history. The attainment of this level of intensity is a recent phenomenon, but it has come out of a long period of agricultural development. To understand the nature of modern farming and its relationship with the environment we need, therefore, however briefly, to consider its history. We cannot, in this book, examine the history of farming in all parts of the temperate world, but we can use as a model the development of agriculture in Britain over the last few hundred years.

This development has not necessarily been continuous and incremental. It has included periods of relatively rapid and abrupt change during which agricultural technology and structure have altered fundamentally. The nature and date of these periods of change have been matters of considerable concern to agricultural historians. Following Ernle's (1961) detailed analysis of agrarian changes in Britain, it was widely accepted that the late eighteenth and early nineteenth centuries saw what amounted to an agricultural revolution. Seebohm (1952) described the period as one in which 'there spread over England the greatest advance in the practice of farming that had taken place since agriculture was first established in this country'. Other historians, however, have argued that the agricultural revolution was a considerably earlier event; Kerridge (1967), for example, places it in the sixteenth and seventeenth centuries. Whichever – if either – of these views is valid, it is clear that the roots of modern farming can be traced back at least to the sixteenth century.

2.1.1 Agriculture in Britain prior to 1500

From Roman times until the early sixteenth century agricultural development was slow. As the population grew, the area of cultivated land expanded by encroachment into the woodlands, but as Hoskins (1976) notes, even in 1500 'England was still a colonial economy with too few people to civilize the whole landscape'. Over the same period, approaches to farming had changed little. A few agricultural treatises appeared, but these drew heavily on Roman and Greek learning. Both the *Book on Agriculture*, written by the Moor Ibn-al Awan (n.d.), and the *Librum Commodorum Ruralium* of Crescentius (*c.* 1300) bear a close similarity to the classical works of Columella (*c.* AD 100) and Varro (35 BC) Virgil's (n.d.) *The Georgics* also had a lasting influence and as late as the mid-eighteenth century agriculture in Britain was still regarded as 'Virgillian' in character (Fussell 1972).

Although agricultural development prior to 1500 had been slow, the agricultural systems that had evolved were far from uniform. Admittedly, the open-field system dominated large areas, but even this showed regional variations, and in many areas open fields were unknown. In much of Sussex, Kent and the Welsh Marches, for example, the land had often been enclosed directly from woodland; according to Leland (1710–1712) 'most part of all Somersetshire is in hedgerows enclosed'. In these areas pastoralism predominated. Elsewhere, in the uplands, infield-outfield systems were found (Kerridge 1967). The infield, normally on the lower land, was cropped every year and fertilised with animal manure; the more distant outfields were given over mainly to rough pasture. On the exposed hill-lands, such as the Chalk plateaux of southern England, convertible husbandry was practised, the grassland being ploughed and cropped for a few years and then allowed to revert to pasture.

It was in the Midlands that the open-field system was most fully developed. Here, the farmsteads tended to be clustered, and around them stretched the large, arable fields, divided into two, or more commonly three, areas of approximately equal size. Each of the open fields was further subdivided into narrow strips which were shared on a more or less random basis between the villagers. The open fields were farmed in strict rotation. In the two-field system, an annual alternation of fallow and corn was employed. In the three-field system, wheat or rye would be followed by barley, oats or beans, and this in turn by a year of fallow. Cattle were grazed on the pasture which was allowed to develop on the fallow land, and were possibly herded across it to spread the manure. Beyond the open fields lay the wasteland, which was used for common grazing, while further pasture was available in riverside meadows and occasional areas of improved wasteland.

The evolution of the open-field system left its mark on the landscape, as Hoskins (1976) has described; it is an impact which is still visible in many areas. It has also been claimed that continuous arable cropping in the open fields, relieved only by periods of fallow, resulted in a significant decline in soil fertility. The extent to which this is true is not easy to judge. Tillage was shallow and crop yields in any case were low, so the rates of nutrient and organic matter loss may have been slow, and losses replenished by natural processes and the limited manuring. The effects of continuous cropping also depend to a great extent upon soil type (see Ch. 8). It is far from certain, therefore, that soil fertility declined sufficiently to cause problems, and it may well be that the eventual demise of the open-field system owed more to climatic vicissitudes – Europe in the sixteenth and seventeenth centuries was in the grip of the 'little ice age' – and to socio-economic changes than to any reduction in soil fertility.

2.1.2   Agricultural development from the sixteenth century

The sixteenth century saw the beginnings of change in farming systems in Britain. The Black Death had caused marked reductions in both rural and urban populations. Wool prices were increasing and encouraging an expansion of pastoralism, which in turn released labour from the land and led to enclosure (see Ch. 11). Rural unemployment became a reality and where they could the landlords evicted the superfluous peasantry. At the same time, a class of yeoman farmers was developing, with a growing influence and authority; agrarian revolt faced the more conservative barons who attempted to exact the customary services from their tenants. By 1500, feudalism was all but dead, and thereafter the agricultural systems associated with it began to decay.

Without doubt, one of the major influences to emerge from the decline of the feudal system was the extension of enclosure. This had been taking place for many centuries, but between the sixteenth and nineteenth centuries it wiped out the open fields and took into cultivation vast areas of heathland, forest and upland moor. Initially, enclosure took place without the aid, and indeed against the will, of Parliament, and Acts were passed and inquiries held to try to prevent the practice. For all that has been written about it, therefore, enclosure was slow and by 1700 about half of the cultivated land in England was still in open fields – a total of some 4.5 million acres (Hoskins 1976). After the Restoration, however, government opposition ceased and from 1750 enclosure was taking place by private Act of Parliament. From then on, the rate of change was more rapid. The open fields were converted into what Marshall (1787) called 'a continuous sheet of greensward', while the heathlands and forests were ploughed up for arable crops. Arthur Young (1804) wrote: 'Half a century ago, Norfolk might be termed a rabbit and rye country. In its northern part wheat was almost unknown, in the whole tract lying between Holkham and Lynn not an ear was to be seen, and it was scarcely believed that an ear could be made to grow. Now the most abundant crops of wheat and barley cover the whole district.'

The changes that came with enclosure did not touch all areas equally (Fig. 2.1). Where enclosure occurred, however, it led to one particularly important development: consolidation of previously disparate holdings. Consolidation was often slow, for it involved major decisions and much manoeuvring; it required outlay on fences and hedges, and it led to inevitable problems regarding the relative worth of the strips of land which were exchanged or sold. Kerridge (1967) records that it took the Nicholas family of Roundway almost a century to consolidate 30 acres (12.1 ha). In time, however, it created an agricultural framework within which innovations could be more rapidly adopted, and in which improvement of land by drainage, manuring or clearance could be more readily contemplated.

Enclosure also affected the landscape. Into the relatively featureless open fields and common land intruded hedgerows; from the moors and fens there emerged pastureland and arable fields; the forests were dissected by clearings or cut down entirely; villages were deserted, while new, isolated farmsteads appeared in their place. The passage of these changes has been described for many areas in the 'county landscape' series (e.g. Rowley 1972; Emery 1974; Bigmore 1979), but to illustrate the effects upon the landscape we will consider two examples.

In Oxfordshire, as elsewhere in Britain, changes prior to 1700 were lim-

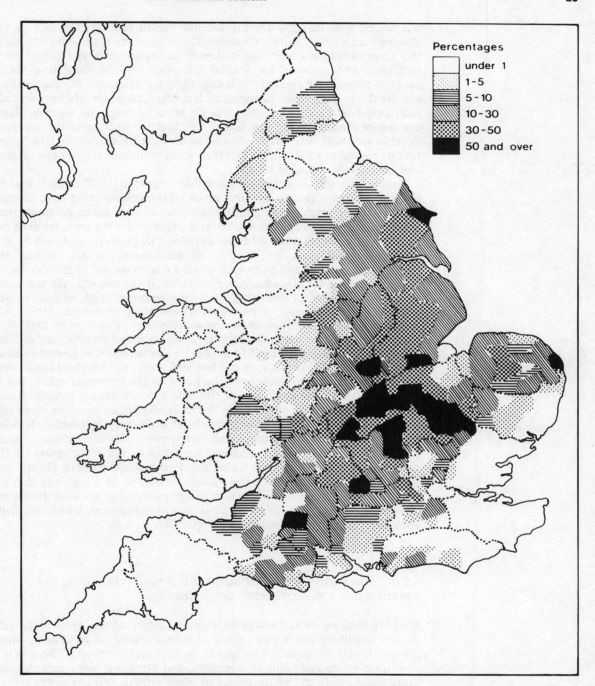

Fig. 2.1 Distribution of parliamentary enclosure in England and Wales during
the eighteenth and nineteenth centuries (from Gonner 1912).

ited, and no more than 30 per cent of the county had been enclosed by
the end of the eighteenth century. Villages, however, were being abandoned as a result of rural depopulation, and Allison *et al.* (1965) have
identified 101 village sites deserted before 1800; to these, Emery (1974)

has added a further 24. During the nineteenth century, the rate of enclosure accelerated. With enclosure came improvements in land drainage; the Thames was dredged and widened and areas such as Otmoor were reclaimed and enclosed for arable cultivation. At the same time, large areas of forest were cleared, including the once extensive Wychwood Forest. By the beginning of the nineteenth century, these forests were already intersected by clearings, but it was illegal coppicing, and equally illegal grazing of the forests by pigs and sheep, that led to their further decline. Wychwood itself was finally cleared as a result of an Act of Disafforestation, passed in 1857; within eighteen months the forest had been cleared and converted to arable land.

Similar changes occurred in Shropshire. Rowley (1972) argues that by the end of the sixteenth century, much of the county must have already been enclosed, and the remaining open fields were often no more than small, scattered patches associated with areas under the ownership of absentee landlords. Almost all these surviving fields were enclosed by private agreement, and not by Act of Parliament, by the end of the eighteenth century; and as in Oxfordshire enclosure led to the decline of many villages and the emergence of isolated farmsteads. At the same time, agriculture was extending on to the common land, in part by the action of squatters and in part by parliamentary enclosure. Often, encroachment of agriculture on to these areas was piecemeal at first; near Much Wenlock, for example, squatters had enclosed about a half of Shirlett Forest by 1725, and in 1814 an Act of Parliament was passed to complete the process (Fig. 2.2). In the lowland areas, the enclosed lands were hedged and fenced to demarcate the fields and to constrain cattle, but in the uplands physical enclosure of the land was not always practised, and in Clee St Margaret for example the upland pasture remains open and undivided. With enclosure came roads 'at least forty feet wide' (Rowley 1972), which had to be completed within two years of enclosure, while improved drainage on the enclosed land led to the reclamation of the many meres and mosses which characterised lowland parts of Shropshire. As in other parts of Britain, the present landscape of the county thus reflects the processes of land division and improvement associated with enclosure. (Further examples of land use development which involved enclosure, are given in section 11.2, p. 310.)

2.1.3 The development of agricultural methods during the seventeenth and eighteenth centuries

Possibly because of the changing social structure of the period, the seventeenth century saw a blossoming of scientific thought in Britain, which ultimately led to marked changes in agricultural methods. There was a new spirit of enquiry among scientists, and attention was being devoted more assiduously to the questions of plant growth and chemistry. Boyle (1661) presaged this new spirit with his book entitled *The Sceptical Chymist* in which he re-examined the alchemic principles of contemporary science. He was followed by Evelyn (1676) and Woodward (1699), both of whom discussed the nature of the food of plants. It was, however, men like Tull, Townshend, Coke and Young who translated these ideas into practical action.

It is often Tull who receives the credit for much of the innovation in

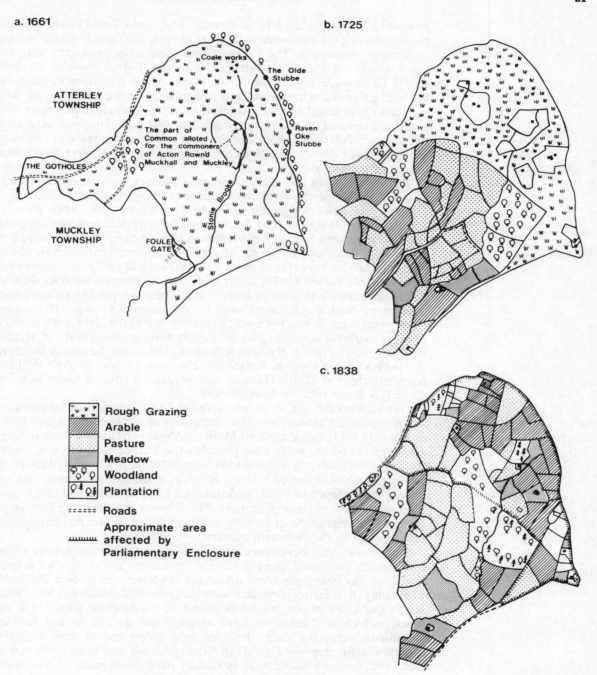

a. 1661

ATTERLEY
TOWNSHIP

Coale works

The Olde
Stubbe

The part of
Common alloted
for the commoners
of Acton Rownd
Muckhall and Muckley

Raven
Oke
Stubbe

THE GOTHOLES

MUCKLEY
TOWNSHIP

Stonie Brooke

FOULE
GATE

b. 1725

c. 1838

Rough Grazing
Arable
Pasture
Meadow
Woodland
Plantation
Roads
Approximate area
affected by
Parliamentary Enclosure

Fig. 2.2   Enclosure of Shirlett Forest, Shropshire, 1661–1838 (from Rowley 1972).

farming methods during the eighteenth century. To some extent this is unjustified, for many of his inventions had been pre-empted by others: by Worlidge and Platte in England, and by Kalm in Sweden. In 1731, however, Tull published a book entitled The New *Horse-Houghing Husbandry*, setting out his concepts of plant growth and their implications for cultivation procedures. The food of plants, he claimed, were the soil particles which the plants devoured through the 'lacteal mouths' of their roots. To increase plant growth, therefore, it was necessary to improve the contact between roots and soil and to make more available the fine particles. This could best be achieved by tillage, which broke down the clods and opened up the pore spaces. Not one ploughing was necessary but several; Tull comments: 'The first and second plowings with common Ploughs scarce deserve the Name of Tillage; they rather serve to prepare the Land for Tillage.

'The third, fourth and every sybsequent Plowing, may be of more Benefit, and less Expence, than any of the preceding ones.'

To this end, Tull designed a range of tillage implements – hoes, ploughs and harrows – as well as seed drills, which he employed with considerable success on his own farm at Crowmarsh in Oxfordshire. His ideas were not universally accepted, however, and he was severely criticised by those who rejected his concepts of plant growth. Nor were his innovations rapidly adopted. Fussell (1973) states that: 'His neighbours were so slow to follow his example that the practice had not been adopted by farmers who lived more than a mile or so away by the reign of George III' – some thirty years later. It was not until the second half of the eighteenth century that his methods began to spread, largely through the efforts of Hamel de Manceau in France, Baldwin in Suffolk, Coke and the Smyth brothers in Norfolk and Knapp in Berkshire. Even so, as late as 1787 William Marshall claimed that: 'There is not perhaps a drill, a horse-hoe, or scarcely a horse rake, in East Norfolk.'

This tardiness is not to be wondered at. The diffusion of information in the eighteenth century was slow, especially in rural communities where, as Fussell (1973) suggests, 'Old Moore's Almanac, Tusser's book of doggerel verse, and the Bible were probably the limit of most farmers' literature'. Additionally, new inventions were often costly to manufacture; a seeding machine invented by John Randall, a somewhat eccentric schoolmaster from Heath in West Yorkshire, cost £20 to make in 1764, a prohibitive amount for many farmers. Nonetheless, by 1800 the first steps towards mechanisation of farming had been taken, a process that continued through to the twentieth century.

Meanwhile other developments were occurring. New crops were being introduced, including clover, turnips, potatoes and ryegrass; animal breeding was being improved under the leadership of Robert Bakewell of Dishley; new farming systems were being devised. Much has been written in particular about the development of the Norfolk four-course rotation, which is claimed to have arrested the decline in soil fertility supposedly occurring under the open-field system and to have been the foundation for improved yields in the eighteenth and nineteenth centuries. The four-course rotation is said to have been initiated by Townshend, following his introduction to Britain of turnips, which formed a vital part of the rotation (Table 2.1). It is also frequently implied that the Norfolk four-course rotation dominated agriculture in Britain during the late eighteenth and nineteenth centuries. We now know that many of these claims are unfounded. Turnips were being grown in Britain well before Townshend's time, and there was never, in any case, a single system of rotation; as Marshall (1790) commented: 'Every occupier, if not under legal restrictions, farms by existing circumstances; crops his lands

Table 2.1 The Norfolk four-course rotation

Year	Crop	Use
1	Turnips or swedes	Folded with sheep in winter
2	Spring barley	Cash crop
3	Red clover	Grazed in spring and summer
4	Winter wheat	Cash crop

Source: After Lockhart and Wiseman 1978

according to the several states in which they are.' Nor were rotations adopted everywhere. Again, Marshall noted that, 'in the far-famed husbandry of East Kent, there is scarcely anything that resembles a "regular rotation of crops"'. The one claim which is valid, however, is that these rotation systems helped to maintain soil fertility and, perhaps more importantly, controlled weeds and reduced the labour demands of agriculture by spreading the workload more evenly through the year.

Further, and possibly even more crucial, innovations came in the mid-nineteenth century with the emergence of agricultural chemistry. Humphry Davy (1813) pointed the way forward, arguing that plant nutrients, instead of being manufactured during growth as was widely believed, were obtained from the soil. Thirty years later, Liebig in Germany, and Lawes at Rothamsted in England, were able to demonstrate the role of specific nutrients in plant growth and to rationalise observations on the effects of manures. Out of their research developed the fertiliser industry which is the foundation of modern agriculture. Other developments followed. Liebig showed that warmth helped animals to reduce their food requirements, so farm buildings were designed to house livestock. High quality pedigree herds of Shorthorn, Hereford and Devon cattle were started, while breeding of sheep improved, producing quality flocks of Lincolns, Southdowns, Border Leicesters and Cheviots. Drainage methods, which had been developed in the previous century by Elkington (Johnstone 1797), were improved and during the late nineteenth century some 4.8 million ha of land were drained (Trafford 1970). In 1838 the Royal Agricultural Society of England was formed, and this provided both a stimulus for research and a means of disseminating ideas.

2.1.4 Nineteenth-century developments

By the beginning of the twentieth century, the scientific basis of modern agriculture had been established. As previously, however, the results were not immediately seen in the fields and farmyards. Adoption of the innovations, though more rapid than in earlier centuries, still took time, and as Street's (1936) description of farming in East Anglia prior to the First World War indicates, farming techniques still had much in common with those which Tull and Townshend had advocated almost two hundred years earlier. Fertilisers were available and widely used, but farmyard manures still provided the main source of nutrients. Cereal harvesters as well as tillage implements had been developed, but they were still horse-drawn; steam power was available but few farmers could afford to own the machines and, at the same time, retain the horses without which they could not farm the land. Cattle and sheep had improved, but few new

breeds had appeared and control of livestock quality for the majority of
farmers was limited so long as farmers used their own bulls for breeding.
During the 1800s the area of arable land had expanded within the sanc-
tuary of the Corn Laws, but in 1869 they were repealed, and between
1871 and 1901 the acreage of corn fell from 8.24 to 5.89 million acres.
Arable land area increased in the First World War, but afterwards acreages declined again.

2.2   THE EMERGENCE OF MODERN FARMING SYSTEMS

2.2.1   Mechanisation

It was not, therefore, until the middle decades of the present century that
agricultural development revived. Much of the stimulus for progress came
from the Second World War, but before then a number of vital innovations had appeared. Possibly the most important was the petrol-driven
tractor. In the late nineteenth century steam ploughs and engines had
been quite common, but they were cumbersome and expensive to run.
The direct effect of the development of the tractor was to reduce greatly
the amount of time and labour needed in agriculture. Kaiser (1930)
showed that to produce an acre of maize by hand took 300 man hours;
with machinery it took only 3.6 hours. Today, it requires less than two
man hours. As a consequence, the decline in the agricultural labour force
has accelerated. Tractors have also increased output by releasing the land
which was previously needed for feeding draft animals. The increased
power available from tractors has changed cultivation methods; implements can now be larger, ploughing can be carried out more deeply and
more effectively, and numerous operations which otherwise would have
to be done by hand can now be mechanised (e.g. drain laying, stacking).
   Harvesting has similarly been mechanised. As Hawkins (1980) notes,
in 1930 the only mechanised harvesters available were for cereal crops and
these were horse- or tractor-drawn. During the 1950s these were replaced
by self-propelled combine harvesters. The history of adoption of these has
been charted by Jones (1965) and was shown to follow the classical pattern of innovation: a slow initial adoption by a small body of leaders, a
period of relatively rapid growth as the innovation was taken up by the
majority of farmers, and then a final, slow phase of adoption by the 'lag-
gers'. The spatial pattern of adoption in England and Wales reflects this
sequence. In accord with diffusion theory, use of the combine harvester
spread concentrically from the nodal area, in the south-west, taking over
ten years to reach full adoption in the north-west of the country. The
effects of this innovation have been far-reaching. Not only has it reduced
labour requirements, but it has also made harvesting much speedier and
therefore less dependent upon weather; yield losses during harvesting
have therefore decreased. Other harvesting implements have developed
similarly. Forage harvesters are now used to collect cut grass for silage
or forage, and balers have eliminated the need for hand-stooking of straw
and hay. Mechanised harvesting of potatoes, sugar beet and peas has also
become possible, contributing to the rapid increase in the acreages devoted to these crops.

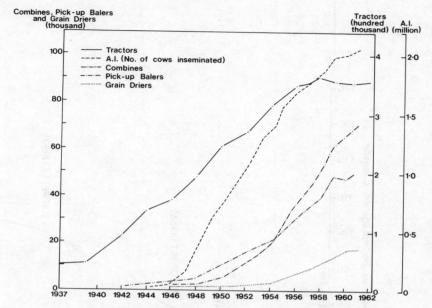

Fig. 2.3 The adoption of agricultural innovations in England and Wales (from Jones 1965).

Over the same period, mechanisation has extended into other areas of agricultural activity. Grain and grass drying systems were developed during the 1950s, while major advances were made in mechanised milking (Fig. 2.3). Today, these and other operations are entering a new phase of development, as they become automated through the use of computers (Hawkins 1980).

2.2.2 Plant and animal breeding

The early post-war period saw comparable advances in plant and animal breeding. In Britain, the National Institute for Agricultural Botany has been responsible for breeding and testing a range of high-yielding and high-quality cereal varieties. These have been developed in part to provide specific grain characteristics, better suited to the needs of the processor and consumer. They have also been bred to cope with specific problems such as fungal disease, pest attack (e.g. by nematodes) and lodging (the laying flat of the cereal crop by wind or rain). Additionally, of course, the aim has been to increase the yield potential of the plants. The importance of these advances can hardly be over-emphasised. Dyke (1968) estimated that improved varieties alone accounted for about 35 per cent of the increase in wheat yields in the 20 years prior to 1963. Austin (1978) shows that new varieties of wheat outyield those widespread in the 1940s by 60 per cent. Riggs *et al.* (1981) compared thirty-seven varieties of spring barley grown in England and Wales between 1880 and 1980. They showed that over the hundred-year period genetic improvement had resulted in annual yield increases of 0.39 per cent; since 1953 yields had risen by 0.84 per cent per year due to improved varieties (see also

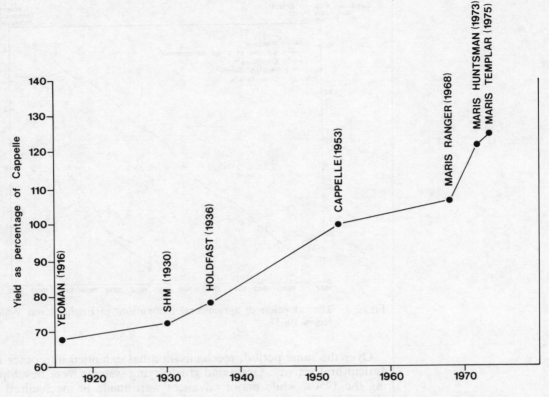

Fig. 2.4   Yields of main wheat varieties introduced in Britain, 1920–75 (after Arnold 1980).

Fig. 2.4). Modern varieties of barley had higher grain yields, shorter straw, higher survival rate of tillers and a better harvest index (i.e. a higher ratio of dry weight of grain to the weight of the above ground dry matter). This last group of factors was believed to be the main reason for increased yield.

Similar developments have been made with other crops. Maize has increased in importance as new varieties have become available, adapted to the short growing season and relatively cool summers found in Britain. Improved varieties have increased the productivity of sugar beet growing and encouraged the more widespread adoption of beet as a break crop in arable systems. Comparable progress has been made in the breeding of grass varieties. These have been developed in an attempt not only to increase yields, but also to improve digestibility, winter hardiness and resistance to trampling. Since 1945, new varieties of ryegrass (*Lolium* spp.) have had particular success and these now form the basis of most improved pastures in Britain.

Of equal importance have been developments in animal breeding. During the 1950s and 1960s, the Friesian rose to pre-eminence as a dairy breed and today accounts for approximately 80 per cent of all dairy cattle in England and Wales. Its success was based on its high yield potential and a decline in the demand for milk, with high butter-fat contents (e.g. for butter or cream production), such as that produced by Channel Island cattle. Dairy herds also provide a large proportion of the beef production, and for this purpose Friesians are crossed with specialist beef cattle such as Herefords, Galloways and Aberdeen Angus. Since 1962, Charolais, Limousin and Simmental cattle have been imported to fulfil the same function, namely for crossing with Friesians for beef production from dairy herds. The last forty years have seen related progress in veterinary science, which has resulted in improved animal health and husbandry and in the widespread adoption of artificial insemination for cattle breeding (Fig. 2.3). This in turn has raised the quality of farm livestock and helped to eliminate weak strains in the breeds.

2.2.3 Fertilisers and pesticides

Between 1939 and 1975, the quantity of fertiliser applied to crops in Britain increased sevenfold. At the same time marked changes in fertiliser practice have occurred, with a change away from farmyard manures and traditional, simple fertilisers (e.g. ammonium nitrate, ammonium phosphate or potassium chloride) to compound fertilisers normally consisting of a mixture of nitrogen, phosphorus and potassium. About 75 per cent of the fertilisers used in Britain are now applied in compound form. Associated with this development has been a change in the relative importance of different fertilisers. Calcium is the only nutrient whose use has declined, but nitrogen has shown the most marked increase; since 1939 there has been a sixteenfold increase in its application. The effect of this increased nitrogen usage has been marked. According to Dyke (1968) it accounted for about 25 per cent of the improvement in yields between 1943 and 1963, while Austin (1978) estimated a 30 per cent increase in yield between 1950 and 1972 due to increased nitrogen application.

Numerous other innovations have appeared. Perhaps the most significant, both agriculturally and environmentally, has been the development of pesticides. In 1931 and 1932 only 120 ha of land were sprayed in Britain; by the late 1970s over 10 million ha were being treated with pesticides (Hawkins 1980). Developments started during the 1940s with the introduction of DDT and MCPA. In the decade following the war they became widely used for control of agricultural pests. In their wake came many similar compounds, such as the herbicides mecoprop (CMPP), dicamba and dichlorprop, and the insecticides aldrin, dieldrin and heptachlor. In time, the potentially harmful effects of persistent residues of these pesticides became apparent and alternatives were sought. As a result, a fundamental change in pesticide usage occurred during the 1960s as organophosphate pesticides were developed (see Ch. 6 and 10). The overall trend has been consistent, however, with the total area of land sprayed increasing by about 5 per cent each year.

2.2.4 Changes in farming practice and structure

Together these developments have had far-reaching implications. Yields have increased markedly. Between 1952 and 1975 agricultural output rose by about 60 per cent (Fig. 2.5). Over the same period of time the total area of agricultural land decreased by about 6 per cent as land was taken over by urban development, mining and afforestation (Best 1981). Thus, increased output has occurred wholly as a result of improvements in yields per unit area; wheat yields, for example, have risen from about 2.7 t ha⁻¹ in the late 1940s to almost 5 t ha⁻¹ in the late 1970s; barley yields from 2.5 to 4.0 t ha⁻¹ (Fig. 2.6). E. W. Cooke (1979) claims that yields of root crops have increased by 50–75 per cent since 1939, while milk yields per cow have increased by 50 per cent.

At the same time, the number of workers employed in agriculture has declined considerably (Fig. 2.7). In 1946 there were about 780,000 people employed in agriculture in the United Kingdom; by 1975 this had fallen to about 315,000 (Hawkins 1980). Associated with this has been a trend towards rationalisation of farms, with small farms being amalgamated. This has led to an increase in the average size of holdings. Between 1949 and 1972, the number of farms less than 20 acres (8.2 ha) in area has declined from 138,654 to 53,879, while the number of holdings more than 300 acres (122 ha) has increased from 12,317 to 16,765. The total number of holdings decreased from 312,364 to 182,853 (Lloyd and Wibberley 1979). At the same time, field sizes have increased (Coppock 1976). Hooper (1970) estimates that hedgerows were removed at a rate of 8000 km y⁻¹ between 1945 and 1970, representing a total loss of about 1 per cent of the country's hedgerows in 25 years. Changes in the size of both farms and fields show marked regional differences, most change occurring in the eastern part of the country. Today, average farm sizes are in the order of 125 ha in the east of Britain, compared to less than 45 ha in the west. Field sizes similarly differ. Coppock (1976) quotes data col-

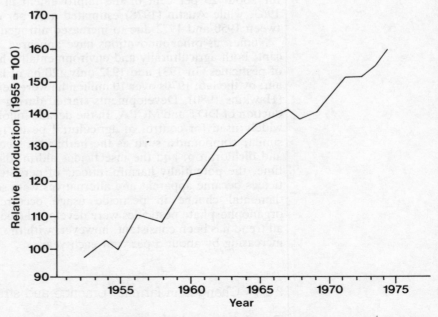

Fig. 2.5   Agricultural food production in the United Kingdom, 1952–75 (after Hawkins 1980).

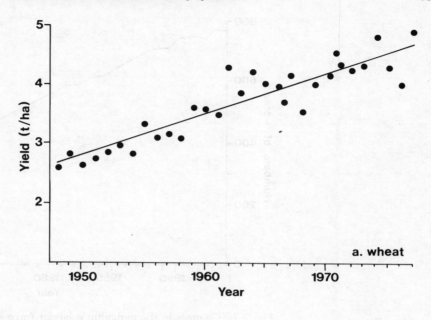

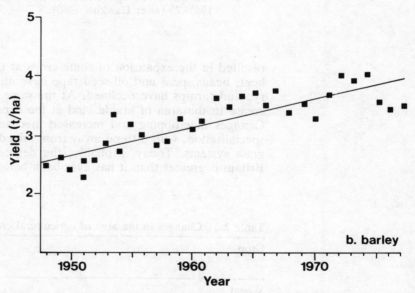

Fig. 2.6 Changes in the yield of wheat and barley in the United Kingdom, 1948–77 (from Austin 1978).

lected by the Forestry Commission, showing that in Devon there are about 20 km of field boundaries per square kilometre, while in Essex there are a little over 12 km per square kilometre.

This pattern reflects the changes in farming practice which have occurred over the last forty years. Improved plant breeding, better crop protection and a fuller understanding of nutrient requirements – as well as external factors, such as changes in demand and market prices – have

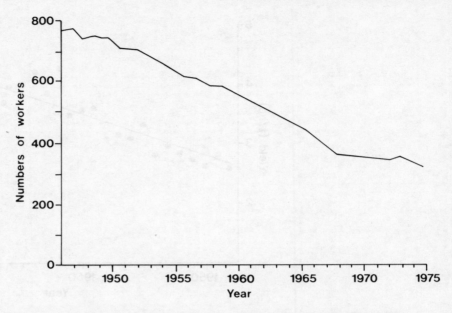

Fig. 2.7   Changes in the agricultural labour force in the United Kingdom,
1945–75 (after Hawkins 1980).

resulted in the expansion of some crops at the expense of others. Sugar
beet, beans, peas and oil-seed rape have all grown in importance; oats,
rye and turnips have declined. At the same time, there has been an increase in the area of arable land at the expense of pasture (Table 2.2).
Changes in cropping and increased mechanisation have also increased
specialisation, with a trend away from mixed farming to all-arable or all-
grass systems. Today, probably, the degree of regional segregation in
Britain is greater than it has ever been before.

Table 2.2   Changes in the area of agricultural crops in Britain

Crop	% of total agricultural area		
	1938	1960	1979
Wheat	4.2	4.5	7.5
Barley	2.2	7.2	12.8
Oats	4.6	3.8	0.9
Potatoes	1.3	1.6	1.1
Sugar beet	0.7	0.9	1.2
Turnips and swedes	1.6	1.0	n.d.
Mangolds	0.5	0.3	n.d.
Oil-seed rape	0.0	0.0	0.4
Temporary grass	7.5	13.6	10.4
Permanent grass	38.4	25.2	27.9
Total agricultural land area (including rough grazing) (million ha)	18.36	18.75	18.36

Source: Compiled from MAFF 1968; Hood 1982

2.2.5 Agricultural development in other temperate areas

To some extent, the pattern of agricultural development experienced in Britain has been repeated elsewhere. After 1750, a relatively fluid exchange of ideas and information took place in Europe through the translation of agricultural texts. Fussell (1972), for example, records that Francis Home's *Principles of agriculture and vegetation*, published in 1757, was translated into French in 1761 and German in 1763. du Hamel de Manceau visited England and was greatly impressed by Tull's farming methods; he took these ideas back to France from where he disseminated them widely, his writings being translated into Spanish, Dutch, German and English between 1751 and 1764. As these examples show, much of the stimulus for agricultural innovation at this time was coming from Britain, and spreading from there across continental Europe. As a result, farming techniques developed in Britain were introduced throughout much of the European mainland before the start of the nineteenth century. The Norfolk four-course rotation was widely adopted in France, for example, and was advocated in Italy by Campini (1774). Similarly, Tull's seed-drill was common in France and Italy, while English advances in drainage methods and improved fertiliser practices were introduced to many parts of north-western Europe. Potato growing, too, spread from England and Switzerland to France, Germany and Denmark (Abel 1978).

In Britain, these technical developments had been associated with social and economic events which had led to a marked expansion in the area of cultivated land and had changed the rural landscape. The same processes were seen in other parts of Europe. Either through the enclosure of the traditional open fields, or through taking wasteland directly into cultivation, agriculture became more intensive and more commercialised in many areas. As in Britain, the landscape changed accordingly. In Denmark in particular the history of agriculture in the eighteenth century reflects that of Britain. Similarly, in the Mass, Scheldt and lower Rhine, and in Flanders, Brabant and Hennegan extensive enclosure of fen, moorland and heathland took place. In France, too, enclosure was common; Toutain (1961) estimated that between 1750 and 1790 the area of arable land in France increased from 19 million ha to 23.9 million ha as cultivation encroached onto what had previously been wasteland.

The pattern of agricultural development in continental Europe did not everywhere mirror that of Britain, however. In some areas English farming methods were improved and developed; in Flanders, the Norfolk four-course rotation evolved into a much more sophisticated twelve-course rotation including crops such as clover, flax, potatoes, rye and turnips (Abel 1978). Elsewhere – especially in Germany and Italy – traditional three-course systems prevailed, and indeed in some localities had not yet even developed. Moreover, when changes did come in many parts of Europe, they took place as a response to different social forces and within different agrarian frameworks. The Napoleonic Code, for example, led not to physical enclosure and consolidation of farmland as had occurred in Britain, but to expropriation of the existing land holdings by the peasants. The result was the fixation of unconsolidated and unenclosed strips in peasant ownership. At the same time, the principle of partible inheritance – the equal division of land amongst heirs – encouraged further fragmentation. Additionally, in many parts of Europe, livestock farming was carried out not as a large, field-based enterprise but as a small, 'stall feeding' system. Thus, the need to enclose fields physically with permanent hedgerows was less important. As a consequence of all these factors, the landscapes inherited by modern farming systems in many parts of Europe retained the basic features of the classical open-field systems. Even today, in much of northern France and Belgium, the landscape bears this imprint.

The effects of agricultural innovation were not restricted to Europe. From England, Spain, the Netherlands and other imperial countries, European crops, livestock and farming methods were being disseminated throughout the New World. Explorers took with them cattle, pigs and horses, and seeds of wheat and barley, which they traded or scattered in the newly discovered lands (e.g. Grigg 1974). Later, settlers and pioneers introduced farming methods learned at home in Europe, and these spread inland as settlement expanded into the interior. In the process, farming methods changed as new crops were discovered, the limitations of traditional crops were recognised, and as experience led to the introduction of new techniques. In the USA, for example, the earliest settlements in New England were like those in Britain: nucleated villages with widely dispersed plots of arable land resembling the open-field strips of Europe. By the time that pioneer movement inland occurred, however, in the late eighteenth century, most of this land was enclosed and isolated farmsteads characterised the landscape. By that time, too, the importance of maize had been established; it gave a higher and more reliable yield, could be harvested over a longer period and was less susceptible to disease. It was the main crop of the pioneers who crossed the Alleghenies into the Ohio valley. But its value-to-weight ratio was low, so as a cash crop it was limited and needed converting into higher-value animal products; in this way it became the basis of the corn-hog economy which dominated the Mid-West prior to 1880. Further north, wheat was preferred as a cash crop because the summers were too short and cool to give good yields of maize.

During the late nineteenth century, agriculture in the USA spread westwards, and as it did so the pattern of cropping changed. The centre of wheat production migrated from Ohio to Chicago to the Dakotas, and, in the present century, became focused on the winter-wheat belt of the Palouse district. In its wake came dairying, in 1869 centred on New York, but by 1899 developed also in Illinois, Iowa and Wisconsin.

Similar changes occurred in Canada and Australia. It was not until the twentieth century that the Canadian prairies were settled, but between 1900 and 1930 wheat farming spread westwards into Alberta and Saskatchewan, though yields generally declined (Ridley and Hedlin 1980). During the 1930s, however, the depression led to falling grain prices and a reduction in the land devoted to wheat. Mixed farming was adopted in much of the Manitoban Park Belt, while cattle ranching re-emerged in the Alberta-Saskatchewan border country. In Australia, sheep dominated the economy until the mid-nineteenth century, but from 1860 to 1890 wheat increased in importance, attaining something close to its present distribution by the end of the century. Since then, yields of wheat have increased, in part due to the use of phosphate fertilisers to counteract the natural phosphorus deficiency in many of the soils, but as in Canada the 1930s saw a period of falling prices and a retreat from the more marginal wheat growing areas. As in Canada, also, the result has been a move back towards more mixed farming in many areas, with wheat and sheep pasture being rotated to help maintain soil fertility and to reduce the economic problems of monoculture.

Elsewhere in the temperate world, the nineteenth and twentieth centuries saw similar expansions of agricultural land, due in many cases to human migration motivated by political factors. During the early part of the nineteenth century, for example, there was a persistent migration of people eastwards and northwards in Russia on to the wooded steppes west of the Urals (Grigg 1974). Between 1800 and 1860, 20 million ha of new land had been brought into cultivation, mainly for wheat and sugar beet, and by 1850 Russia had become a major grain exporting country. After 1880, this movement extended further eastwards, across the Urals. The process was halted by the upheaval of the First World War, and practically no new settlement occurred in the troubled decades that followed. In the 1950s, however, an unprecedented expansion into the Siberian plains took place as part of the 'virgin and idle lands' programme, and 36 million ha of new land were ploughed up for grain crops. The areas being settled at this stage were more marginal for agriculture, with poorer soils than the lands to the west and with a growing season of little more than 130 days. As a result, yields have fluctuated widely, whilst problems were encountered due to the use of intensive and mechanised farming practices in soils sensitive to erosion and exhaustion. Nevertheless, since 1960 dryland farming methods have increasingly been introduced, and the area now produces 50 per cent of Russia's grain, and about 70 per cent of its wheat.

2.3 CONCLUSIONS

It is apparent that the agricultural history of different parts of the temperate world varies in detail, though the broad trends have been similar: increased intensification and the expansion of cultivation on to new lands. It is also clear that modern farming practices owe much to the legacy of previous generations. Past activities have moulded not only farming principles and experience but also much of the environment in which modern agriculture takes place. For these reasons it is not always easy to isolate the effects upon the environment of present-day practices.

PART 2

AGRICULTURAL PRACTICES

3 TILLAGE

3.1 INTRODUCTION

Tillage is one of the fundamental practices of agricultural management. It is the procedure by which man disturbs, overturns and rearranges the soil, and is aimed at creating favourable physical conditions for crop growth. As such, it has a long history. Simple tillage tools date back to the initiation of settled agriculture, while ploughs incorporating a wooden shaft and a single wooden share were used by the Romans (Simonson 1968). Until the Middle Ages, however, tillage practices changed little, and the manner of their effect was only poorly understood. It was only in the eighteenth century, in fact, that clear concepts of tillage began to emerge and radical developments in tillage methods were introduced. Much of the credit for these developments is attributed to Jethro Tull. In his book *The New Horse-Houghing Husbandry* referred to earlier (section 2.1.3), Tull (1731) set out the principles and practice of what he claimed to be a new method of tillage, based upon the regular and repeated ploughing of the land. Tull's ideas of why tillage was beneficial have not survived the test of science, and it was in any case many years before his methods diffused throughout the agricultural world and were widely adopted. To claim as Pereira (1975) did, therefore, that after Tull 'for the next two centuries farmers in Britain and elsewhere stirred their soils like Christmas puddings' is a considerable overstatement. Nevertheless, by 1866 Copland was able to state that the drill and hoe were common throughout England, and since then tillage has become an integral part of almost all agricultural systems.

Even so, the practice is not without its problems. Increasingly in recent years it has been realised that soil compaction caused by tillage may have adverse effects upon crop yields (Table 3.1). Moreover, as fuel costs have risen the economic viability of repeated tillage has been called into question. As a result, various alternative techniques – such as reduced cultivation and direct drilling – are gradually being introduced. The purpose of this chapter, therefore, is to examine the principles and practice of conventional tillage techniques, to assess their impact upon soil conditions in both the short and long term, and to compare them with alternative procedures such as direct drilling.

Table 3.1 Effects of structural damage of the soil on crop yield

Crop	Soil	Yield on control site (t ha^{-1})	Yield on compacted site (t ha^{-1})	Reduction in yield %	Source
Winter wheat	Clay	5.77	4.09	29.6	Wilkinson 1975
	Sand	4.57	2.23	51.1	Wilkinson 1975
	Loamy sand	5.27	1.38	73.8	Batey and Davies 1971
Spring barley	Clay	5.32	2.96	44.3	Wilkinson 1975
	Sand	3.49	1.77	49.1	Wilkinson 1975
	Sandy loam	3.14	1.51	51.9	Batey and Davies 1971
Spring oats	Sandy loam	3.64	1.26	65.4	Batey and Davies 1971
Sugar beet	Clay	46.44	19.08	58.9	Wilkinson 1975
	Sand	44.68	22.84	48.8	Wilkinson 1975
	Loamy sand	57.19	23.59	58.8	Batey and Davies 1971
	Sandy	12.0	9.0	25.0	Harrod 1975
	Sandy	46.1	27.9	50.3	Harrod 1975
	Sandy loam	34.14	20.08	41.2	Batey and Davies 1971
	Silty clay loam	40.75	17.32	57.5	Batey and Davies 1971
Potatoes	Clay	26.36	14.61	45.7	Wilkinson 1975
	Silt-loam	26.61	11.42	57.1	Batey and Davies 1971
Peas	Clay	2.64	0.83	67.6	Wilkinson 1975
	Sand	2.62	0.83	67.5	Wilkinson 1975
	Sandy loam	2.64	0.88	66.7	Batey and Davies 1971
Onions	Silty clay-loam	27.00	15.80	41.5	Davies 1975
	Loamy peat	31.63	22.59	28.6	Batey and Davies 1971

3.2   THE FUNCTIONS OF TILLAGE

Tillage is carried out for three main reasons:

1. Physical preparation of the seedbed, for example by ploughing, harrowing and rolling the soil to break up the aggregates;
2. Removal of crop residues and weeds, by chopping them up with discs and burying them by ploughing;
3. Improvement of rooting and drainage conditions, for example by deep ploughing.

3.2.1   Physical preparation of the seedbed

In many agricultural systems, crops need to be sown annually as seeds. The germination and development of these seeds, and thus the ultimate

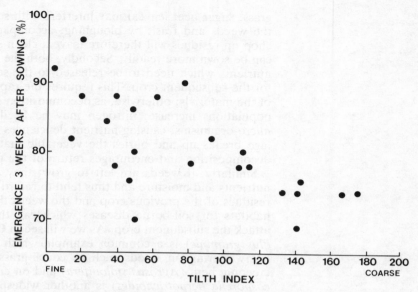

Fig. 3.1   The relationship between soil structure and the emergency of seedlings of oats. The tilth index is a measure of the coarseness of the soil aggregates (from Thow 1963).

yield of the crop, depend to a great extent upon the physical character of the seedbed – upon what the farmer refers to as the tilth of the soil. Thow (1963), for example, has shown that the emergence of oat seedlings is related to the size of the soil aggregates (Fig. 3.1), while many workers have demonstrated that the depth of seed emplacement greatly affects rates of seedling emergence and subsequent crop growth (e.g. Room Singh and Ghildyal 1977). One of the main aims of tillage, therefore, is to create a suitable seedbed, into which seeds can be sown at a uniform rate and at a constant depth, and which allows rapid germination and vigorous seedling development.

Accurate and controlled seed emplacement during sowing is facilitated by the presence of a fine, even surface tilth. Moreover, as we will see in Chapter 8, germination of crop seeds is governed by moisture, aeration and temperature conditions in the soil, and these conditions tend to be optimised by the existence of fine and well developed soil aggregates. Large clods or compact surface layers hinder sowing operations and inhibit emergence, so tillage is often necessary to break up the surface aggregates, to destroy crusts and to loosen compacted zones in the upper layers of the soil. This may require several different tillage operations, including primary tillage with a conventional mould-board plough, chisel plough or rotary cultivator, followed by one or more secondary cultivations with harrows to break up the clods and create a smooth, fine surface tilth.

3.2.2   Removal of crop residues and weeds

The removal of weeds and crop residues serves a number of purposes. In the first place, weeds and trash from previous crops (e.g. corn stubble,

grass, sugar beet leaves) may interfere with sowing operations. Burial of
the weeds and trash by ploughing, accompanied perhaps by discing to
chop up residues will therefore leave a clean surface into which the seed
can be sown more readily. Secondly, both the weeds and residues contain
nutrients which need to be released to the soil to make them available
for the subsequent crop. This requires the rapid microbial decomposition
of the materials; otherwise, as decomposition proceeds and micro-organic
populations increase, nitrogen may be fixed within the bodies of the
micro-organisms, causing nutrient deficiences for the following crop. Tillage breaks up and buries the vegetable matter and thereby speeds up
decomposition and encourages return of the plant nutrients to the soil.

Similarly, if weeds are left to grow, they compete with the crop for
nutrients and moisture and thus tend to retard seedling development. The
residues of the previous crop and the weeds themselves may also provide
habitats for soil-borne diseases which can then survive the winter and
attack the subsequent crop. As we will see in Chapter 6, take-all (*Ophiobolus graminis*) is a common example which attacks wheat and barley,
overwintering on weeds such as couch-grass (*Agropyrons repens*) and
creeping bent (*Agrostis stolonifera*) and on crop residues. Eyespot (*Cercosporella herpotrichoides*) is another widespread fungal parasite which
may also survive on weeds and crop residues, and consequently tends to
build up under continuous arable cultivation unless host material is removed by effective tillage.

3.2.3 Improvement of rooting and drainage conditions

Growing plants require physical support, water, nutrients and oxygen
from the soil, all of which are provided via the plant roots. The ability
of the roots to meet these requirements, however, is dependent upon the
volume of soil they are able to exploit. In general, root extension is limited by lack of suitable pore spaces in the soil (for example in compact
or poorly structured layers) and by excess moisture. One of the aims of
tillage, therefore, is to create a suitable rooting environment by loosening
the soil and by helping to remove excess water.

Under natural conditions, the majority of soil pores are small; few are
greater than 2 mm in diameter, and the majority are less than 60–100 μm.
In contrast, the primary roots of many agricultural crops are in the order
of 400 μm or more in diameter; even the laterals may be coarser than
300 μm. Consequently, the amount of soil volume immediately available
to roots is limited, and roots need to be able to force their way between
the soil particles.

They do so by exerting pressures on the walls of the pore spaces, causing them to expand. While the finer root hairs provide anchorage, the root
tip is pushed into the pore space by elongation of the meristem cells. So
long as the soil is sufficiently compressible or friable, the pore space is
opened up and the root extends. The maximum force that can be exerted
by growing roots is limited, however, (Fig. 3.2) so that where the soil is
hard or compact, extension may be severely curtailed. Ploughing opens
up the pore spaces, breaks compact layers and disturbs the aggregates so
that they pack less tightly together. In these ways, ploughing facilitates
root extension.

Similarly, tillage improves drainage and aeration in the soil. We will

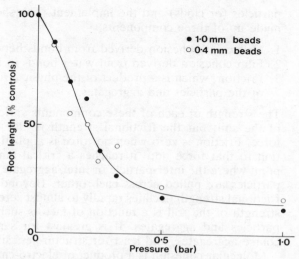

Fig. 3.2   The effect of external pressure on the elongation of barley seminal
root axes (from Cannell 1975).

examine the effects of tillage upon drainage more fully in the next chapter. It is clear, however, that by increasing porosity, tillage allows the freer entry and egress of air and water, thus ensuring a constant supply to the roots. At the same time excess water is removed by drainage and toxic gases are lost by diffusion to the atmosphere.

3.3   TILLAGE PROCESSES

At a detailed level, the processes involved in tillage vary according to the nature of the operation and the implement used. Moreover, many other factors may influence the interactions with the soil: the speed at which the operation is carried out, the condition of the implement, and above all the state of the soil (see Spoor 1975, 1979 and Baver *et al.* 1976, on which much of the following discussion is based). Some indication of the general processes involved can be provided by examining the reactions engendered by pulling a simple tine (e.g. the tine of a harrow or chisel plough) through the soil.

3.3.1   Shear force and shear strength

When a tine is drawn through the soil it sets up shear stresses within the soil, operating at right angles to the tine surface and in the direction of travel. These stresses encourage deformation and shearing of the soil. As the applied force of the tine increases so does the shear strength until, at some point, the soil fails. Failure can occur in one of two ways: either the soil will shear within itself, as particles or clods move past each other, or the tine will slide through the soil and shear will occur between the soil particles (or clods) and the implement. The shear strength of the soil is made up of three components:

1. Molecular cohesion derived from bonds between the particles;
2. Film cohesion derived from water bonds in the soil;
3. Friction, which is a product of the physical roughness and interlocking of the particles and aggregates.

The strength of each of these components varies according to the nature of the soil, but the frictional strength is also a product of the applied force. Friction is zero when no force is applied, but increases in proportion to that force until it reaches a critical limit. This will occur at the point where the inter-particle or inter-aggregate links are destroyed as the particles are pulled across each other. Beyond this limit, therefore, the frictional strength declines rapidly to almost zero. The maximum frictional strength of the soil is a function of factors such as the size and shape of particles and aggregates. It is greatest for sands or soils composed of coarse aggregates, and least for structureless silts and clays.

Molecular cohesion is a product of electro-chemical bonds between the clay and organic colloids within the soil. Ionic attraction between positively charged ions (cations) and negatively charged sites on the surfaces of the colloidal micelles accounts for some of this bonding, but also important are the negative–negative bonds created by van der Waal's forces. These are electro-chemical forces which occur between colloidal surfaces and they tend to be at a maximum when the soil is dry, for in this state the particles are in close contact, allowing molecular attractions to operate between a large number of colloidal surfaces. As the soil is wetted, water enters the spaces between the particles and forces them apart, thus reducing the molecular bonds. Other soil properties may also affect molecular cohesion, however, for at least part of the colloidal bonding is pH dependent. Under acid conditions the surface charge of the colloidal micelles is reduced. Characteristics such as particle size, the types of clay present and the organic matter content are also significant in determining the strength of molecular cohesion. In general fine-grained soils have a higher molecular cohesion than coarse (silty or sandy) soils; 2 : 1 clays such as montmorillonite and hydrated clays such as vermiculite increase cohesion; soils rich in organic matter also have higher colloidal charges and therefore a greater source of molecular cohesion.

Film cohesion arises from the air–water and water–water bonds within the soil. Compared to the molecular bonds these are relatively weak, and they increase initially as the soil is wetted. At a point where the soil pores become saturated, however, the strength of film cohesion is reduced, for the only source of bonding is the surface tension effect. This is reduced as the thickness of the water films increases and, at thicknesses of about 60 μm, is almost negligible. Indeed, when the soil is fully saturated the water may exert a positive pressure on the particles, forcing them apart. The strength of film cohesion is dependent, therefore, upon the physical properties of the soil, for factors such as particle size, aggregation and pore size all influence the amount of water which the soil can hold and the thickness of the water films at specific moisture contents. Clays, containing mainly fine pore spaces, become saturated relatively easily and retain the water for long periods. As a result they tend to exhibit relatively high levels of film cohesion, for even when saturated the tensions binding the water to the soil are high. Coarse-grained, porous soils, on the other hand, have relatively low levels of film cohesion.

In addition to these generally dynamic sources of shear strength, soils also derive a degree of strength from chemical and organic cementation. Calcium, iron and aluminium compounds may all act as cements, although these are to a greater or lesser degree soluble and their effect may be reduced as the soil is wetted. Organic cements, from the gums and slimes produced by soil organisms and roots, are more stable and provide a major source of almost permanent strength in organically rich soils. Many of these organic compounds, such as the humic acids and polysaccharides are water-repellent and provide the aggregates with a stability against disaggregation by wetting. Physical bonding by fungi and plant roots may also give a considerable degree of strength, particularly in the surface layers.

The distribution of these sources of shear strength through the soil is not uniform, largely because the soil does not consist of a heterogeneous mix of unconsolidated particles, but of a closely interacting arrangement of aggregates. These aggregates or clods are internally relatively cohesive; they derive most of their strength from the molecular and film cohesion between the particles of which they are built. On the other hand, the links between the aggregates are relatively weak, for the fissures and pores which divide them are large (often greater than 60 μm), and molecular cohesion cannot provide an effective bond.

It is therefore possible to identify two components of the total shear strength of the soil: that provided by the individual particles (known as the intra-aggregate or clod shear strength), and that provided by the aggregates (the inter-aggregate or bulk shear strength). These are comprised as follows:

1. Clod shear strength: consisting mainly of molecular and, to a lesser extent, film cohesion;
2. Bulk shear strength: consisting of friction and, to a small degree, film cohesion.

The relative importance of these two sources of shear strength varies according to the nature of the soil. In general, well structured soils composed of small, friable aggregates exhibit high levels of both clod and bulk shear strength. Heavy clays, which may be either structureless or nearly so, owe their strength mainly to clod shear strength; coarser sandy soils in which the aggregates are weak have a low total shear strength, and most of this is derived from inter-aggregate bonds.

The shear strength cannot be considered as a fundamental property of the soil; instead, it varies temporally according to moisture content. As the moisture content increases, there is a rapid decline in molecular cohesion, as the particles are pushed apart. Film cohesion, however, increases initially as the water films between the particles grow to bridge the pore spaces and provide a bond between the grains (Fig. 3.3A).

Ultimately, the water films thicken to such an extent that the surface tension effect is reduced; film cohesion, too, declines and the soil becomes increasingly fluid. Due to the different characteristics of shear strength within and between the aggregates, increasing moisture content produces contrasting effects on the various components of the soil. Clod shear strength, which is largely derived from molecular cohesion, declines with increasing moisture content in a similar fashion to molecular cohesion. On the other hand bulk shear strength owes more to frictional effects, which themselves do not change significantly with moisture content, and film cohesion. As a result the bulk shear strength of the soil follows a pattern

A.

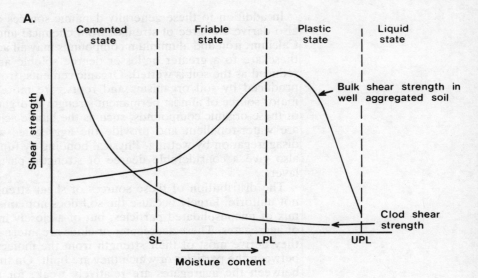

B.

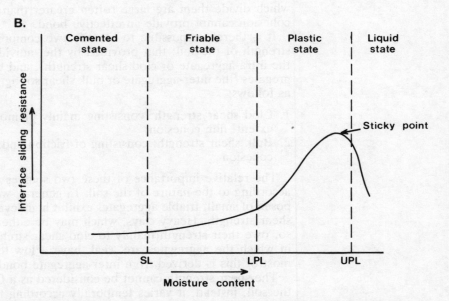

Fig. 3.3   The effect of moisture content on (a) the bulk and clod shear
strength of a structured soil, and (b) the interface sliding resistance of
a cultivation tine (after Spoor 1975).

generally similar to film cohesion (Fig. 3.3A).

It was noted earlier that failure could occur either through shearing
within the soil body or through shearing at the interface of the soil and
tillage implement. The latter depends upon the sliding resistance of the
soil and tine. This, in turn, is a function of two main components: the
frictional strength derived from roughness at the interface, and the adhesive strength provided by the soil 'sticking' to the tine. This can be expressed as follows:

$$R_s = C_a A + W \tan \theta \qquad\qquad [3.1]$$

where R_s is the sliding resistance
        C_a is the cohesion operating over
        A the surface area of contact
        $W \tan \theta$ is the coefficient of friction.

With a polished tine surface, the frictional effect is minimal, and most of the sliding resistance is derived from adhesion, which is a function of moisture content. Like the film cohesion existing within the soil, adhesion is a product of the surface tension effect; thin films of water which bridge the spaces between the soil particles and the tine face provide an adhesive bond. As has been shown, the strength of this bond initially increases as the moisture content increases and more bonds are created, but ultimately declines as the water films become so thick that the capillary attraction is negligible. At high moisture contents, therefore, the water may lubricate the interface, reducing both the adhesion and frictional resistance and allowing the tine to move freely through the soil (Fig. 3.3B).

3.3.2   Soil failure under tillage

It is clear from the preceding discussion that so many variables are involved in the response of the soil to tillage that generalisations are difficult. What is valid for one soil at one moisture content may not be true for other conditions. To illustrate the effects of moisture content, we can take as an example a well-structured, loamy soil. At low moisture contents the tendency will be for the tine to cause inter-aggregate failure (i.e. the aggregates are rearranged without being broken) as the clods slide over the tine face due to the combination of low sliding resistance, low bulk shear strength and high clod shear strength.

At higher moisture contents, with the increase in bulk shear strength and, to a lesser extent, sliding resistance, the rearrangement of the aggregates is accompanied by internal shearing. This leads to breakage of the clods. It is in this state that the soil is most workable, for the total shear strength is relatively low, and the effect of tillage is often that most desired.

Further increases in moisture content lead to a rise in bulk shear strength to a maximum, and also to an increase in the sliding resistance. Tillage under these conditions requires higher forces to be applied to the soil, due to the higher total shear strength, and thus a higher work load is involved. Moreover, the soil aggregates are relatively unstable, and subject to both destruction and compaction. Thus damage to the soil by compaction, smearing and puddling becomes a danger.

At yet higher moisture contents, the sliding resistance reaches a maximum while the soil resistance declines. It is in this state that the soil tends to stick to the tine, while tillage results in loss of structure due to the breakdown and recompaction of the aggregates. Eventually, as the soil becomes fluid the soil simply flows around the implement. No significant rearrangement of the soil is achieved by the implement, but reorientation of the particles at the interface of the tine and soil may cause smearing, while the tractor may seriously compact the soil. Essentially, the soil is in an unworkable state.

3.3.3 The significance of tine shape

The design of the implement is of fundamental importance to the tillage process, although the range of designs and methods of use make meaningful summary difficult.

The processes outlined previously are valid for most simple tine implements (e.g. harrows and chisel ploughs), but differences in the angle of incision of the tine affect the operation. At low angles (i.e. where the tine is pointing forward), the tendency is to lift the soil. This loosens and rearranges the aggregates, often bringing larger clods to the surface (Figure 3.4). At an angle of about 45° the applied forces operate horizontally in the soil, and the result is general rearrangement of the soil. At higher angles (including tines which slope backwards) the forces operate downwards into the soil. This means that most of the force is expended in breaking aggregates, since there is little scope for rearrangement of the clods; when the soil is in a plastic state this will also lead to compaction and clod formation.

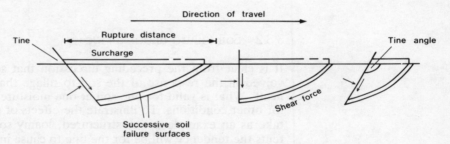

Fig. 3.4 The effect of tine angle on soil failure (after Spoor 1975).

Tine width and depth are also important. Narrow tines confine the forces to a small surface area, and thus exert a high, localised stress in the soil. Shallow working also leads to a narrow zone of disturbance. With wider tines, or with deeper penetration, the zone of disturbance is increased, and more general rearrangement and clod disintegration occurs.

As this indicates, the effects of tillage vary with the type of implement and the manner of its use (e.g. tine angle, speed of operation) as well as with the soil conditions. If the optimum structure is to be achieved, these interactions must be taken into account when selecting tillage methods (Table 3.2). By the same token, the effects of tillage on the soil depend upon soil type. As we have seen, factors such as texture, organic matter content and drainage status all determine the response of the soil to tillage, and Jones (1979) has classified soils in the west Midlands of England in terms of their suitability for tillage on this basis (Table 3.3).

3.3.4 The mould-board plough

One particularly significant variation to the general themes presented above should be considered. This is the case of the mould-board plough, possibly one of the most widely used and efficient tillage implements to have been evolved. Even today the detailed processes involved in using the mould-board plough are not clearly understood, although its main

Table 3.2   The effects of soil consistence on the operation of different tillage methods

Operation	Bulk shear strength	Clod shear strength	Soil consistency	Direction of resultant implement force on soil	Conventional implement
	Relative values				
Disintegration	High	Low	Friable	Downwards	Roll. rotary cultivator
Clod formation	Low	Low	Plastic	Downwards or sidewards	Mould-board plough
Rearrangement	Low	High	Friable	Sideways	Spike or oscillating harrow
Puddling	High	Very low	Plastic	Downwards or sideways	Disc harrow
	Very low	Very low	Liquid	Any direction	Rotary cultivation
Loosening	Low	High or low	Friable Cemented	Upwards	45° rake angle tine
Inversion	High	High or low	Friable Plastic		Mould-board plough
Mixing	Low	High or low	Friable Liquid		Rotary cultivation Disc harrow
Smoothing	Low	High or low	Friable		45° tine Leveller
Moling	High Low	Low High or low	Plastic (below) Friable (above)	Upwards	Mole plough 45° leg
Smearing	High	Low	Plastic		Slipping wheel
Cutting and movement					

Source: From Spoor 1975

function when compared to other implements, is the complete overturning of the soil and burial of the original surface material.

Attempts to identify these processes more closely were made by Nichols (1929). He showed that, contrary to established opinion, the soil disintegrates due to the operation of forces acting at right angles to the share and mould-board. The advancing plough share apparently catches material which is compressed in a direction normal to the share curve. Due to the resistance exerted by the soil mass ahead of this block of compressed soil, the soil block shears and moves up the share. Because the curve of the share is constantly increasing, however, the angle of the applied force changes as the block is raised until, as can be deduced from earlier discussions, it is acting downwards into the soil block. This leads to a pulverising effect, and the characteristic combination of granulation, lifting and overturning produced by the mould-board plough.

Table 3.3 Classification of the soils of the west Midlands of England in terms of their ease of tillage

		A horizons (0–25 cm)				E and/or B horizons (25–50 cm)				B and/or C horizons (50–100 cm)			
		C	L_{dt}	C_{at}	θ_{vt}	C	L_{dt}	C_{at}	θ_{vt}	C	L_{dt}	C_{at}	θ_{vt}
Group I	$\overline{x}$	9.3	1.44	20.4	28.6	7.3	1.51	24.8	20.6	5.5	1.55	24.9	18.4
	SD (x)	4.5	0.13	7.7	7.2	3.9	0.12	7.5	5.2	3.5	0.15	7.4	4.9
	SE $(\overline{x})$	0.9	0.03	1.7	1.5	0.8	0.03	1.6	1.1	0.8	0.03	1.6	1.1
	n	22	22	21	22	22	22	22	22	21	20	20	20
Group II	$\overline{x}$	16.8	1.40	13.8	38.7	13.9	1.60	18.0	27.1	22.7	1.79	8.9	30.9
	SD (x)	3.9	0.16	4.7	6.6	4.6	0.12	4.7	3.7	9.5	0.12	4.0	5.4
	SE $(\overline{x})$	1.1	0.04	1.3	1.8	1.3	0.03	1.4	1.0	2.9	0.04	1.2	1.6
	n	13	13	13	13	13	13	12	13	11	11	11	11
Group III	$\overline{x}$	24.9	1.34	11.5	45.4	23.4	1.59	13.7	34.1	33.0	1.79	8.0	35.2
	SD (x)	7.2	0.17	4.9	6.0	6.3	0.23	4.7	4.5	8.3	0.17	4.3	4.5
	SE $(\overline{x})$	2.4	0.06	1.6	2.0	2.1	0.07	1.6	1.6	2.9	0.06	1.6	1.7
	n	9	9	9	9	9	9	9	9	8	7	7	7
Group IV	$\overline{x}$	28.0	1.34	9.7	48.1	32.7	1.72	9.3	36.8	44.0	1.89	3.8	40.2
	SD (x)	6.8	0.16	2.5	6.4	11.2	0.16	4.7	4.9	7.8	0.16	2.9	5.2
	SE $(\overline{x})$	1.6	0.04	0.6	1.5	2.6	0.04	1.1	1.1	1.9	0.04	0.8	1.3
	n	19	19	19	19	19	19	19	19	16	16	16	16
Group V	$\overline{x}$	38.3	1.38	10.2	49.4	47.7	1.70	7.7	42.8	55.7	1.80	4.2	44.7
	SD (x)	7.3	0.14	2.4	5.4	11.7	0.16	3.6	6.1	11.8	0.16	3.4	7.3
	SE $(\overline{x})$	2.2	0.04	0.7	1.6	3.5	0.05	1.1	1.8	3.7	0.05	1.1	2.3
	n	11	11	11	11	11	11	11	11	10	10	10	10
Group VI	$\overline{x}$	36.4	1.14	11.1	55.4	39.5	1.66	9.1	41.5	32.0	1.68	6.3	41.2
	SD (x)	11.7	0.15	3.2	3.3	18.5	0.10	1.9	5.0	5.0	0.16	3.8	5.1
	SE $(\overline{x})$	4.8	0.06	1.3	1.3	7.6	0.4	0.8	2.0	2.5	0.08	1.9	2.5
	n	6	6	6	6	6	6	6	6	4	4	4	4

Physical properties
C   clay (gravimetric %)
L_{dt} packing density (gcm^{-3})
C_{at} air capacity (vol. %)
θ_{vt} retained water capacity (vol. %)

$\overline{x}$ Mean
SD (x) Standard deviation of population x
SE $(\overline{x})$ Standard error of mean x
n No. of samples

Source: From Jones 1979

3.4   LONG-TERM EFFECTS UPON THE SOIL

The changes that take place in the soil during tillage arise from both the effect of the tillage implement and the tractor that is pulling it. In general, while the tillage implement is loosening the soil, the tractor is causing compression. In neither case, however, is the effect permanent. Following tillage, the soil structure tends to readjust back towards its original state as rainfall, the burrowing of worms, pressures exerted by roots, the action of frost and wetting-and-drying combine to cause changes in the arrangement of the particles. When recuperation of the soil structure is complete, and the soil returns to its original state before the next tillage operation, there is unlikely to be any cumulative effect of tillage upon soil conditions. When these readjustments are incomplete, however, repeated tillage may result in a progressive change in soil structure. In some cases this leads to long-term damage to the soil (Greenland 1977). Moreover, the change in structural conditions has implications for other soil properties, such as the chemical status and the organic activity. In the long term these effects may significantly influence crop yields (see Table 3.1, p. 44).

3.4.1   Bulk density and aeration

Most tillage operations such as ploughing and harrowing result in some
rearrangement of the aggregates, which causes an opening up of pore
spaces and fissures. As a result, the ploughed layer of regularly tilled soils

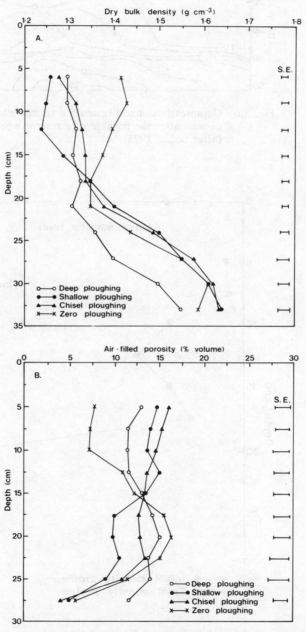

Fig. 3.5   Bulk density and air-filled porosity with depth in soils under different
tillage treatments (a) dry bulk density during the growing season for
barley (results = average of eleven sampling occasions, 1968–70) (b)
air-filled porosity at mid-growth stage of barley (from Soane 1975).

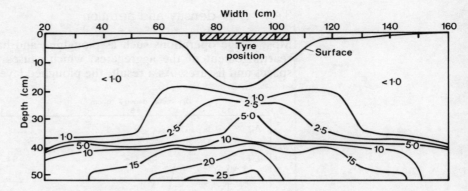

Fig. 3.6 Compression of soil beneath a tractor wheel. Cone resistance (bars) of soils after the passage of a tractor wheel with a conventional tyre (after Soane 1973).

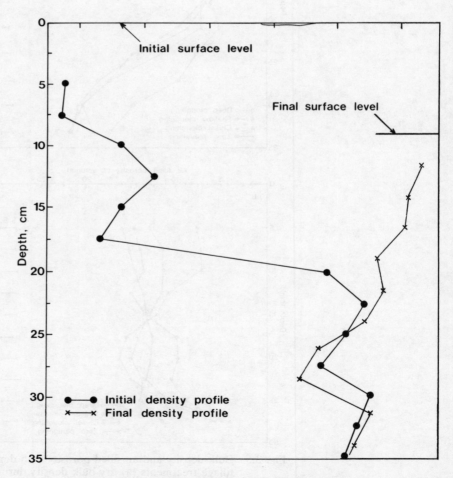

Fig. 3.7 Changes in bulk density with depth following the passage of a tractor on loose soil (from Soane 1975).

tends to be more porous, looser and less dense than that of untilled soils. We will consider these effects more fully when we examine the influence of direct drilling on the soil (section 3.5.2). As the data in Fig. 3.5 indicate, however, reducing the intensity of tillage (for example by using shallow chisel ploughing rather than deep mould-board ploughing) results in both higher bulk densities and reduced air porosities.

There are, however, exceptions, and in some cases bulk densities of tilled soils increase significantly. One possible reason for this is that tillage under unsuitable soil moisture conditions may lead to marked compaction of the surface soil. This is due mainly to the compressive effect of the tractor and implement wheels (Figs. 3.6 and 3.7). When it is realised that conventional tillage processes may require as many as six or seven separate passes over the land with different implements, it is apparent that a large proportion of the field area may be affected by wheel-compaction; indeed Soane (1975) has shown that up to 90 per cent of the surface may be crossed by 'wheelings' (Fig. 3.8) during a single year.

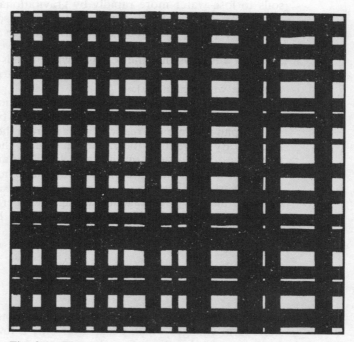

Fig. 3.8 The extent of field coverage by vehicle wheels during a typical annual sequence of cultivation operations (from Soane 1975).

While tillage commonly loosens the upper layers of the soil, at depth the reverse may occur, and bulk densities commonly increase in the lower part of the plough layer. In part this is due, again, to compression by vehicle wheels, but in addition it may result from the development of a plough-pan. Repeated tillage, at a similar depth, especially in unfavourable conditions or with a blunt ploughshare, can cause smearing and re-alignment of the soil particles, creating a dense, platy layer. This inhibits water and air movement and prevents root extension. In wet periods plants may suffer from waterlogging in the root zone; in dry periods the shallow depth of the roots may mean that they are unable to exploit water reserves deeper in the soil and may suffer from moisture deficiencies. As a consequence, yields of crops may be significantly reduced (Adams *et al.* 1960; ASAE 1971).

3.4.2   Infiltration capacity and water retention

The structural changes wrought by tillage also affect the hydrological properties of the soil in a variety of ways. Increased surface roughness following primary tillage operations leads to greater surface storage of rainfall following a storm, and hence reduces the probability of overland flow (e.g. Reid, 1979). It also enables infiltration to continue for longer. The loosening influence of tillage similarly encourages infiltration by providing a surface zone containing larger pores which can accept and transport water more rapidly.

Related to these changes in water acceptance, is a tendency for tilled soils to lose water more rapidly by gravitational drainage, at least from the upper layers. This has two important consequences. First, it means that there is less likelihood of waterlogging in the surface horizon, and the field capacity of the soil occurs at a relatively low moisture content. Second, it follows that the soil as a whole tends to retain less water after rainfall, and thus has a smaller capacity to store water (i.e. a reduced available water capacity).

Again it is necessary to emphasise the exceptions to this general pattern which can arise under conditions of untimely tillage. Development of a plough-pan may greatly reduce drainage from the topsoil and cause waterlogging. By preventing the rapid removal of the gravitational water this may have repercussions on the rate of infiltration; during prolonged rainfall, saturation of the topsoil may occur due to impeded drainage and saturated overland flow may take place. Paradoxically, this impedance to drainage may be advantageous during drier periods for it encourages retention of plant-available water (i.e. moisture held at tensions of about 0.05 to 15 bars).

3.4.3   Organic matter

The changes in physical aspects of the soils under tillage are essentially short-term changes which occur after only a few tillage operations, or a few years of cultivation. Over longer periods more fundamental, and less readily reversible changes may occur in the soil. One of these effects is seen in the organic matter content. As Clement and Williams (1964), Jenkinson and Johnston (1977) and Russell (1977) have shown, tillage of grassland soils leads to a slow but persistent decline in the content of organic matter. Compared to organic carbon contents of 5–10 per cent in pasture soils, arable soils often contain no more than 1–2 per cent (Fig. 3.9).

Again it is necessary to treat these differences with caution, for not all of the changes can be attributed to the direct action of tillage. Loss of organic matter content under arable cultivation may be a product of removal of organic residues in the harvest rather than of the direct influence of tillage. Data on the quantities of organic matter returned to the soil

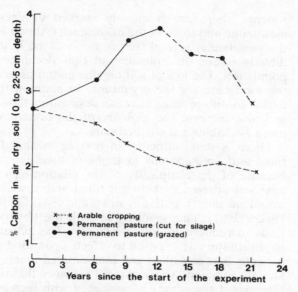

Fig. 3.9 Changes in the organic carbon content of arable and pasture soils in the Highfield ley-arable experiment, Rothamsted, 1949–72 (after Johnston 1973).

by different crops are difficult to obtain, for root decay may be occurring throughout the year, but it is clear that the weight of roots in cereal crops is often less than that under grass (see section 8.3.5). The reduction in organic carbon under continuous arable cultivation is therefore attributable mainly to the lower returns of plant material from cereal or root crops compared to grass (see Ch. 8).

Nevertheless, there is evidence that tillage does help to diminish the contents of organic matter. The improved aeration of tilled soil seems to encourage chemical oxidation of the organic residues which are not replenished by annual returns from the crop. To some extent this may seem surprising; as we will see (section 3.4.4) tillage also causes a decline in soil organisms and this might be expected to result in a slower rate of decomposition. It is mainly the mesofauna which are affected by tillage, however, and their job of breaking down the larger plant residues is probably filled, at least in part, by tillage processes such as discing and ploughing of the stubble. Thus micro-organic decomposition is able to proceed and the improved accessibility provided by tillage allows a greater proportion of the organic matter to be attacked. As a consequence, the organic content of tilled soils declines and the age of the residual material increases (Jenkinson 1968). The speed at which these changes occur and the rapidity of the decline in organic matter contents attest to the fact that they are not merely a product of reduced inputs, but are a function of accelerated decomposition (Kononova 1966).

3.4.4 Soil organisms

Many of the organisms which inhabit the soil are severely affected by tillage. Apart from the obvious effects upon the larger animals, such as the rodents, there are frequently marked declines in the activity of the mesofauna and some of the microfauna (Abbott *et al*. 1979). In the case of the mesofauna, several factors are involved in these effects. Tillage may directly injure the animals and thus decimate (almost literally!) their population. The loss of soil organic matter associated with tillage reduces the food base for the organisms. In addition the structural changes induced by tillage may increase desiccation through improving drainage, and also increase the mobility of the organisms, making some species more available to their predators.

There is some difficulty in making meaningful comparisons between tilled and untilled soils in terms of their faunal populations, however, because of the complexity of the relationships. Numerous studies have analysed differences between tilled arable soils and untilled pasture or woodland soils (e.g. Evans and Guild 1947; Guild 1951, 1952; Abbott and Parker 1981); almost all agree with the obvious conclusion that tillage leads to reduced biological activity. The extent to which this is due to physical injury as opposed to effects upon food supply is not clear. Comparisons between tilled and long-untilled arable soils or ley pasture soils are less common. There is some evidence that tillage of pastureland (and subsequent reseeding) is associated with increased populations of many organisms, but this is more likely to be a product of the improved grass sward and the establishment of more nutritious organic products, than of changes in aeration or moisture conditions afforded by tillage (Ch. 7). Maltby (1975), for example, in comparing reclaimed and unreclaimed moorlands, noted a major increase in fungi and yeasts in the reclaimed soils. He attributed this to the establishment of vegetation which decomposed more readily than the sward of the unreclaimed pasture. More recently comparisons of earthworms on tilled and untilled arable soils showed the surprising fact that populations in both increased over time, although the increase was more marked where no tillage was carried out (Gerard and Hay 1979). Comparisons of earthworm numbers in direct drilled and conventionally ploughed soils on experimental plots in Devon, England, are shown in Table 3.4.

Table 3.4 Soil conditions under direct drilled and ploughed plots at Bigbury, Devon

	Ploughed		Direct drilled	
	$\bar{x}$	SD	$\bar{x}$	SD
Earthworms (no. m^{-3})	208.3	372.2	637.3	400.0
Infiltration capacity (l h^{-1})	5.9	3.3	2.8	3.4
Cone index	87.3	17.9	102.3	21.8

Source: Field survey by D. J. Briggs, 1982

3.4.5 Structural stability

Changes in the organic matter content and microbial activity combine to affect the structural stability of the soil. It has been mentioned that aggregation of soil particles is achieved largely through the action of cohesion and cementation, and in both cases organic matter plays a major part. Organic colloids, for example, have a high surface charge that enables them to retain cations through adsorption, and thus they encourage cation-bonding. Humic acids and polysaccharides, produced either by organic decomposition or by excretion from living organisms, are also important cementing agents. They form water-resistant films around the aggregates which provide protection from disruption by wetting.

Destruction of organic compounds by tillage, therefore, reduces soil structural stability. Low (1955), for example, comparing structural stability under permanent grassland and arable cropping, noted a decline in stability from 73 to 35 per cent after only a single year of arable cultivation. Similarly, Dettman and Emerson (1959) found structural stability levels of as low as 5 per cent in unmanured arable land, compared with 50 per cent in land that had been down to grass for only 4 years. Many other studies have shown similar patterns (see, for example, Low *et al.* 1963; Tanchandrphongs and Davidson 1970; Cooke and Williams 1971; Williams 1978).

The implications of this loss in stability are many. One inevitable consequence is that the soil becomes more prone to capping and crusting under the impact of rainfall (MacIntyre 1958), and this may lead to reduced seedling emergence and yield (Russell 1971a; Low 1973). Additionally, the reduced stability makes the soil susceptible to erosion by both wind and water (Chepil 1953, 1954, 1955; Johnson *et al.* 1979); indeed, structural stability is one of the main controls upon soil erodibility, as recognised in the soil loss equations developed for wind (Woodruff and Siddoway 1965) and rainfall (Smith and Wischmeier 1962).

More immediate are the effects upon the ease of tillage. Reduced stability of the aggregates is reflected by lower molecular cohesion and, as a consequence, the aggregates may be liable to shearing and pulverisation even at low moisture contents.

Structural deterioration as a result of excessive tillage therefore triggers off positive feedback processes that make the soil progressively more prone to damage and progressively more difficult to cultivate (Fig. 3.10).

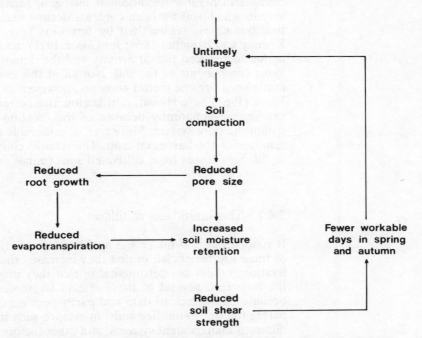

Fig. 3.10 Feedback effects of soil compaction due to untimely tillage.

In soils of low structural stability, timeliness of cultivation is consequently of paramount importance if a downward spiral in the structural conditions of the soil is to be avoided.

3.4.6   Chemical properties

Less attention has been given to the consequences of tillage for soil chemical properties, but effects undoubtedly occur. The improved aeration of soils encourages oxidation and possibly results in the conversion of compounds to less soluble, oxidised forms (for example, Fe^{2+} to Fe^{3+}). On the other hand, the better drainage may facilitate leaching by speeding up rates of water movement through the soil. We will discuss these aspects more fully in Chapter 4. More permanently, the loss of organic compounds leads to reduced cation retention and a diminished cation exchange capacity (Kulkarni and Savant 1977). This may lead to a marked reduction in the quantity of plant-available nutrients in the soil. Williams and Lipsett (1961), for example, comparing soils under pasture and arable rotations, noted a 30 per cent decline in organic matter content after 20 rotations, and an associated fall of 30 per cent in nitrogen and sulphur contents, a 14 per cent reduction in sodium, calcium, magnesium and potassium, and a 17 per cent decline in phosphorus content. They argued that this could not be attributed entirely to nutrient removal in the harvested crop – no more than a third of the magnesium lost could be accounted for in this way, for example – and suggested that it was partly a product of the falling nutrient reserves in, and cation retention on, the organic residues.

As this indicates, one of the nutrients most severely affected by these changes in organic conditions is nitrogen. Many studies have shown the way in which soil nitrogen contents decline under continuous arable cultivation, unless replenished by fertilisers (e.g. Salter and Green 1933; Keeney and Bremner 1964; Jenkinson 1971), and undoubtedly part of this is due to reduced faunal activity and the diminished retention in the organic components of the soil. Not all of this change is a result simply of diminished organic matter contents, however; in addition, as Whisler and Klute (1965) have shown, nitrification may be inhibited by compaction by machinery, apparently because of the creation of anaerobic conditions within the pore system. Moreover, considerable quantities of nitrogen are removed in the harvested crop. The relative contributions of these effects to nitrogen losses from cultivated soils cannot easily be assessed.

3.4.7   The timeliness of tillage

It is apparent that tillage has a wide range of effects upon the soil. Some of these are beneficial, in that they increase crop yields and facilitate cultivation; others are detrimental in that they may reduce soil fertility. In the long run, several of these effects are somewhat tendentious, partly because of the lack of data and partly because of the difficulties of comparing tilled with untilled soils; in essence such studies are comparing two different management systems, and other factors such as harvesting methods, sowing dates, fertiliser practices and crop types are rarely constant.

Nevertheless, it is clear that many of the detrimental effects of tillage arise mainly when the soil is cultivated under unsuitable conditions (i.e. when it is too wet). This has led to the concept of the timeliness of tillage. It is widely argued that many of the problems of tillage could be avoided if tillage operations were timed to coincide with optimum soil conditions.

Timeliness of tillage is an accepted aim of most farmers, but it is apparent that in practical terms it is an objective that is often difficult to attain. As the Strutt Report (MAFF 1970a) observed, there are often too few 'workable days', especially in wet seasons, and the farmer is then faced with a serious conflict of policy. He may need to cultivate large areas of land in time to sow the crops at their optimum time; delay in sowing may significantly reduce yields (e.g. Smith 1972). He has available only limited labour and machinery. Rather than defer cultivation, on the off chance that conditions may improve and with the serious risk of hinder-

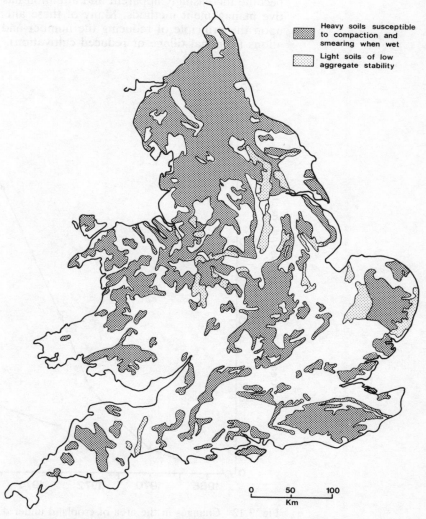

Heavy soils susceptible to compaction and smearing when wet

Light soils of low aggregate stability

Fig. 3.11 Distribution of soils susceptible to structural damage in England and Wales.

ing sowing, he may well choose to carry out tillage when the soil is wet. Simply, he may not have the opportunity to carry out 'timely' tillage.

These problems are most acute on heavy soils and in wetter areas, for there the number of workable days is most restricted. Difficulties are also increased on large, highly mechanised farms, for then large areas of land must be cultivated within a short period of time, and the flexibility to adjust cropping plans is often limited. As a result, problems of soil damage in Britain tend to be most widespread in the intensive arable areas of the clay lowlands (Fig. 3.11).

3.5 DIRECT DRILLING

In recent decades, the possible adverse effects of conventional tillage have become increasingly apparent and attention has been devoted to alternative management methods. Many of these alternatives have been based upon the principle of reducing the number and intensity of tillage operations (minimal tillage or reduced cultivation). In its extreme form, this

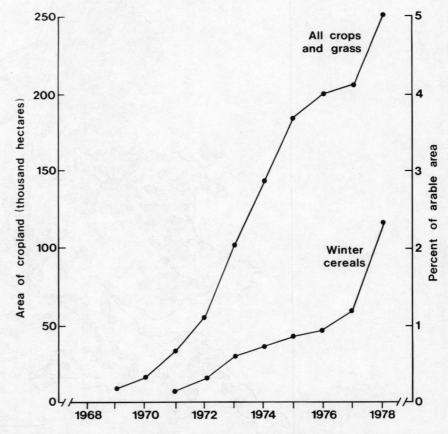

Fig. 3.12 Changes in the area of cropland under direct drilling in the United Kingdom, 1968–78 (after Cannell 1981).

involves no seedbed preparation at all, and crops are sown into the untilled soil by a machine that cuts a narrow seed-slot. Crop residues are
normally allowed to decompose *in situ*, herbicides are used to control
weeds and pests, and rooting and drainage conditions are maintained by
encouraging earthworm activity. This system is known as direct drilling
in Britain, or as zero tillage in the USA.

The concept of direct drilling originated in North America, largely in
semi-arid areas where it was seen as a means of conserving soil and moisture and reducing planting costs (Baeumer and Bakermans 1973). By not
using tillage, it was possible to leave a mulch of crop residues on the surface to protect the soil from erosion and to reduce evapotranspiration.
During the 1960s use of direct drilling and other methods of minimal cultivation was extended into the humid temperate areas, partly in an attempt to avoid structural problems in unstable soils and partly to reduce
energy costs. Between 1969 and 1978, for example, the area of direct
drilled crops in Britain increased from less than 10,000 ha to almost
250,000 ha (Fig. 3.12), whilst in a survey of Warwickshire, Phillips (1978)
found that 81.5 per cent of farmers had used a system of minimal cultivation. In the USA, the area of direct drilling has risen less rapidly, but
use of minimal cultivation increased more than threefold between 1972
and 1980 (Fig. 3.13).

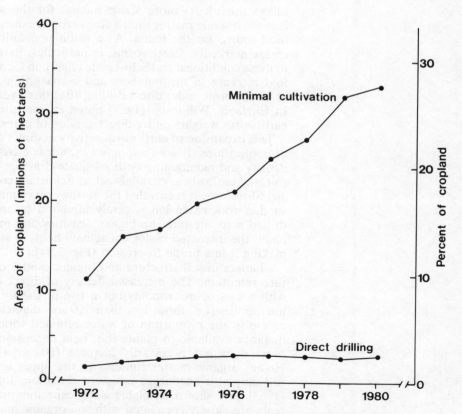

Fig. 3.13 Changes in the area of cropland under direct drilling and minimal
cultivation in the USA, 1972–80 (after Cannell 1981).

3.5.1   Effects on the soil

With this increase in the use of direct drilling, there has come a growing interest in the effects of these methods upon soil conditions. Numerous studies have been carried out comparing direct drilling with various types of conventional tillage, and it is clear that structural, organic and chemical properties may all be modified. Many of these changes, of course, are the reverse of those seen in conventional tillage systems. As can be anticipated, therefore, bulk densities tend to increase in the topsoil under zero tillage, while air porosities and permeability decline due to the lack of loosening by ploughing (Fig. 3.5, page 55). Baeumer and Bakermans (1973), for example, quote average reductions in porosity of 0–6 per cent (volume for volume) under direct drilling, while Soane et al. (1975) show increases in bulk densities from about 1.2 to 1.4 g cm^{-3} after direct drilling of barley for 6 years. Associated with this increased density there occurs a marked rise in the mechanical resistance of the soil (as measured by a cone penetrometer). Soane et al. (1975) recorded a fivefold increase in penetrometer resistance.

At greater depth this pattern is often reversed and bulk densities in the subsoil fall under direct drilling (see Fig. 3.5) partly because of the lack of mechanical compaction by wheels or smearing by ploughs, and partly because of increased organic activity under direct drilling. The lack of tillage provides a more stable habitat for the soil organisms, while increased organic matter inputs from crop residues supply a more constant food source for the fauna. As a result populations of soil organisms increase markedly. Earthworms, in particular, have been found to respond to these conditions, and Schwerdtle (1969), in Germany, recorded a twelvefold increase in the numbers and a sixteenfold increase in the weights of earthworms under direct drilling after three years' cropping with maize. In England, Wilkinson (1967) noted an increase of over 50 per cent in earthworm weights under direct drilling of cereals (see also Table 3.4).

This expansion of earthworm activity in direct drilled soils has a number of implications. It goes a long way towards offsetting the rise in soil bulk density and maintaining both adequate drainage and aeration in the topsoil; the earthworm channels act as important conduits for rainwater and air. Moreover, it seems that the earthworm channels also facilitate rooting so that root extension is rarely inhibited to any great extent in direct drilled soils, despite the higher densities and mechanical resistance. Finally, the increased biological activity helps to stabilise the soil structure, making it less prone to erosion (Fig. 3.14).

Changes in soil structure and organic matter contents also affect moisture retention. The increased density of direct drilled soils is associated with a loss of macropores but a rise in the total volume of meso- and micropores (i.e. those less than 100 μm diameter). This leads to an increase in the proportion of water retained within the soil, although the quantity available to plants (i.e. held at tensions from about 0.05 to 15 bars) does not necessarily increase (Baeumer and Bakermans 1973). Higher organic matter contents in the upper layers of the direct drilled soil may also encourage moisture retention, for as Salter and Williams (1963) have shown, available water capacities and retained moisture contents are closely correlated with soil organic matter contents.

In turn, the adjustments in soil structure, organic matter content and moisture content may affect soil temperature under direct drilling. Untilled soils are generally darker than tilled soils, so they absorb greater

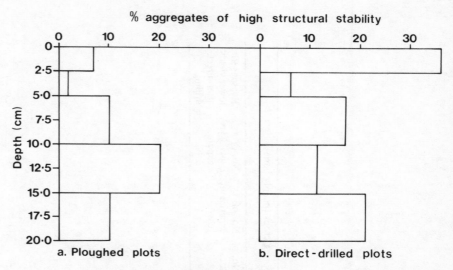

Fig. 3.14 Structural stability of sandy loam soil after two years of conventional ploughing or direct drilling (data from Cannell and Finney 1975).

proportions of incoming energy. The higher bulk density also means that their thermal conductivity is greater, so heat is transmitted more readily to lower layers. On the other hand, the greater moisture content increases the heat capacity of direct drilled soils, more energy being required to heat the water than the soil particles. As a consequence, soil temperatures under direct drilling tend to be lower in the topsoil during the summer, but slightly warmer at depth. Conversely, in the winter, surface temperatures may be marginally higher, while direct drilled soils tend to warm up rather more slowly than tilled soils during the spring (Hay 1977; Hay *et al.* 1978). This may delay germination and seedling emergence.

3.5.2 Effects on crop yields

It might be anticipated that the combined effects of these changes in soil conditions under direct drilling would be to increase soil fertility and raise crop yields. In reality this is far from generally the case. Davies and Cannell (1975) analysed results from direct drilling experiments in the UK prior to 1974 and found that yields under direct drilling were on average from 0–13 per cent lower than under conventional tillage, depending upon the crop, the previous cropping history and the time when the experiment was carried out. They attributed this partly to inexperience in the application of direct drilling methods. Since then Cannell *et al.* (1979) have showed that in more recent experiments yields were generally comparable (Table 3.5). In particular, it seems that problems with weeds (especially *Agropyron repens*) and pests (especially slugs) prevented the establishment of an optimum plant population under direct drilling, and it is now appreciated that these need to be controlled by more effective use of herbicides and insecticides. Similarly, it has become apparent that direct drilled crops respond to higher levels of nitrogen fertiliser than do crops grown under conventional tillage (Bakermans and de Wit 1970; Davies

Table 3.5 Yields of arable crops in direct drilling experiments

Dominant soil group	Proportion of cereal growing areas		Winter crops (mainly wheat)				Spring crops (mainly barley)				
	England and Wales	Scotland	DD yield % of ploughing	No. of sites	No. of experiment years	Reasons for lower yields after direct drilling	DD yield % of ploughing	No. of sites	No. of experiment years	No. of expts 3 yrs or more	Reasons for lower yields after direct drilling
Brown alluvial and brown calcareous alluvial soils	1.0	1.0									
Alluvial gley soils	5.0	2.0	93	2	5	Surface waterlogging, compaction					
Humic alluvial gley soils	2.0	2.0	103	2	3						
Earthy peat soils	2.0	2.0									
Rendzinas	11.0		100	8	13	Couch-grasses, N level too low	99	5	14	3	Couch-grasses
Brown sands	4.0	15.0	95	1	2		81	3	6	1	Compaction, restricted rooting
Brown calcareous earths	4.0	4.0	98	2	8						
Brown earths	10.0	22.0	100	2	7		96 (98)	4 / 4	20 / 19	3 / 3	Drill problems (Excluding result with drill problem)

Soil series	(%)	(%)	No.	Yield (% of ploughed)	No.	Main limitations (experiments)	No.	Yield (% of ploughed)	No.	No.	No.	Main limitations
Argillic brown earths	9.0		3	103	7	Surface waterlogging	95		4	11	1	Surface waterlogging, restricted rooting, low plant population, N level too low
Paleo-argillic brown earths	4.0		3 3	94 (97)	9	Crop residues *Excluding results where crop residues were not removed*						
Brown podzolic soils	22.0											
Gley podzols	1.0											
Calcareous pelosols	8.0			96 (100)	31 29	Couch-grasses, annual grass weeds compaction wetness *Excluding 1 unsuitable site with severe grass weeds*	93		4	12	3	Compaction, wetness crop residues
Sandy gley soils	1.5	2.0										
Cambic gley soils	0.5	4.0										
Argillic gley soils	0.5	0.5										
Stagnogley soils	36.5	30.0	17	101	44	Compaction, waterlogging, black-grass	91		8	22	3	Waterlogging, grass weeds, restricted rooting
Total			49		129				28	85	14	

Source: From Cannell *et al.* 1979

and Cannell 1975). This is apparently because more of the soil nitrogen is bound up within the crop residues and soil fauna, while mineralisation is inhibited in the relatively compact direct drilled soils (Dowdell and Cannell 1975). Thus to optimise yields under direct drilling, high rates of fertiliser application are required.

Numerous other factors influence the performance of direct drilled crops. From the results of experiments to date, it seems that winter-sown crops (e.g. winter wheat) perform better than spring-sown crops (e.g. spring barley), possibly because direct drilling in spring is often delayed by wet soil conditions or because spring drilled crops experience moisture stress in autumn (Hodgson *et al*. 1977; Ellis *et al*. 1979). Previous cropping practice is also important, and in many of the early experiments it seems that yields of direct drilled barley crops were reduced by competition from grass weeds inherited from the preceding pasture crop. Finally, of course, the efficiency of the drilling operation itself may affect yields and in several experiments poor performance of direct drilled crops can be attributed to drilling under unsuitable soil conditions (Davies and Cannell 1975).

3.5.3   Suitability of soils for direct drilling

As experience has accumulated from different areas, it has become apparent that direct drilling is not equally successful in all areas. Numerous factors influence the suitability of the land for direct drilling, including soil conditions, climate and topography. One of the main soil factors is drainage. Direct drilling in wet soils may cause smearing of the seed-slot, with the result that root development of the seedling is restricted and growth retarded. In addition, slow surface drainage may allow the seed-slot to become water logged so that seeds are damaged before germination. As a result, yields on gley soils are often lower under direct drilling than under conventional tillage techniques. Conversely, excessive drainage may also limit yields, for leaching of nitrates may cause severe N deficiencies in direct drilled crops.

Soil structure and texture impose similar constraints upon direct drilling. Smearing, lack of seed-slot closure and inadequate coverage of the seeds may all reduce the rate of germination in poorly structured or heavy-textured soils. Moreover, surface compaction results in the need for higher power rates to draw the drill through the soil so that wheel-slippage and uneven sowing rates may create serious problems. Consequently, the performance of direct drilled crops is often poor relative to those grown under conventional tillage systems on clay soils (Table 3.5).

The main climatic factor that determines land suitability for direct drilling is rainfall. High rainfall areas are characterised by a greater risk of soil damage during drilling and by problems of waterlogging within the seed-slot. High rainfall, particularly in the autumn, may also delay drilling and prevent the sowing of winter cereals, while pests (such as slugs) and weeds are encouraged in wet conditions. The acreage of land under direct drilling therefore shows a close correlation with the amount of autumn rainfall (Cannell *et al*. 1979). As a result, the distribution of land suitable for direct drilling in Britain shows a marked concentration in the south and east (Fig. 3.15).

In general, direct drilling is possible in a wide variety of topographic

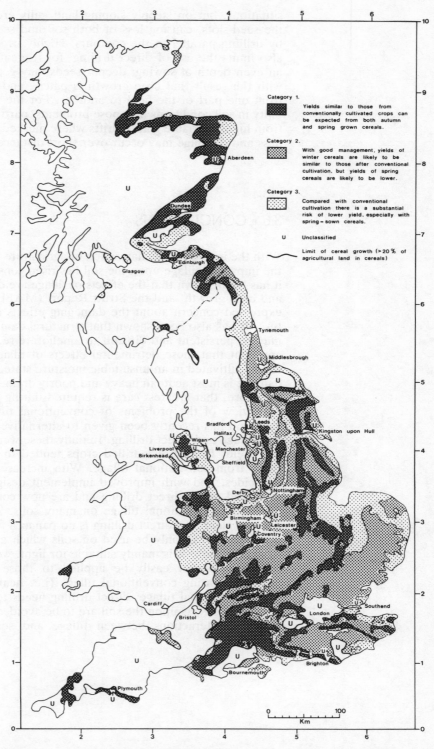

Category 1.

Yields similar to those from conventionally cultivated crops can be expected from both autumn and spring grown cereals.

Category 2.

With good management, yields of winter cereals are likely to be similar to those after conventional cultivation, but yields of spring cereals are likely to be lower.

Category 3.

Compared with conventional cultivation there is a substantial risk of lower yield, especially with spring - sown cereals.

U   Unclassified

Limit of cereal growth (> 20 % of agricultural land in cereals)

Fig. 3.15   The suitability of soils for direct drilling in Great Britain (after Cannell *et al.* 1979).

situations, but on steeply sloping land gully erosion may be initiated in the seed-slots, causing loss of both soil and seeds. This can be avoided by drilling parallel to the contours. Highly irregular surface topography also limits the use of direct drilling, for it creates difficulties in ensuring an even depth of sowing; deeper seeds may germinate and emerge later with the result that crop growth is patchy and ripening dates may vary from one part of the field to another. For the same reason, local variability in soil conditions may pose problems, particularly on soils developed from highly variable glacial drifts where marked changes in texture, stoniness and drainage may occur over short distances.

3.6   CONCLUSIONS

With the increased mechanisation of agriculture over the last few decades, the impact of tillage upon the soil has risen considerably. In several cases it has been shown that the effects of tillage are detrimental to soil fertility and crop growth, and the Strutt Report (MAFF 1970a) for example, has expressed concern about the damaging effects of tillage upon soil structure. It has also been shown that structural damage by tillage implements may be persistent and difficult to ameliorate (e.g. Greenland 1977). It is apparent that these detrimental effects of tillage arise mainly when the soil is cultivated in an unsuitable moisture state, and as a result structural damage is most acute in heavy and poorly drained soils. It is in these soils, therefore, that greatest care is required during tillage.

In view of the problems of conventional tillage practices, increasing attention has recently been given to alternative techniques such as minimal tillage and direct drilling. Initially these were of limited applicability, and yields from direct drilled crops tended to be significantly less than those from conventional tillage. With increased use of N fertiliser and pesticides, and with improved implement design, however, yields have increased under direct drilling and are now comparable with those produced by conventional tillage on many soils. At the same time, it has become clear that direct drilling is no panacea for the problems of tillage practices. It can only be used on soils which are already in good structural condition. It is mainly suitable for light, well-drained soils, and consequently cannot easily be applied to those soils that give greatest problems during conventional tillage (i.e. heavy clay soils). Moreover, like conventional tillage, direct drilling needs to be employed with care. If adverse effects upon the soil are to be avoided, therefore, it is essential that the interactions between tillage and soil conditions are clearly understood.

4 DRAINAGE AND IRRIGATION

4.1 THE WATER NEEDS OF PLANTS

Water availability represents one of the main controls upon crop growth. Plants require water for two main reasons: to maintain turgor and to supply nutrients. Water deficiencies lead, under extreme conditions, to loss of turgidity and wilting of the plant; under less extreme conditions, growth is inhibited by lack of nutrients and by reduced assimilation of energy. Excesses of water may also occur, however. The plants obtain most of their water from the soil, and here the water competes with air for a place in the pore spaces. Where water is present in abundance, air may be excluded from the pores and plants may suffer from lack of oxygen and related problems of anaerobis.

Water is extracted from the soil mainly as a result of transpiration. This is a process controlled partly by physiological factors, such as leaf area and the size, density and orientation of the stomata; it is also in large part a climatological phenomenon. The rate of *potential transpiration* is a function of atmospheric conditions (wind speed, humidity and turbulence) and energy inputs from solar radiation. Transpiration results in the evaporation of water from the stomata in the plant leaves, and this creates a disequilibrium between the moisture status in the leaf and that in the lower parts of the plant. Water therefore moves through a process of mass flow towards the leaves. The effect is thereby transmitted through the membranes of the roots into the soil and a related movement of water occurs through the pores surrounding the root.

There are various factors limiting the uptake of water by the plant in response to transpiration losses. Resistances occur within the plant stem and roots; the permeability of the root membranes may exert a considerable limitation upon the rate at which water enters the plant, whilst the length and size of the transmitting pores within the plant may be significant. The volume of roots is equally important because this determines the area of contact between the plant and the soil water. Plants with a more extensive root system can take up water more readily than plants with a restricted system.

The rate of water supply is also governed by soil factors. In particular, as the soil dries out and the water films within the soil become discontinuous, the rate at which mass flow takes place through the pore system to the root declines. Ultimately, the demand for water by the plant ceases to be met by the soil. When this happens, there occurs a phase during which the transpiration rate exceeds the rate of water uptake, and the plant experiences a net loss of moisture. The leaves start to dry out and the stomata close up. This regulates transpiration to some extent, but it does not prevent it altogether; prolonged drought therefore leads to continued loss of turgor, reduced nutrient inputs and diminished growth. In agricultural crops one of the consequences is loss of yield. Ultimately, if water is not supplied to the plant in sufficient quantities by renewed rainfall or irrigation, the plant wilts and dies.

The effect of water deficiencies upon crop growth and yield depend very much upon the timing and duration of drought. Plants are at their most susceptible during flower formation and fertilisation. Lack of moisture may also be important during the germination period, since seedling emergence may be delayed and thus the duration of the growing season curtailed. At these critical times relatively brief periods of water shortage may seriously reduce yield. Throughout the rest of the growth period, it is the intensity and duration of the drought which is more critical. As has been shown, the rate of transpiration falls when water is scarce, and this leads to reduced nutrient uptake. Thus, prolonged water deficiencies may cause an overall reduction in dry matter production by the plant.

Except at extreme conditions when permanent wilting occurs, the critical effect of moisture stress tends to be the lack of nutrients; thus plants which are supplied with an abundance of, in particular, nitrogen are able to withstand the effects of drought better than those receiving smaller quantities of nutrients. The application of fertilisers acts to some extent to counteract the effects of water shortage, therefore, though this is only effective if the fertilisers are supplied to the moist zone of the soil, from which the plants are obtaining their water.

4.2   THE OXYGEN NEEDS OF PLANTS

As has been mentioned, plants require oxygen as well as water. They obtain their oxygen from two main sources: from the soil air and from the open atmosphere. Oxygen supply through the leaves and thence down through the plant to the terminal oxidases in the roots is sufficient to maintain growth in plants adapted to aquatic conditions (e.g. rice), and to support at least the upper 2 cm of roots in many cereal seedlings (Jensen *et al.* 1964; Greenwood and Goodman 1971). Nevertheless, it is rarely adequate to satisfy requirements in more active and mature arable crops. Consequently, conditions affecting the supply of oxygen from the soil air are critical.

Movement of oxygen through the pore system of the soil to the plant roots is only indirectly a function of the size of the pores. In general, diffusion rates approximate to the transport equation:

$$\rho \, \frac{dc}{dt} = D_x \, \frac{d^2c}{dx^2} + D_y \, \frac{d^2c}{dy^2} + D_z \, \frac{d^2c}{dz^2} \pm S$$

[4.1]

where c is the concentration of oxygen at time t

x, y and z are the spatial coordinates of the soil

ρ is the porosity;
D is the effective coefficient of oxygen diffusion, about 1×10^{-2} cm^2 s^{-1} (Russell 1973)
S is the rate of respiration per unit volume of soil.

In air-filled pores, oxygen diffusion is therefore rapid and oxygen deficiencies are rare. In saturated pores, however, D is much lower, possibly only 1/1000th or less of the rate in free air. As a result, oxygen diffusion is unable to sustain root or microbial requirements for any length of time. In the absence of sufficient oxygen, substances such as alcohol and cyanide may be formed in the plant tissues, and plant growth severely curtailed (Rose 1968; Smith and Scott Russell 1969).

4.3 AGRICULTURAL DRAINAGE SYSTEMS

4.3.1 The effects of waterlogging

Because air and water compete for the same position in the soil, it is clear that periods of excess water lead to reductions in oxygen supply. Thus the effects of waterlogging on the soil are felt mainly through the control upon the soil air. It is, however, not only the direct consequence of a reduced supply of oxygen which inhibits plant growth in waterlogged soils. Under certain circumstances, toxic compounds may build up as a result of oxygen deficiencies.

One of the compounds affected in this way is carbon dioxide. The content of carbon dioxide in the soil atmosphere is closely related to that of oxygen, for CO_2 is given off during respiration by plants and micro-organisms. The concentration of carbon dioxide depends upon the balance between this rate of supply and the rate of diffusion to the surface. The rate of diffusion can again be described in terms of the transport equation, but in this case the value of D for water is relatively high. Thus, harmful concentrations of carbon dioxide accrue less readily than do inhibitory deficiencies of oxygen under saturated conditions. Indeed, Greenwood (1967) argues that toxic levels of CO_2 are unlikely in arable soils.

Other gaseous products of anaerobis are more significant. Smith and Russell (1969) demonstrated that ethylene could reach toxic levels in the soil air or soil water under anaerobic conditions. Butyric acid, acetic acid, lactic acid and proprionic acid may also be produced (Stevenson 1967), although their effects on most agricultural crops are less marked (Russell 1973). Inorganic compounds may similarly accumulate in waterlogged soils, and some of these, such as manganous ions, nitrite, sulphite and sulphide may be harmful to plants.

The creation of these inorganic substances is a function of reduction processes. Under aerobic conditions, oxygen in the soil air acts as a sink for electrons released during microbial respiration. In the absence of this oxygen, other compounds are exploited, and in the process are converted from a higher to a lower valency state. This leads to a loss of oxygen and the formation of reduced compounds. For example:

$$NO_3^- + 2H^+ + 2e \rightarrow NO_2^- + H_2O \qquad [4.2]$$

$$Fe_2O_3 + 2H^+ + 2e \rightarrow 2FeO + H_2O \qquad [4.3]$$

Reduction of sulphates in this way is particularly important because it commonly produces hydrogen sulphide through the action of the bacteria *Desulphvibrio*. Hydrogen sulphide is toxic to many plants at concentrations as low as 10^{-6} M.

The organic compounds are a by-product of organic decomposition under anaerobic conditions. Instead of being oxidised to carbon dioxide and water, carbohydrates are only partially decomposed to simple fatty acids and various hydroxy acids (Russell 1973). These are then rapidly decomposed to gaseous compounds including methane, propane and ethylene. The exact means by which these compounds are formed are not yet known, although it seems clear that enzymic activity is fundamental to the processes. It is also apparent that the processes occur early in the period of saturation and take place very quickly. Ethylene, for example, is produced within hours of the oxygen content falling below 1 per cent and, in the absence of free passage to the soil surface, may remain at toxic levels for many weeks thereafter (Dowdell *et al.* 1972). In the presence of ethylene, root growth is inhibited (Fig. 4.1). Root cells may in fact be

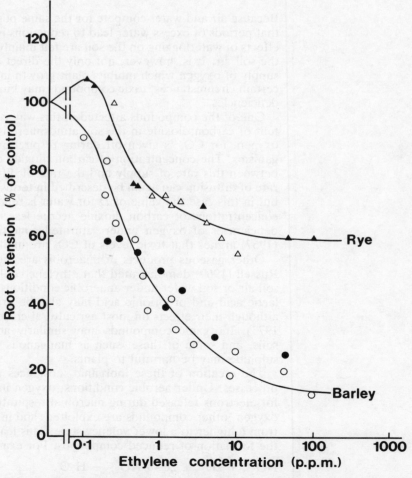

Fig. 4.1   The effect of ethylene concentrations in the soil on rates of root extension (after Smith and Robertson 1971).

severely damaged, with the result that there is a lasting effect upon plant growth.

Anaerobic conditions might, at first sight, seem to be inimical to leaching and nutrient loss, for it is generally argued that leaching is encouraged by conditions favouring rapid percolation of water. Many compounds, however, are relatively insoluble in aerobic environments, and are only subject to leaching to any great extent in oxygen-deficient conditions. Thus, Russell (1973) states that 'slow percolation through the soil when waterlogged may impoverish it faster than normal percolation under aerobic conditions.'

Nutrient losses in these circumstances are attributable, to a major degree, to the reduction of manganese dioxide and ferric oxide. The manganous and ferrous ions which are thereby produced tend to displace other cations from the exchange sites, leading to increased quantities of exchangeable ions such as calcium, magnesium and potassium in the soil solution. In addition, several nutrients become subject to leaching through the direct effects of reduction. Iron and manganese are clearly affected in this way, but so is nitrogen which is commonly reduced to highly soluble nitrite or nitrous oxide.

Phosphate, also, may be released through reduction. Much of the inorganic phosphorus in the soil is held as ferric compounds coating clay particles. Reduction of these compounds results in the gradual leaching of the phosphate. Indeed, all the nutrients brought into solution may be washed downwards through the soil, the more soluble of them being removed completely, whilst the others are precipitated at various depths according to their oxidation potentials. Nitrogen, and to a lesser extent, calcium, magnesium and potassium are thus lost in considerable quantities, whereas manganous ions are often precipitated in the subsoil, and the ferrous ions, which are readily re-oxidised, are deposited only a short distance beneath the actively reducing conditions associated with the microbial population of the topsoil. Whether removed completely or merely translocated to greater depths, these nutrients all tend to be less readily available to plants as they are washed from the main root zone in the topsoil. In this way, waterlogging may ultimately produce severe nutrient shortages.

These various effects of waterlogging combine to influence crop yield to a marked extent. It is difficult to isolate the contribution of individual processes, since the factors tend to interact, but numerous studies have shown the combined effects of waterlogging. It is clear from these studies that even short periods of saturation at key times in the growth cycle may significantly diminish yields. Indeed, the effect of a single day's water logging on crop yield (Table 4.1) implies that it is not the direct effect, but the indirect and longer-term consequences of ethylene production and

Table 4.1   The effect of seven days' waterlogging on crop yields in Hungary

	J	F	M	A	M	J	J	A	S	O	N	D
Month of waterlogging (% of reduction in yield)												
Grass	—	—	—	10	15	20	20	10	—	—	—	—
Sugar beet	—	—	50	50	50	40	40	40	40	10	—	—
Potatoes	—	—	80	80	90	100	100	100	40	—	—	—
Winter wheat	5	5	15	25	40	50	—	—	—	—	—	—

Source: From Trafford 1970

other toxic substances that are most critical.

It is also apparent that the timing of waterlogged conditions is important in determining the effects upon yields. Currie (1975) demonstrated that saturation of the soil during the seedling stage has little influence on yield because – despite the adverse effects upon oxygen supply, nutrient supply and soil temperature – the requirements of the young plant are limited. As has been mentioned, atmospheric supplies of oxygen may meet the plant's needs, whilst the shallowness of the rooting means that escape of toxic compounds to the surface, or their leaching beyond the reach of the plant, are favoured. In contrast, the same conditions during the periods of vigorous growth and root extension have far more serious consequences for yield.

There are a number of more indirect effects upon crop growth. Under waterlogged conditions, the incidence of disease often increases, for many agricultural pests favour moist conditions, particularly during the summer period. Poor soil drainage also tends to reduce the length of the growing season, for wet soils characteristically warm up more slowly in the spring. Three factors are important in this respect: the albedo of a dry soil is often lower than that of a wet surface, so less energy is reflected from the soil; evaporation is reduced so less energy is lost in the process; and, most importantly, the specific heat of water is greater than that of the air which it displaces; consequently more energy is required to raise the temperature of the soil mass when it is wet. These relationships are not entirely simple, for below certain moisture contents (often about 20 per cent by volume), the effects may be reversed. Beyond that point, however, drainage increases the thermal diffusivity of the soil and result in a more rapid rise in soil temperature. The advantages of this to crop growth are considerable. It allows earlier germination, a more effective use of solar radiation throughout the growing season, and earlier flowering. There are related practical benefits, for the flexibility of management procedures is greatly increased.

4.3.2 The need for agricultural drainage

Drainage problems arise because water cannot escape rapidly enough from the soil profile. This may occur because the gravitational potential is too low (as in the case of an area with a high regional water table) or because the negative pressure potentials are too great. This latter condition is characteristic of many fine-grained or compact soils, dominated by fine pore spaces. Most of the water infiltrating into these soils is affected by the strong matric forces which hold the water against the surfaces of the particles and retard its movement through the soil pores. As a consequence, such soils have a very low saturated hydraulic conductivity which prevents the rapid removal of rainwater.

On the basis of these conditions, a major distinction has often been made between groundwater gley soils and surface-water gley soils (e.g. Clarke 1940). The former are defined as soils in which gleying is a product of a high water table produced by external factors such as geology or topography; surface-water gleys are considered to be soils in which drainage is impeded by internal conditions, such as a heavy texture, lack of natural structure or compaction. More recently, Avery (1980) has introduced a further distinction, recognising pelosols – that is, non-alluvial and noncolluvial soils which have clayey horizons within 30 cm of the surface and gleyed, slowly permeable subsoils.

In terms of their drainage management, these three major soil groups show different properties. Surface-water gleys can generally be improved by encouraging the vertical movement of water through the profile; groundwater gleys require facilities to remove the water from the area; pelosols typically need both kinds of treatment.

The purpose of agricultural drainage is to control water movement and retention in order to provide optimum conditions for plant growth. This, of course, does not mean removing water from the soil entirely; it is essential to retain sufficient water, held at suitable tensions, to allow an adequate supply for plant requirements. Thus, drainage should be seen as a means of water control rather than water removal.

Where excess water exists, however, any or all of three general objectives need to be attained:

1. Improved vertical movement of water;
2. Improved lateral movement;
3. Lowering of the regional water table.

4.3.3 Improving vertical movement of water

One of the main reasons for inadequate vertical movement of water through the soil profile is the presence of too fine or too tortuous a pore system. This problem can be overcome by structural modifications to the soil, aimed at opening up the pores and thereby creating a larger volume of continuous macropores.

These objectives can be met by overturning and breaking the soil through, for example, normal tillage operations. Since the macropores occur mainly between the aggregates, their number is best increased by the creation of more, smaller and more stable aggregates. In the short term ploughing can usually achieve this, although, in the longer term, it may be necessary to stabilise the structure by increasing the organic matter content of the soil or by using various soil conditioners (e.g. Emerson *et al.* 1978). Conventional tillage is of limited value, however, where impedance occurs at greater depth, and is only effective where the need is to break up surface crusts or shallow pans.

In clay soils, or in soils containing deeper pans, the internal drainage can be improved by subsoiling. This involves the use of a deep, plough-like implement which shatters the impeded horizon and creates a system of coarse, far-reaching fissures down which the water can drain. The use of the subsoiler is limited by soil conditions, and is only suitable for relatively heavy and dry soils which will break under the impact of the implement (Swain 1975). If the soil is too wet, it merely deforms plastically before the subsoiler, and smearing at the interface may seal any pores which do exist. If the soil is too coarse, then the fissures created by the subsoiler may close up as the blocks break down and resettle.

In recent years the efficacy of subsoiling has been criticised and the process has been said to lead to the creation of a deep channel in the soil but little shattering or fissuring, much as in moling (Trafford 1978); indeed, it has been referred to as square moling (Trafford 1975). It has been shown by Spoor (1975), however, that the lack of shattering often results from the use of the subsoiler below its critical depth. As with any tine, the forces created as it moves through the soil are translated into stresses running horizontally and upwards into the soil. At shallow depths the effect is to produce lifting of the soil, with associated fracturing of the clods. At greater depths, the load of the overlying soil prevents active lifting and, instead, the soil is consolidated around the implement. The critical depth for the subsoiler appears to be in the range of 20–40 cm; below this depth the beneficial shattering effect is lost.

Numerous experiments with subsoiling have also been troubled by the application of the method under unsuitable moisture conditions. Where care has been taken, the effects upon soil drainage and crop yields are marked (Table 4.2). Improved aeration, at least in the upper 30–40 cm, allows more extensive root development and increased growth rates. These improvements in drainage also lead to an increased loss of nitrogen, however, and yields are optimised only if additional N fertilisers are applied (Davies 1978).

Table 4.2   Effects of subsoiling on crop yields

| Crop | Yields (t ha^{-1}) | | % | Soil type |
	Subsoiled	Control	gain	
Grass[a]	1.28	0.98	30.6	Surface-water gley (Cottam series)
Grass[a]	2.62	1.42	84.5	Surface-water gley (Cottam series)
Barley[b]	4.75	4.24	12.0	Rendzina over Cornbrash
Barley[b]	3.41	3.51	−2.8	Rendzina over Cornbrash
Barley[b]	4.52	3.90	15.9	Sandy loam over indurated gravel
Barley[b]	3.82	3.60	6.1	Sandy loam over indurated gravel
Winter wheat[b]	5.56	5.35	3.9	Argillic gley soil (Worcester series)
Winter wheat[b]	4.31	3.90	10.5	Argillic gley soil (Worcester series)
Spring barley[b]	4.17	4.03	3.4	Argillic gley soil (Worcester series)
Winter oats[b]	3.41	3.03	12.5	Argillic gley soil (Worcester series)
Winter oats[b]	4.16	4.18	−0.5	Surface-water gley (Charlton Bank)
Winter wheat[b]	3.84	2.23	72.2	Surface-water gley (Charlton Bank)
Winter beans[b]	3.51	3.78	−7.1	Surface-water gley (Charlton Bank)

[a] From Davies (1978): data for experimental plots in North Wales
[b] From Swain (1975): data for experimental plots in the English Midlands

4.3.4   Improving lateral movement of water

In many circumstances, slow vertical movement of water is associated with impeded horizontal flow. This is particularly the case in clay soils, where even the presence of vertical fissuring does little to allow the lateral escape of the water at depth. Without the facility for horizontal movement in the subsoil, improved percolation has little effect, for ponding will occur within the fissures and macropores.

Methods to improve horizontal movement are varied. Tillage and subsoiling will help to some degree, although in neither case is the lateral continuity of the fissures adequate to ensure free flow of water. In addition, neither method ensures a consistent gradient for the water within the subsoil. Therefore, it is often necessary to install some form of drainage system to help remove excess water. Drains may take the form of moles, pipe drains or open ditches.

Mole drains

In most cases, attempts to improve the lateral flow of water involve the creation of a network of large channels, tunnels or ditches which feed water from the soil into suitable arterial drainage systems. Mole draining represents one of the simplest forms of such methods. Using a bullet-shaped implement – the mole – dragged by a tractor, a continuous tunnel (<20 cm diameter) is cut within the soil. By aligning this mole along the appropriate gradient it is possible to provide a large, continuous, artificial pore within the soil. The gradient of this mole tunnel is critical; if it is too steep channelling and erosion may occur, whereas if it is too shallow ponding may take place in depressions caused by slight irregularities in the mole drain. A gradient of 0.5–3.0 per cent is normally considered optimal.

The procedure is confined mainly to clayey soils, for the mole tunnels collapse in coarser materials. It can also be carried out only within a restricted range of moisture conditions. As with subsoiling, problems of smearing may arise if the soil is too wet (i.e. beyond its lower plastic limit), and shattering occurs if the soil is too dry.

Moling can be carried out at close lateral intervals – often 2–5 m – and it is relatively cheap. For these reasons it is commonly used as a secondary drainage system, feeding water into primary tile or ditch systems. The life-span of mole drains is rarely long, however, and the procedure needs to be repeated every five to seven years as the moles collapse under the effects of tillage, consolidation by machinery, natural settling of the soil and structural disintegration.

The depth of moling varies, although traditionally a depth of 50–60 cm is used. More recently, so-called 'mini-moling' has been adopted, with a small-bore implement (c.5 cm diameter) and at depths of 30–40 cm or less. These may be spaced at lateral intervals of as little as 1 m.

Pipe drains

More permanent subsoil drainage is achieved using tile or plastic pipes. Typically, these are 10–30 cm diameter and are inserted within trenches at depths of about 0.5–2.0 m. In Britain, most systems are within the range of 70–120 cm depth. The trenches are usually back-filled with gravel, polystyrene or fuel ash, to encourage water inflow to the pipes, and to provide a connection area between the permanent pipes and secondary systems such as mole drains. It has also been argued that the back-fill acts as a filter, trapping sediment washed into the drain. As Thomasson (1975) states, however, this is rarely effective, since the permeable fill is too coarse. Moreover, the creation of an active filter is, to some extent, incompatible with the other objectives of tile drainage, since it may impede flow into the drain.

It follows that one of the problems encountered by pipe drainage is siltation. In the case of tile drains, water enters the system through narrow gaps between the pipes, and over time these tend to become clogged with silt and clay. More modern plastic pipes include perforations or slits, but these too may become silted up.

Open ditches

Open ditches are probably the most ancient form of drainage improvement and they were certainly widely used in Roman times in many parts of Europe. They have several advantages over other systems, for they are cheap to install, relatively easy to maintain (it is possible to see their condition directly) and under certain conditions they are highly effective. Their efficiency is related to the fact that they have a larger perimeter than any other system, and thus seepage into the ditch occurs over a larger boundary area. This may result in a greater reduction in the height of the water table.

On the other hand, open ditches are wasteful of land and they cannot be closely spaced for they inhibit field operations, particularly in arable farming. Even in grazing systems the danger for animals may deter their use, although in some areas, such as the marshlands of East Anglia, the ditches are used as field boundaries to constrain livestock.

4.3.5 Arterial drainage

Where the cause of poor drainage is a regional phenomenon, such as a high water table produced by the regional topography or proximity of the land to sea-level, improvement of soil drainage requires a different approach. The need is to lower the water table generally. This may be achieved in a variety of ways, such as:

(a) straightening rivers to increase their gradient and improve channel flow, preventing flooding etc.;
(b) deepening or dredging rivers to lower the mean water level;
(c) pumping water from agricultural land into the rivers.

Arterial drainage schemes have a long history. In the East Anglian Fens, for example, drainage schemes date back to Roman times, and extensive reclamation was carried out in the seventeenth century under the patronage of the Earl of Bedford. In 1631, however, the Dutch engineer Sir Cornelius Vermuyden was invited to supervise the project, and during the next twenty years he established the system which is the basis of the present drainage network of the Fens (see section 11.6.1).

Similar developments have taken place in other fenland areas in Britain, including the Somerset Levels, the Cuckmere and Ouse valleys in Sussex and the lower Trent valley (e.g. Curtis *et al.* 1976). During the present century, the reclamation of new land by such schemes has gradually declined in Britain as attention has become focused more on improving existing systems, but elsewhere extensive projects have been carried out. Probably the most impressive has been the Dutch polder

Table 4.3 Reclamation of the Zuyder Zee, Netherlands

Name of polder	Area (ha)	Period dikes built	Period polder developed	Farmland (%)	Woods and reserves (%)
Wieringermeer	20,000	1927–9	1930–40	87	3
North East Polder	48,000	1936–40	1942–62	87	5
Eastern Flevoland	54,000	1950–56	1957–present	75	11
South Flevoland	43,000	1959–67	1968–present	50	18

scheme. Since the Zuyder Zee (Reclamation) Act of 14 June, 1918, four polders have been reclaimed, giving a total of some 165,000 ha of new land (Table 4.3). As the polders are constructed, dikes are built, fed by lateral drains spaced at intervals of 1.6–2.0 km. These, in turn, are supplied by main drains some 300 m apart, and a close network of field drains spaced at distances of 8–24 m (Glopper and Smits 1974).

Extensive drainage systems of this type have provided some of the world's most fertile farmland. The East Anglian Fens include the largest area of grade 1 agricultural soils in Britain (MAFF 1966), while grass and arable yields on the polders are among the highest not only in the Netherlands but in the whole of Europe. Arterial drainage schemes are not without their problems, however, as we will discuss in Chapter 13.

4.4   VARIABLES IN DRAINAGE DESIGN

The selection of a specific drainage system involves a number of decisions. These relate not only to the general type of the drainage system, but also to variables such as:

(a) The spacing of the drains;
(b) the depth of the drains;
(c) the size of the drains;
(d) whether to use single or multiple drainage systems.

These variables apply to all forms of drain. A full discussion of drainage design is provided by Bailey *et al.* (1980), who also supply design charts for field use.

4.4.1   Spacing

Up to a spacing of about 6 m, the distance between drains acts as an important control upon the rate of inflow and thus upon the time it takes for the soil to regain its equilibrium moisture content. It does not, however, affect the moisture content at equilibrium, and beyond 6 m the rate of inflow is independent of spacing (Baver *et al.* 1976). Close spacing of drains therefore encourages rapid removal of water, but is obviously expensive. Distant spacing is cheaper but less effective, and failure of a drain may have more widespread implications for the field. Moreover, wide spacing results in the development of a water table with a markedly humped profile, particularly in soils with a slow hydraulic conductivity. As a consequence, drainage conditions vary considerably in the inter-drain area and crop growth may be affected.

4.4.2 Depth

The depth of the drain controls the height of the water table when drain-age ceases. Deep drains result in a lower water table, and thus increase the volume of unsaturated soil available to roots. Drain depth also con-trols the equilibrium moisture content (i.e. the moisture content when drainage is complete) of the soil above the water table, for this depends upon the gravitational potential which is a function of height above datum. Deep drains therefore lead to lower equilibrium moisture contents at the surface, and a consequent increase in the bearing capacity of the surface soil. It is, however, possible to over-drain some soils, with the result that they become too dry and lose cohesion; peats are particularly prone to this problem, and under these conditions they become suscep-tible to erosion and experience a reduction in their shear strength. As is apparent from the discussion on tillage processes, this effect is related to the interaction between cohesive and adhesive forces (section 3.3). Ex-cessive drainage may also result in problems of droughtiness during drier parts of the season. Similarly, over-draining of peats may lead to marked lowering of the soil surface due to drying and contraction of the peat. This may bring tile drains within reach of tillage implements, and, more gen-erally, cause problems of flooding. In parts of the East Anglian Fens, for example, over-draining has lowered the surface by as much as 5 m, bring-ing it below the water level in the main arterial waterways. As a conse-quence, severe flooding took place during 1948 and peat shrinkage has also led to undermining of buildings and roads (see Ch. 13).

In addition to influencing the conditions at equilibrium, drain depth affects the time it takes for equilibrium to be reached. In soils with a low hydraulic conductivity this can be a serious consideration, for the water may pond within the surface layers of the soil during storms and cause waterlogging in the root zone. Where vertical conductivity is low, there-fore, relatively shallow drains are often used; it is in these soils that the problems of water table curvature arise, so close spacing is also essential. Conversely, in more permeable soils, deeper drains can be installed. One additional advantage of deep drains is that they are less prone to damage by machinery or livestock. In the case of open ditch drainage, the rate of inflow is increased by cutting deeper ditches, for this increases the area through which seepage may occur.

4.4.3 Size

The ability of a drain to carry water is related to the cross-sectional area of the drains. In the case of pipe drains, the general relationship is ex-pressed by the function:

$$Q_{max} \sim d^{2.67} \tag{4.4}$$

Thus, a small change in pipe diameter (d) permits a large increase in flow capacity (Q_{max}). With the 75 mm clay pipes traditionally used for agricultural drainage in Britain, under-capacity of drains is rarely a problem, but the smaller (50 mm) plastic drains which have been developed in recent years may limit flow and impair drainage efficiency. (More recently, larger diameter plastic drains have been accepted as meeting design standards and these are now being widely used). Methods for assessing the critical drain diameter are provided by Trafford and Dennis (1974) and Bailey *et al.* (1980).

4.4.4 Choice of drainage system

The selection of a particular drainage system is made in the light of a number of considerations. These include:

(a) economic factors (e.g. cost, expected yield on investment, subsidies);
(b) management factors (e.g. requirements of crops, tillage requirements);
(c) soil factors (e.g. texture, hydraulic conductivity, the specific type of drainage problem);
(d) climatic factors (e.g. rainfall intensities, frequency, distribution).

A general scheme for incorporating these factors in a decision program is given in Fig. 4.2.

Many of these factors are difficult to assess, so decisions often involve a large element of subjective judgement. The prediction of the cost-effectiveness of any scheme, for example, requires assumptions about future – often long-term – trends in prices and interest rates, and these are often unreliable. Nor is information always available on the specific drainage requirements of particular crops, and so the prediction of yields may be uncertain. This problem is exacerbated by the fact that responses to drainage frequently depend upon other management factors, such as rates of fertiliser application, and timing of sowing and cultivation.

In the case of the soil and climatic factors more quantitative approaches can be used. Two main approaches to determining drainage spacing, depth and, where relevant, size have been developed, one based on steady-state theory, and the other on more dynamic models of water flow.

Many of the steady-state equations attempt to simplify the drainage system by assuming one-dimensional flow. They thus follow the premises originally proposed by Dupuit (1863), namely:

1. That flow streamlines in the soil are horizontal;
2. That the velocity of flow along these streamlines is proportional to the slope of the free water surface but independent of depth.

On the basis of these assumptions the horizontal flux of water towards the drain approximates to the relationship

$$q = ky\frac{dy}{dx} \tag{4.5}$$

where q is the horizontal flux
y is the height of the water table above the impermeable layer
x is the horizontal distance from the drain
k is the hydraulic conductivity.

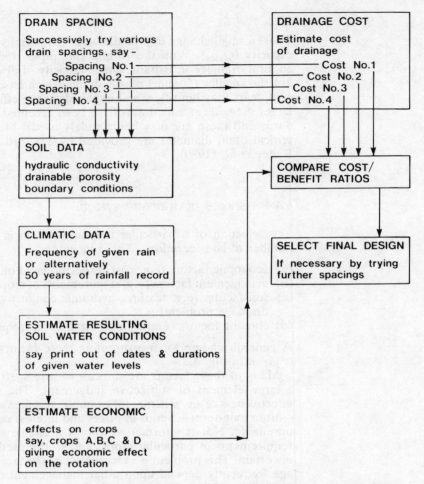

Fig. 4.2   A decision program for drainage design (from Trafford 1975).

This equation can then be used to determine the drain spacing necessary to keep the water table below a predetermined height under conditions of prolonged rainfall. Numerous refinements to this approach have, in fact, been proposed (e.g. Hooghoudt 1940; Aronovici and Donnan 1946; Schilfgaarde 1963) although all involve somewhat unrealistic assumptions about water flow in the soil. Despite the theoretical limitations of these equations, however, they have been successfully used in many parts of the world and Hooghoudt's method in particular has been widely adopted in Holland and Germany (Figure 4.3, Table 4.4).

Hydrodynamic approaches are intended to provide a more realistic method of assessing drainage requirements, by modelling changes in water table height in response to given rainfall inputs. Non-steady-state formulae are used to assess two- and three-dimensional flow in the soil. These are generally based on Laplace's equation which states that:

$$\frac{\delta^2 H}{x^2} + \frac{\delta^2 H}{y^2} = 0$$
[4.6]

where H is the height of the water table above the base of the drain

a.

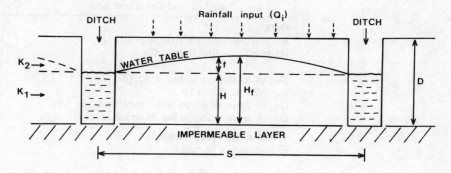

b.

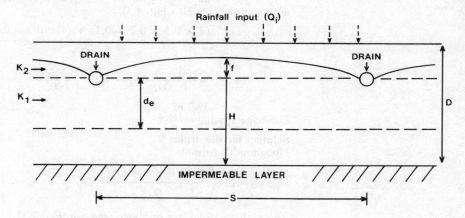

c.

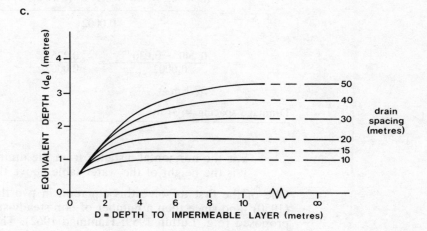

Fig. 4.3 Parameters in Hooghoudt's formula for drainage design (a) parameters for open ditches (b) parameters for tile drains (c) determination of equivalent depth (see Table 4.4 for a worked example).

Table 4.4   Solution of Hooghoudt's formula: worked examples for (a) open
            ditches and (b) tile drains

Data

Rainfall: maximum design event (Q_i)	0.0002 m³ m⁻²
Hydraulic conductivity (K_1)	0.5 m d⁻¹
(K_2)	0.9 m d⁻¹
Depth to impermeable layer (D)	1.2 m
Equivalent depth (d_e)	0.6 m
Depth from drain to impermeable layer (H)	0.7 m
Depth from water table to drain level (f)	0.1 m

Solution for open ditches
Hooghoudt's formula:

$$S = \left(\frac{8\,K_1\,H\,f + 4\,K_2\,f^2}{Q_0} \right)^{0.5}$$

where Q_0 = *drain outflow (m³ m⁻²)* = Q_i at equilibrium; and all other variables
are as above (see also Fig. 4.5)

$$S = \frac{(8 \times 0.5 \times 0.7 \times 0.1) + (4 \times 0.9 \times 0.1^2)^{0.5}}{0.0002}$$

$$= \frac{0.280 + 0.036^{\,0.5}}{0.0002} = \frac{0.316^{\,0.5}}{0.0002}$$

$$= \quad 39.7 \text{ m}$$

Drainage spacing = 40 m

Solution for tile drains
Hooghoudt's formula:

$$S = \left(\frac{8\,K_1\,d_e\,f + 4\,K_2\,f^2}{Q_o} \right)^{0.5}$$

where all variables are as above (see also Fig. 4.5).

$$S = \frac{(8 \times 0.5 \times 0.6 \times 0.1) + (4 \times 0.9 \times 0.1^2)^{0.5}}{0.0002}$$

$$= \frac{0.240 + 0.036^{\,0.5}}{0.0002} = \frac{0.276^{\,0.5}}{0.0002}$$

$$= \quad 37.1 \text{ m}$$

Drainage spacing = 37 m

---

x is the horizontal distance from the drain
y is the height of the water table above the impermeable layer.

One of the first to solve this equation in two dimensions was Kirkham
(1950), and since then a number of non-steady-state formulae have been
proposed (e.g. Luthin 1959; Hammad 1962). The more recent developments of this approach have attempted to take account of factors such as
convergence of drains and non-homogeneity of the soil.

Both steady-state and hydrodynamic methods have a major disadvantage in that they require a great many data which are not readily available. All the procedures need information on hydraulic conductivity, and

non-steady-state equations also use estimates of the drainable porosity (i.e. the pore volume which will drain by gravity). The cost of obtaining these data may be prohibitive, while as Bouma *et al*. (1980) have shown, estimation of information on these properties from soil survey maps can lead to major errors in predictions of water table levels. The major attractions of the hydrodynamic approaches are that they are theoretically satisfying and that they allow models of water table response to be based on long-term (normally at least ten years) records of rainfall characteristics. They thus take account of the average distribution of rainfall through the year, and provide an indication of the frequency with which the water table will reach a given height under specified drainage conditions. In this way realistic estimates of the economic cost of different drainage designs should be possible in terms of both installation costs and potential crop losses. Nevertheless, lacking detailed data, not only on soil conditions but also on crop responses, such estimates are not always reliable. Moreover, as comparisons of different quantitative approaches have shown (e.g. Wesseling 1964; Nwa and Twocock 1969), there is often little to choose between the various methods. Commonly, therefore, drainage design is based upon field experience, individual preference and the accumulated knowledge of previous work, rather than on mathematical models (Bailey *et al*. 1980).

It is also important to appreciate that no drainage design can be perfect; it is not possible to cater for the most extreme events. In general, an attempt is made to guard against the more frequent events, occurring every five to ten years. The magnitude of these can be determined from rainfall recurrence curves.

4.5 EFFECTS OF AGRICULTURAL DRAINAGE ON SOIL HYDROLOGY AND CROP YIELD

Rather surprisingly, there is no great wealth of data on the effects of drainage upon either hydrological conditions or crop yields. This is largely because of the difficulties of carrying out controlled, representative experiments under field conditions, and, of course, the economic disincentive to do so for systems of low-cost technology. Useful reviews of experiments have been made by Trafford (1970, 1977, 1978) and Trafford and Walpole (1975).

The results from different studies are often somewhat contradictory. Many of the apparent contradictions relate to inherent differences in soil conditions, so that it is difficult to compare results from different experiments directly. Moreover, it is not always clear that the optimum drainage schemes have been applied, with the result that, over the range of conditions studied, hydrological responses have not always been as significant as might be anticipated. Another problem has been the immense range of variables to be controlled and replicated; depth, spacing, pipe size, layout, drainage system may all be varied. Furthermore, each design may be used singly or in combination. Consequently, general patterns of results from drainage experiments are difficult to discern.

Several more recent studies have indicated that the most marked improvements in drainage, particularly on surface-water gleys, are obtained

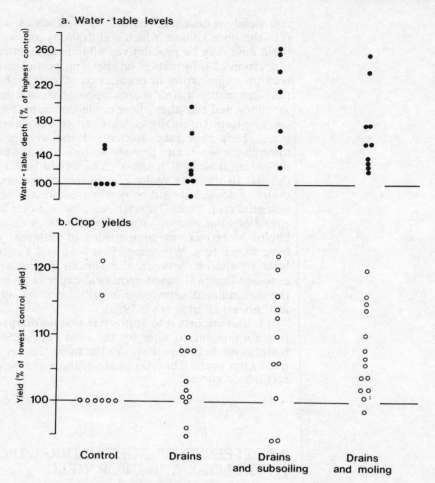

Fig. 4.4　Effects of different drainage systems on water tables and crop yields on clay soils (from Trafford and Walpole 1975).

by using several systems of drainage in combination. Experiments at Drayton Experimental Husbandry Farm (Warwickshire), in soils of the Denchworth series for example, showed that pipe draining alone had little effect upon waterlogging, but that when used in combination with either moling or subsoiling, the water table level fell and the number of days of waterlogging declined significantly (Fig. 4.4, Table 4.5). These data seem to indicate that in many soils the need is to improve both vertical percolation and lateral movement of the water; moreover, lateral movement requires the use both of main drains to remove the water from the area and of a close network of feeders to transport the water to these main conduits. It seems likely that future schemes will increasingly use combined pipe and mole or subsoiling methods.

The fact that water tables fall is no guarantee that crop yields will increase. In some circumstances, reductions in yield may occur. On the one hand this may be due to over-drainage, so that plants suffer from water deficiencies; on the other hand yields may fall because of inadequate applications of nitrogen. Trafford (1972) reviewed results from clay soils of the Denchworth, Ragdale and Windsor series and showed that marked

Table 4.5 The effects of agricultural drainage on water table levels at Drayton Experimental Husbandry Farm, Warwickshire

Year	Average median winter water table level (cm below surface)			
	Control	Pipes only	Pipes + Moled	Pipes + Subsoiled
1970–71	18	27	39	48
1971–72	18	14	25	26
1972–73	39	37	53	59
Number of days within 30 cm (1971–72)	107	106	—	47

Source: After Trafford and Walpole. 1975

improvements in yield were obtained by using combined drainage systems; tile drains alone had little effect upon yields (Fig.4.4). Davies (1978), and many others, however, have found that yields increased most dramatically only when drainage improvements were supplemented by increased nitrogen applications (Fig. 4.5). Results indicating effects of drainage upon crop yields are shown in Table 4.6.

Yield, of course, is one of the crucial factors in the equation to determine the success of drainage schemes, but it should not be seen in isolation. Even if yield does not rise, there may be significant economic or management benefits associated with greater flexibility of cropping and tillage, longer growing or grazing seasons and reduced structural damage (e.g. Kellett 1978); The benefits of drainage for grazing systems are often due to reduced susceptibility to poaching, increased accessibility to the

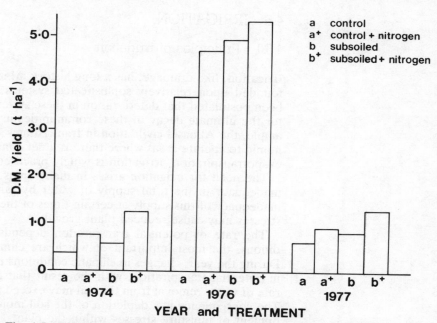

Fig. 4.5 Effects of nitrogen fertilisers on response of grass to subsoiling at Waen, Clwyd, Wales (all yields measured in spring) (data from Davies 1978).

Table 4.6 The effects of agricultural drainage upon crop yields

Location and crop	Spacing (m) and yield (t ha^{-1})				Source
Drayton	Control	60 m	30 m	15 m	Trafford 1977
Winter wheat	3.69	4.42	4.29	4.45	
Cambridge	Control	27 m	10 m	5 m	Barker 1963
Winter wheat	2.51	3.56	3.98	3.54	
Zahati, Moravia	Control	14 m	10 m	6 m	Trafford 1970
Barley	3.60	3.48	3.90	3.97	
Brooksby	Control	80 m[a]	40 m[a]	20 m	Wilkinson 1975
Winter wheat	2.24	2.35	2.27	1.93	

Note: a = tiles plus porous fill and moling

land and improved animal cleanliness and health rather than to direct increases in yields. Similarly, consequent improvements to the quality rather than the quantity of the herbage may be a major benefit.

A further problem with defining the success of drainage schemes purely in terms of yield relates to the difficulty of obtaining meaningful comparative data on yield. Comparisons of productivity on drained and undrained plots tell only part of the story, for drainage allows a change in the whole farming system. It is the relative financial profitability, plus, perhaps, the gains in the ease of management and the long-term benefits of soil conservation that determine the overall effectiveness of agricultural drainage.

4.6 IRRIGATION

4.6.1 Principles of irrigation

Irrigation, like drainage, has a long history. Many early civilisations were founded upon relatively sophisticated systems of irrigation, and it has been postulated that deterioration in these systems was partly responsible for the ultimate decay of these communities in some instances – for example, the Abbasid civilisation in Iran (Adams 1962). Today, even in the humid temperate areas where there is a net annual excess of rainfall over evapotranspiration, irrigation is widely practised.

The need for irrigation arises in many areas, therefore, not from any inadequacy in the total supply of water by rainfall, but because of the inadequacy of this supply at certain times of the year. Resulting moisture stresses may cause reduced plant growth.

The rate of potential transpiration depends upon a variety of conditions, the most important of which are climatic (e.g. Penman 1963). During the year, changes in climatic conditions result in marked variations in potential transpiration, with the result that for significant periods the rate of water removal from the soil may exceed the rate of replenishment. The consequence is a depletion of the soil moisture store and the development of moisture stresses within the plant. These cause physiological adjustments, such as the closure of stomata in the daytime, which lead to reduced transpiration and growth. The exact relationship between soil moisture conditions, potential and actual evaporation is complex and is not known in detail; it varies from one crop to another, but it is clear that as moisture stress increases most crops experience a rapid (possibly exponential) decline in transpiration rates and growth may eventually cease. The effects of such stress upon yields may be dramatic (e.g. Day *et al.* 1978).

It is in order to combat these periods of moisture stress that irrigation is used. The practice is an expensive one, and in many cases can only be applied with profit to high value crops; it is also a procedure which requires careful control if adverse effects such as salinisation or soil erosion are to be avoided.

The main purpose of irrigation is normally to replenish the soil moisture store so that water is made available to plants. Under certain circumstances – although less commonly in temperate areas – irrigation is also used to provide a means of leaching salts from the rooting zone in order to prevent salinity problems and improve soil structure. In both cases, the effectiveness of the procedure depends upon a number of factors specifically:

1. The method of application;
2. The timing of irrigation;
3. The amount of water supplied;
4. The quality of the applied water.

4.6.2 Method of application

Three general methods of irrigation may be identified: surface methods, overhead methods and subsurface methods. By far the most widely used are surface methods of application, for in general these involve relatively low-cost technology and are easy to apply over a wide area. They include various methods of surface flooding, either by damming and diverting main streams, or by distributing water from streams via a network of canals or ditches which are subsequently dammed and breached to flood small parcels of land. None of these methods is widely used in temperate areas, although because of their simplicity and cheapness they are often employed in developing countries.

In contrast, furrow irrigation and corrugation irrigation are more common in temperate agricultural systems. The former involves the construction of parallel furrows down which water is allowed to flow. Seepage occurs through the base and side of the furrow into the ridges within which the crops are planted. It is a system mainly used with row crops such as vegetables or root crops, and its application is strictly controlled by slope conditions; erosion occurs if flow is too rapid, so it is normally applied only on slopes of 0.2–0.6 per cent (0.1–0.3°). Corrugation irrigation is similar in principle, but in this case the furrows are much shallower and smaller (often about 10 cm deep and 50–150 cm apart) and consequently very small streams of water are used. Because of this it can be used on steeper slopes, with more closely spaced crops.

Overhead methods of application are varied. Water is normally sprinkled onto the soil surface from sprays or pipes. The sprays may be static, rotating or travelling, and they have the advantage of being largely independent of surface topography. Moreover, they are easier to control, so that rates of water application can more readily be adjusted to local requirements. They are also less wasteful of land than surface methods, and can be used with almost any crop, including grass, cereals or root crops. Against these advantages, however, must be set a number of potential disadvantages. The method may be relatively expensive and it requires a constant supply of high pressure water. Overhead irrigation may also create a degree of soil structural damage, for the water droplets are large, and capping, structural disaggregation and erosion may occur. In addition, water loss through evaporation and dispersal by wind may be considerable, so the method is unsuitable in many cases for very arid or very windy areas.

Without doubt, the most effective method of irrigation is through subsurface application (sub-irrigation). This is often achieved by the use of ditches or underground drains. Through damming of the drains, water may be held in the soil so that a constant supply can be provided to the plant roots. This system is particularly effective since it reduces evaporation losses during application, avoids erosion or structural damage, and supplies water where it is most needed (below rather than at the surface). It also has a number of important effects upon plant growth, for it encourages deeper rooting and thereby ensures a better supply of nutrients. The major disadvantage, however, is clearly that of cost for it is often more expensive than other methods. As a result subsurface irrigation is used in relatively few situations, mainly in association with high cost crops such as vegetables.

4.6.3 Timing and rates of application

With all methods of irrigation one of the critical factors is the rate of water application. Numerous problems can arise if the water is applied either too liberally or too sparsely.

Excessive application not only wastes water, but may also cause soil erosion, structural damage and excessive leaching of nutrients. In addition, the rise in the water table may bring soluble salts such as sodium and calcium to the surface and encourage salinity problems, while saturation of the soil within the rooting zone for long periods reduces aeration and inhibits plant growth. Inadequate water supplies may also exacerbate problems of salinity by encouraging precipitation of dissolved salts at the surface. Moreover, only the upper layers of the soil will contain plant-available water, so that a concentration of roots at or near the surface may develop. This reduces the ability of the plant to withstand periods of moisture deficiency (e.g. when irrigation ceases), reduces nutrient uptake and thereby diminishes yields.

Various methods have been devised for calculating the water requirements of irrigated crops (e.g. Olivier 1961; Hanks and Hill 1980), although their application in the field is often inhibited by lack of information on soil conditions and plant requirements. One approach involves the use of measured soil moisture contents and estimations of irrigation efficiency to assess the amount of water necessary to return the soil to the required moisture state (Arnon 1972):

$$d = \frac{10\ (W_f - W_o)SD}{E} \qquad\qquad [4.7]$$

where d is the amount of water to be applied (mm)

W_f is the soil moisture content at field capacity (%)

W_o is the existing soil moisture content (%)

S is the specific gravity of the soil (kg m^{-3})

D is the depth of soil under consideration (mm)

E is the irrigation efficiency (that is the percentage of applied water that remains in the soil).

Tensiometers or regular field sampling to allow gravimetric analysis may be used to determine the moisture content and predict when to apply irrigation.

Meteorological methods of predicting irrigation requirements are also employed. These involve estimating rates of transpiration and determining the potential soil moisture deficit by taking into account rainfall, drainage and antecedent moisture conditions. The method is most relevant at a regional level, and is used, for example, to give general guidelines for irrigation requirements in Britain (L. P. Smith 1976).

4.6.4 Water quality

The quality of applied water rarely represents a problem in temperate areas, where salinity of the water supply is normally limited. In more arid areas, the presence of high concentrations of sodium and calcium salts within the water may encourage salinisation of the soil, and result in the development of pans at or near the surface. Trace elements such as molybdenum, boron and cadmium may also create difficulties in some circumstances, particularly where water is being obtained from industrial areas or from groundwaters associated with ore-bearing rocks. The accumulation of these over time can lead to dangerous levels of heavy metals in the soil, with consequent inhibition of plant growth and possibly deleterious effects upon man.

4.7 EFFECTS OF IRRIGATION UPON SOIL CONDITIONS AND CROP YIELDS

As with agricultural drainage, irrigation leads to significant changes in soil conditions, some of which are long-lasting, and not all of which may be beneficial in agricultural terms.

Few studies have been made of long-term changes in soils under irrigation. Studies of changes in soil aeration have been made by Willey and Tanner (1963), Williamson (1964) and Williamson and Willey (1964) and these indicate that it takes some days for the oxygen diffusion rate of the surface soil to recover following irrigation. They also show that surface applications of water by flooding saturate the topsoil and therefore impair aeration more than sprinkler methods. However, there is no firm evidence that long-term reductions in aeration occur, except where water is over-applied and the water table is maintained close to the surface for long periods of time.

A more serious effect upon the soil relates to erosion (Mech and Smith 1967). As has been mentioned, most methods of irrigation may encourage soil loss by increasing surface runoff and by causing disintegration of aggregates. Erosion is most severe, however, where water is applied by surface methods such as furrow or flood irrigation. It is also most intense early in the irrigation process. Mech (1949), for example, analysed soil loss under maize during 24 hours of continuous furrow irrigation. In total, 50.9 t ha^{-1} soil were lost, but 39.9 t ha^{-1} of this were eroded during the first 30 minutes, and erosion ceased after 4 hours. Again, few quantitative studies have been carried out to analyse the importance of erosion losses in the long term.

Chemical and organic changes in soil conditions might also be anticipated under irrigation. In arid areas, or where water quality is poor, accumulation of salts and trace elements may present a serious problem. Bicarbonate, chloride, sulphate, sodium and boron may all reach toxic levels, and careful control of application rates, moisture conditions and water quality is essential to remove these by leaching. One of the main indications of soil chemical conditions under irrigation is the exchangeable sodium percentage (the concentration of sodium cations as a percentage of the cation exchange capacity), and Anderson *et al*. (1972) found that levels of this rose progressively during fourteen to seventeen years of irrigation with sodium-rich water in south-west Mexico. At the same time organic matter content fell. The problems of such changes are not only nutritional; sodium tends to destabilise the soil structure, blocking the soil pores and inhibiting water movement. In many temperate areas, however, these problems are less severe, and the overall effect of irrigation may be to reduce nutrient availability by encouraging downward movement of nutrients as the water percolates into the soil (e.g. Wali *et al*. 1980). There are also likely to be effects upon organic activity; prolonged irrigation may lead to reduced activity due to lack of aeration, but shorter-term irrigation may encourage microbial processes by reducing moisture stress during the dry periods of the year.

The effects of irrigation upon yields are more widely appreciated, and many studies have shown the benefits of reducing moisture stress. (e.g. Feddes and van Wijk 1970; Evans and Neild 1981) The benefits probably derive not only from the increased supply of water and increased transpiration rates, but also from the added nutrient uptake. Experiments at Gleadthorpe Experimental Husbandry Farm, Nottinghamshire, show consistent increases in potato yields (Fig. 4.6), particularly in relatively dry years, results repeated by Evans and Neild (1981). It is, of course, in these years that irrigation shows its greatest financial benefits, for market prices are generally high and any increase in yield is at a premium.

In general, maximum yields under irrigation require large inputs of additional water, and the marginal increases in output are often low. In addition, irrigated crops tend to respond to increased inputs of fertiliser, especially nitrogen. Even at relatively low levels of soil moisture, there may be marked increases in yields when fertilisers are applied, largely because the availability of moisture facilitates the uptake of nutrients. Thus, there are often benefits in irrigating in association with fertiliser applications, particularly where soils are liable to be dry (e.g. Garwood 1979).

These responses seem to apply to a wide variety of crops. Similar results have been found by many workers, not only for cereals, but also for root

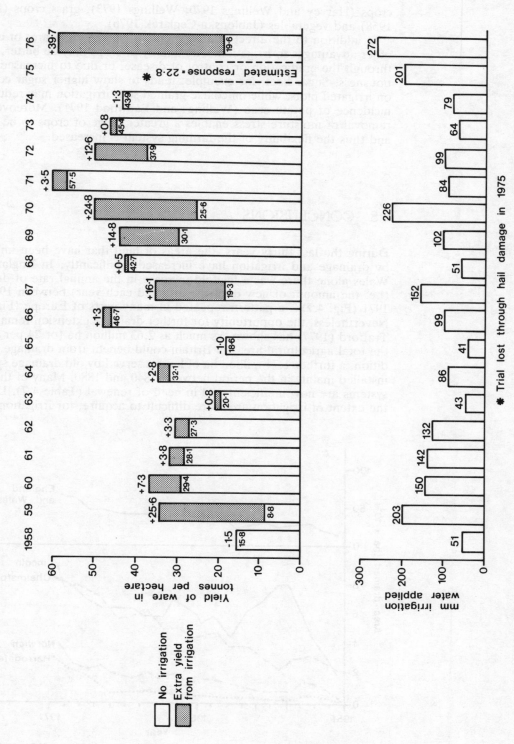

Fig. 4.6   Yields of maincrop potatoes in relation to irrigation of sand soils at Gleadthorpe Experimental Husbandry Farm, Nottinghamshire (from MAFF 1977).

crops (Harvey and Wellings 1970; Wellings 1973), grass crops (Munro 1958) and vegetables (Jablonska-Ceglarek 1976).

In addition to the direct improvements in yields, there are, of course, other advantages to irrigation. The quality of crops may be better, either through the eradication or control of disease, or due to increased nutritiousness. Sugar beet, for example, tends to show higher sugar contents on irrigated plots, while on coarse-grained soils irrigation may reduce the incidence of potato scab (Wellings and Lapwood 1971). Moreover, the removal of moisture stress enables a greater range of crops to be grown and thus the flexibility of the farming system is increased.

4.8   CONCLUSIONS

During the last thirty years, the areas of land that have been improved by drainage and irrigation have increased significantly. In England and Wales alone there was a fivefold increase in the annual rate of drainage (i.e. the amount of new drainage installed each year) between 1951 and 1971 (Fig. 4.7) – a pattern repeated in many parts of Europe (Fig. 4.8). Nevertheless, the opportunity for further drainage extension remains. As Trafford (1977) has shown, as much as 2.63 million ha (or 24 per cent of the total agricultural area) in Britain could benefit from drainage. In addition, a further 1.74 million ha of land is served by old drainage systems, installed mainly in the period between 1840 and 1880. Many of these old systems are now inefficient and in need of renewal (Table 4.7). Data on the extent of irrigation are more difficult to acquire, for irrigation instal-

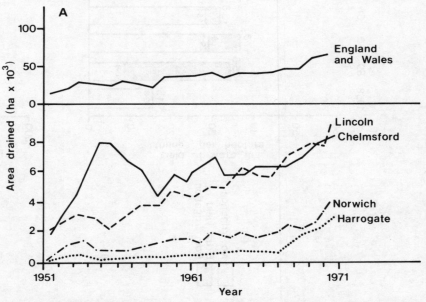

Fig. 4.7   Rates of agricultural drainage in England and Wales 1951–71 (after Green 1973).

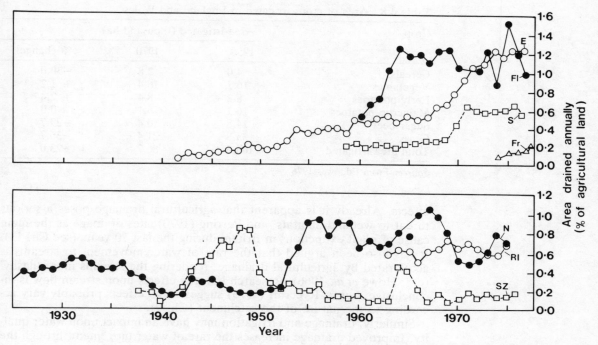

Fig. 4.8 Rates of agricultural drainage in Europe, 1926–76. E = England; FI = Finland; S = Sweden; Fr = France; N = Norway; RI = Republic of Ireland; SZ = Switzerland (from Green 1980).

Table 4.7 The need for agricultural drainage in England and Wales

Classification	Area (million ha)	% of agricultural land
Land drained by post-1940 grant-aided work	1.34	12.2
Land drained by old (mainly 19th century) systems	1.74	15.8
Land naturally freely drained	4.25	38.6
Land needing improvement and likely to be economic	2.63	23.9
Land needing improvement but not likely to be economic	1.05	9.5

Source: After Trafford 1977

lations are not permanent and their use varies greatly in response to climatic conditions. Between 1956 and 1970 the area of land under irrigation in England and Wales fell from about 107,000 ha to about 87,000 ha, largely because of increased costs of irrigation at a time when the market price of irrigated crops was falling (Table 4.8). Since then, however, there has probably been a reversal of this pattern, and the extent of irrigation has again increased.

Both agricultural drainage and irrigation increase crop yields, improve crop quality and make for more flexible cropping systems. Nevertheless, the extension of drainage and irrigation is not without its environmental

Table 4.8   Area of crops irrigated in England and Wales

Crop	Area irrigated (thousand ha)		
	1956	1970	% change
Cereals	4.0	2.8	−30.0
Vegetables	16.2	16.4	+ 1.2
Early potatoes	8.2	8.4	+ 2.4
Maincrop potatoes	16.1	16.1	0.0
Sugar beet	14.8	10.4	−29.7
Grass	33.8	21.4	−36.7
Total area irrigated	106.6	86.7	−23.0

Source: From Eddowes 1976

effects. Already it is apparent that agricultural drainage poses a serious threat to wetland habitats, and Perring (1970) cites drainage as the main cause of plant extinctions in Britain during the last 70 years (see Ch. 13). It has also been argued that the rate of water movement to streams is accelerated by agricultural drainage, rendering the streams more 'flashy' (e.g. Howe *et al*. 1966). Research into the effects upon stream flow is inconclusive, but as Ryecroft (1975) suggests, the effects probably vary according to the nature of the catchment (Ch. 9).

Similarly, drainage and irrigation may have an impact upon water quality. Improved drainage increases the rate of water movement through the soil and may facilitate leaching of nitrate, leading to greater eutrophication of waterways (Ch. 9). Trafford (1978), however, claims that nitrogen losses are maximised when water flows on or through the surface layers of the soil, for it is in these layers that the nitrates are concentrated. Thus, he argues, improved drainage may diminish leaching losses. The effects of irrigation are equally controversial. On the basis of theoretical considerations it might be anticipated that irrigation would lead to greater water movement through the soil, more intense leaching and increased pollution of groundwaters. As with the effects of drainage, however, definitive data are lacking and the impact upon water quality remains largely conjectural. These and other environmental implications of drainage and irrigation will be considered in more detail in Chapters 9 and 13.

5 FERTILISERS AND MANURES

5.1 HISTORICAL ASPECTS

Over the last hundred years, the consumption of fertilisers in the United
Kingdom has risen almost 25-fold (Green 1973). Since 1939 alone, use of
nitrogen fertilisers has increased 16-fold, potassium 5-fold and phosphorus 3-fold (Fig. 5.1). Today, as Table 5.1 indicates, the input of fertiliser
represents one of the main costs in most intensive farming systems in
Europe. Without doubt, the development of the use and understanding

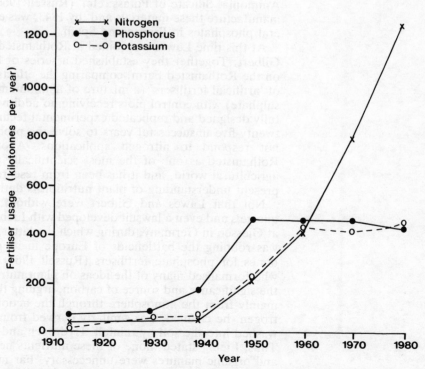

Fig. 5.1 Changes in fertiliser usage in the United Kingdom (data for 1913 and
1929 from Cooke 1974; data for 1939–80 from Hood 1982).

Table 5.1   Costs of fertiliser inputs to different farming systems in Britain

Crop	1967		1975	
	Cost (£ ha^{-1})	% of total costs	Cost (£ ha^{-1})	% of total costs
Winter wheat	7.50	9.6	24.00	11.9
Early potatoes	35.00	9.3	100.00	13.2
Maincrop potatoes	42.50	11.3	80.00	9.0
Sugar beet	31.00*	14.1*	66.00	16.4

Source: Data from Eddowes 1976
* 1968 data

of fertilisers has been one of the major contributions to the evolution of modern farming systems, and it is a development which has had far-reaching implications for environmental conditions.

The development in fertiliser usage owes much to the work of three men – Lawes, Gilbert and Liebig – during the mid-nineteenth century. Following on the recognition by Sir Humphry Davy (1813) of the importance of soil nutrients to plant growth, Lawes established a number of long-term experiments to test the manurial value of bones. The significance of bones as manure had long been appreciated, but their influence was unpredictable and on many soils they seemed to have no detectable effect. Lawes showed that treatment of the bones with acid guaranteed their effect, and he ultimately applied for patents to develop 'patent manures composed of Super Phosphate of Lime, Phosphate of Ammonia, Silicate of Potass: etc.' (Russell 1966). He set up a factory to manufacture these materials and, by 1843, was exploiting reserves of mineral phosphates in France and Spain.

At this time Lawes was joined at Rothamsted in Herfordshire by J. H. Gilbert. Together they established a series of field and laboratory trials on the Rothamsted Farm, comparing the effects of farmyard manure and of 'artificial fertilisers' (a mixture of ammonium chloride and ammonium sulphate) with control plots receiving no additives. They introduced carefully designed and replicated experimental techniques, and they tried, for twenty-five unsuccessful years to solve the problem of why clover would not respond to nitrogen applications. Above all, they established Rothamsted as one of the most scientifically productive centres of the agricultural world, and it has been from research there that much of our present understanding of plant nutrition is derived.

Not that Lawes and Gilbert were without their competitors. Bitter quarrels and even a lawsuit developed with Liebig, Professor of Chemistry at Giessen in Germany, during which the latter complained that 'England was robbing the battlefields of Europe in its ghoulish search for more' bones for phosphate fertilisers (Russell 1966). It was, however, Liebig who formalised many of the ideas on plant nutrition. In 1842, he outlined the significance and source of carbon, arguing that plants obtained carbon mainly from the atmosphere through the action of the green leaves. Nitrogen, he claimed, was similarly derived from the atmosphere, but was washed into the soil by rainfall. Hydrogen and oxygen came from water. Thus, Liebig stated, none of these nutrients need be added as fertilisers, and organic manures were unnecessary. Far more critical in his opinion were the mineral salts, potassium, sodium, calcium, silica and phosphate. Comparisons of plant requirements calculated from analyses of vegetable ash with the soil contents of these minerals, allowed the fertiliser requirements to be exactly determined. He was, therefore, an early protagonist of the value of artificial as opposed to organic fertilisers, a controversy which has lasted to the present day.

5.2 MODERN CONCEPTS OF PLANT NUTRITION

Today the need for inputs of plant nutrients to replenish losses provoked by farming is undisputed. Arguments still exist about the ideal form of these inputs, but it is apparent that in almost all intensive agricultural systems a potential drain of nutrient reserves occurs due to the removal of nutrients during harvesting. The extent of these nutrient losses varies markedly, according to soil fertility, management methods, crop type and yields, but the magnitude of the losses is almost invariably well in excess of the natural inputs from rainfall and weathering (Fig. 1.4 p. 13). Continued cultivation is therefore liable to lead to a gradual depletion of soil nutrient reserves and, ultimately, reduced yields. Such, in fact, is the conclusion drawn both from experiments carried out by Daubeny (1845) and from the long-term continuous cereal experiments on Stackyard Field, Woburn (Fig. 5.2).

As has been noted, the rate at which nutrient depletion occurs depends to a great extent upon crop type and management methods. The data in Table 5.2 indicate the magnitude of nutrient contents in different crops under conditions of moderate yield. It is immediately clear that harvesting procedures are of vital importance. If, in the case of cereal crops, the straw is also harvested, then losses of nutrients are considerably increased; if the straw is ploughed back into the soil or if it is burned, nutrient returns are greater although, in the latter case, losses may occur in the smoke and through leaching (see Ch. 8). Similarly, in the case of grass crops, much depends upon whether the pasture is grazed. Returns of urine and dung ensure an almost complete recycling of most of the major nutrients, while the nutrients returned in these forms tend to be more readily available than those following what Floate (1970) calls the 'plant cycle'. Thus grazing also speeds up the rate of nutrient cycling. In con-

Table 5.2 Nutrient contents of agricultural crops

Crop	Nutrients in harvested crop (kg ha^{-1})						Yield (t ha^{-1})
	N	P	K	Ca	Mg	S	
Wheat (grain + straw)	120	25	80	20	15	25	6
Barley (grain + straw)	100	18	60	15	8	20	5
Sugar beet (roots + tops)	250	30	240	80	30	35	50
Beans (grain only)	200	35	100	40	15	35	5
Kale (whole crop)	200	25	180	200	20	100	50
Grass (dry crop)	250	30	250	70	20	20	10
Clover (dry crop)	220	20	200	150	25	20	8
Potatoes (tubers)	180	25	200	10	20	25	50
Lucerne (dry crop)	280	30	200	250	20	30	10

Source: From G. W. Cooke 1982

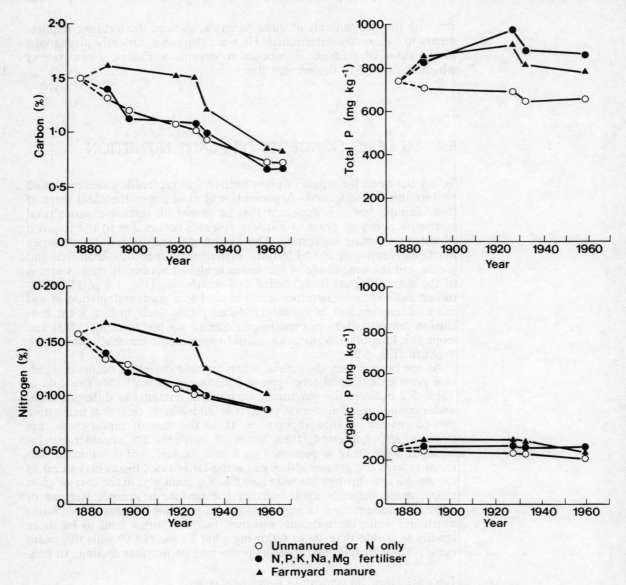

Fig. 5.2   Changes in carbon, nitrogen, total phosphorus and organic
phosphorus contents in the continuous wheat and barley experiments
on Stackyard Field, Woburn (after Mattingly *et al*. 1975).

trast, where the grass is removed for silage or hay, losses may be considerable, particularly when it is appreciated that several 'cuts' may be made each year.

Inputs of nutrients are not only made to replace year to year losses in harvesting. It is also apparent that greatly increased yields may be achieved by raising nutrient reserves of the soil above their 'natural' level by high rates of fertiliser application. This is perhaps nowhere more obvious than in relation to grass crops, where large increases in yield have recently been achieved through liberal applications of nitrogen (see section 7.2.2). Moreover, it is clear that the benefits are seen not only in increased yields; as Armitage and Templeman (1964) conclude, yields are also more consistent from season to season and the growing season may be extended by several weeks.

5.3 THE NATURE OF FERTILISERS

Although Liebig (1842) claimed that nitrogen need not be added as a fertiliser to agricultural crops, it is now recognised that this, together with phosphorus and potassium, represents the main fertiliser requirement in most farming systems. Other nutrients, particularly calcium and some of the micronutrients, may be required under certain conditions but are applied less generally.

In the past, before the development of reasonably precise quantitative concepts of plant nutrition – also before the advent of accurate soil testing and the emergence of an organised fertiliser industry – the main means of fertility control was through the addition of organic manures and fertilisers, in particular farmyard manure (FYM). In recent decades, however, the use of farmyard manure has declined considerably, for a variety of reasons. Inorganic fertilisers became cheaper as mass production became possible and as transport facilities improved. The advantages of ease of handling and storage of artificial fertilisers made them more convenient. Meanwhile, the gradual specialisation of farming systems, and the decline of mixed farming, meant that home production of FYM ceased on many farms, while difficulties of transporting bulky manure deterred its import from livestock areas. Conversely, in intensive grazing areas, the need to store manure and apply it on grassland which is being grazed almost continuously throughout the summer and autumn has created difficulties in utilisation of FYM.

5.3.1 Inorganic fertilisers

The growth in the use of inorganic fertilisers during the last century and a half, and particularly during the last fifty years, is a phenomenon common to most of the developed world. The growth is typical of almost all farming systems with the exception of sugar beet production, and is demonstrated in particular by trends in nitrogen usage (see, also, Hood 1982). Probably the only nutrient to show an overall reduction in use is calcium, though there are still many areas which would benefit from liming (Table 5.3).

Table 5.3 Lime requirements in England and Wales

	Arable land	Temporary grass	Permanent grass	Total agricultural land
% of area needing lime	27	29	44	34
Average lime requirement (t $CaCO_3$ ha^{-1})	3.75	3.65	4.90	4.40

Source: From G. W. Cooke 1975

Inorganic fertilisers are available in a variety of forms, both solid and liquid, single or compound (Table 5.4). Solid fertilisers make up about 95 per cent of the total applications in the United Kingdom and are commonly provided in the form of granules or pellets. Nitrogen, however, is also applied as a liquid fertiliser; in Britain this accounts for only about 6 per cent of total N usage, but elsewhere it is more important – about 30 per cent in Denmark and over 35 per cent in the United States. The

Table 5.4 Common types of inorganic fertilisers

(a) NITROGEN

Fertiliser	Chemical composition	% N	Form	Comment
Ammonium nitrate	NH_4NO_3	34	Prilled or granular	Becoming more widely used; cheap
Ammonium sulphate	$(NH_4)_2SO_4$	21	Crystalline	Now rarely used
Sodium nitrate	$NaNO_3$	16	Granular	Expensive; rarely used
Calcium nitrate	$Ca(NO_3)_2$	15.5	Prilled	Widely used in continental Europe
Urea	$CO(NH_2)_2$	45	Prilled	Expensive; may cause scorch
Anhydrous ammonia	$NH_3.H_2O$	82	Liquid	Cheap; readily lost by leaching
Aqueous ammonia	NH_4OH	21–29	Liquid	Cheap; less susceptible to loss

(b) PHOSPHORUS

Fertiliser	Chemical composition	% P_2O_5	Form	Comment
Ground rock phosphate	Variable	25–30	Powder	Often highly insoluble
Superphosphate	$Ca(H_2PO_4)$	18–21	Powder	*c.* 30% of world P fertiliser
Triple superphosphate	CaH_2PO_4	47	Powder	*c.* 20% of world P fertiliser
Basic slag	Variable	8–22	Powder	By-product of steel–making
Dicalcium phosphate	$CaHPO_4$	40	Granular	Water insoluble
Superphosphoric acid	$H_4P_2O_7$ or H_3PO_4	60	Liquid	Becoming more widely used

(c) POTASSIUM

Fertiliser	Chemical composition	% K_2O Form		Comment
Potassium chloride	KCl	60	Granular	Most widely used K fertiliser
Potassium sulphate	K_2SO_4	48–50	Powder	Expensive; sulphur has some fertiliser value
Kainit	Variable	12–16	Powder	Contains Mg or Na
Potassium nitrate	K_2NO_3	37	Powder	Expensive; mainly used for vegetables

use of liquid fertilisers is growing, largely because of their cheapness and ease of handling.

Another trend in fertiliser practice is the increased use of compound fertilisers. These consist of mixes of nitrogen, phosphorus and potassium, occasionally with small quantities of micronutrients. The composition is referred to by a code representing the percentages of N, P_2O_5 and K_2O; thus a 20 : 10 : 10 fertiliser consists of 20 per cent nitrogen, 10 per cent phosphate and 10 per cent potassium oxide (although in most cases the potassium exists in the form of potassium chloride, KCl), the remainder consisting of the carrier materials (eg. calcium, chloride, sulphate). In 1973 when the last general survey was carried out, 51 per cent of the nitrogen and 92 per cent of the potassium applied in Britain was in the form of compound fertilisers. In Europe as a whole 'complex' fertilisers, including only fertilisers blended during manufacture and excluding later, mechanical mixes by the distributor, accounted for 33 per cent of nitrogen, 40 per cent of phosphorus and 34 per cent of potassium in 1968–69 (FAO 1970). Since then, the proportions have almost certainly risen markedly.

5.3.2   Organic manures

Organic manures comprise a wide range of products, including farmyard manure, composts and urban wastes. Unlike inorganic fertilisers the composition of these varies considerably. FYM, for example, varies according to the quantity of straw or other plant material included, the type, diet and physiological condition of the animal from which the waste products (urea and faeces) were derived, and even the weather. Similarly, the composts vary according to the nature of the plants, their degree of decomposition and their original nutrient content, whilst urban wastes (e.g. sewage sludge) vary in relation to the industrial character of the source area and the degree of treatment. Moreover, in almost all cases, the composition changes during storage due to organic processes. Consequently, the compositions quoted in Table 5.5 are highly approximate. It should also be noted that, in contrast to inorganic fertilisers, manures are chemically impure. Thus, they often contain significant quantities of trace elements such as zinc, copper, nickel, chromium and lead.

The quantities of organic manures applied to the land are not known in detail, for much of the manure is traditionally home produced and does not pass through a commercial market. The most recent full survey of fertiliser practice in England and Wales, in 1957, showed that 50 million

Table 5.5   The chemical composition of farmyard manures and slurries

Manure	Moisture (%)		N (%)		P (%)		K (%)	
	x	range	x	range	x	range	x	range
Turkey manure	55	10–81	1.2	0.4–5.7	0.6	0.22–1.9	0.7	0.08–1.4
Cattle manure	76	8–86	0.6	0.3–2.2	0.1	0.04–0.9	0.5	0.4–1.2
Pig slurry	97	85–99	0.2	0.02–1.0	0.1	0.01–0.35	0.2	0.08–0.33
Broiler chicken manure	32	9–75	2.3	0.4–3.6	0.9	0.09–1.7	1.1	0.25–2.0
Battery chicken manure	66	12–88	1.5	0.5–4.5	0.5	0.13–2.1	0.6	0.17–3.3
Sewage sludge	55	5–94	1.0	0.1–2.7	0.3	0.04–2.1	0.2	0.01–0.7

Source: Data from Berryman 1965

tonnes of manure were produced, and this accounted for about 20 per cent of N applications, 25 per cent of P and 40 per cent of K. Since then the amount of FYM used has undoubtedly dropped considerably, although in other parts of Europe the use of liquid manure (gülle) remains important. Urea and organic rich drainage waters are mixed with faeces and litter, matured, then diluted with water and sprayed on the land. This procedure is used widely in upland, grassland areas such as the mountain regions of Switzerland and Germany, most farms being equipped with underground tanks to enable the manure to be stored until spraying is possible.

5.3.3   Inorganic fertilisers versus organic manures

The changing pattern of fertiliser usage has not occurred without controversy. There has, in particular, been considerable debate about the relative merits of organic manures and inorganic fertilisers, much of it, it might be said, somewhat ill-informed and prejudiced. Clearly, inorganic fertilisers have a wide range of benefits; it is because of these that they have been so widely adopted. Some of these benefits have already been mentioned: their cheapness, their cleanliness and ease of handling, their ease of storage and transport. In addition, the guaranteed composition of inorganic fertilisers means that it is easier to determine rates of application and to predict effects upon yields. It is also easier, particularly with liquid fertilisers, to ensure an even and precise distribution of the fertiliser within the field and, because they can often be applied during tillage or sowing, application of inorganic fertilisers often limits the number of passes necessary over the land.

On the other hand, inorganic fertilisers lack several of the important properties of organic manures. In particular, FYM and composts provide an important input of organic matter to the soil which improves soil structure (Table 5.6). It is difficult to assess the effects of these structural improvements upon yields, for many other factors come into play when FYM is added (including increased nutrient supplies and possibly improved water retention). It is clear from long-term experiments, however, that even occasional dressings of FYM may have a lasting effect upon soil structure, although there is some evidence that the effect is less substantial than that provided by residues from grass crops (e.g. Cooke 1974). What is also apparent is that regular dressings of FYM *can* result in crop yields equivalent to those attained by inorganic fertilisers (Table 5.7). In areas where structural deterioration, and consequent soil erosion, is likely

Table 5.6   Effects of farmyard manure on soil organic matter (organic carbon, kg ha⁻¹) contents in the Hoosfield Continuous Barley experiment

Year	Organic carbon (kg ha⁻¹)		
	No FYM	FYM 1852–71	FYM 1852–1975
1852	30.7	30.7	30.7
1872	—	52.5	—
1882	26.7	51.4	59.9
1913	26.8	43.2	75.5
1946	25.3	38.8	76.3
1975	26.5	41.4	86.8

Source: Data from Jenkinson and Johnston 1977

Table 5.7   Yields of crops under long-term applications of farmyard manure and inorganic fertiliser

Site	Crop	Control (no fertiliser)	FYM	NPK	FYM as % of NPK
Broadbalk[a]	Wheat	2.08	3.50	3.11	112
Saxmundham[b]	Wheat	1.28	2.38	2.43	98
Saxmundham[b]	Barley	1.03	2.03	2.26	90
Hoosfield[c]	Barley	1.59	3.03	2.87	106
Barnfield[d]	Sugar beet	3.80	15.60	15.60	100
Barnfield[d]	Mangolds	3.80	22.30	30.90	72

[a] Data from Dyke (1964): FYM = 35 t ha^{-1}; NPK = 144 kg ha^{-1} N, 32 kg ha^{-1} P, 112 kg ha^{-1} K.
[b] Data from Trist and Boyd (1966): FYM = 15 t ha^{-1}; NPK = 37 kg ha^{-1} N, 16 kg ha^{-1} P, 51.5 kg ha^{-1} K.
[c] Data from Johnston and Poulton (1977): FYM = 35 t ha^{-1}; NPK = 96 kg ha^{-1} N, 32 kg ha^{-1} P, 112 kg ha^{-1} K.
[d] Data from Warren and Johnston (1962): FYM = 35 t ha^{-1}; NPK = 96 kg ha^{-1} N, 34 kg ha^{-1} P; 229 kg ha^{-1} K.

to be a problem, therefore, there may be significant advantages in applying organic manures. Such soils include the fine sands and silts of the arable areas in eastern Britain, though, unfortunately, it is these areas which are furthest removed from large supplies of FYM.

A number of other factors need to be taken into account in assessing the value of organic manures. As has been indicated, one of the main problems is in predicting the effect of FYM upon crop yield. In part this difficulty arises from the natural variability of manure (Table 5.5 p. 107); in part, also, it is due to the unreliable availability of the nutrients contained within the material. Thus, even if the composition is known, it is not easy to determine the quantity of nutrients that will be released to plants. As much as two-thirds of the nitrogen and one-half of the phosphorus, for example, may not be immediately available. Although this delay in nutrient release has the advantage of providing large residual effects and of preventing leaching, it means that short-term crop responses are often less than anticipated. Thus, Herriott et al. (1965), found that gülle liquid manure supplied only about 60 per cent of the nitrogen obtained from equivalent dressings of inorganic fertiliser. This discrepancy is exacerbated by the fact that storage of organic manures results in marked losses in nitrogen through volatilisation.

A further difficulty in predicting the effects of organic manures arises from their slow decomposition. Decomposition occurs largely through microbial activity and is accompanied by a marked increase in microbial populations. These raise the demand for nitrogen and may lead to temporary N deficiencies. Crops growing in the soil at this stage may therefore be hindered by lack of nitrogen.

For this reason, timing of applications of FYM is critical. On grasslands timing is also important because grazing animals reject recently treated grass. Norman and Green (1958), for example, found that dung patches were neglected for one to one and a half years, although there is no evidence to show that the effects of manurial treatment persist for so long. Moreover, as grazing intensities are increased it seems that the fastidiousness of the animals tends to break down (see Ch. 7).

It is clear, however, that given the present structure of farming in many parts of the temperate world and the existing economic conditions, there is little incentive for the farmer to increase the use of organic manures at the expense of inorganic fertilisers. Whether this situation will persist is doubtful. Controls upon the disposal of waste manures are already making it more attractive to return FYM to the land. Furthermore, increases in the costs of petroleum products, of which nitrogen fertilisers are a by-product, may well mean that the price advantage of inorganic fertilisers is lost. The use of manures at least to supplement inorganic fertilisers may therefore become increasingly worthwhile.

5.4   PRINCIPLES OF FERTILISER APPLICATION

Successful fertiliser application requires the supply of fertilisers:

(a) in the right form;
(b) in the correct amount;
(c) at the right time;
(d) in the right place (i.e. correct placement).

5.4.1   The form and amount of fertiliser

Given the wide range of fertilisers available to the farmer, the question of the form of fertiliser may seem to be one of the most difficult decisions. In the past, experiments have implied that there is relatively little to choose between the various forms of inorganic fertiliser and similar crop responses can be obtained whatever form is used so long as the same amount of nutrient is applied (e.g. G. W. Cooke 1975). The choice between the various types of fertiliser, therefore, has often been made upon the basis of cost at the farm gate, cost and ease of storage and application, and availability. More recently, however, evidence has emerged to suggest that crops respond differently to the different forms of fertiliser. Yields of grass, for example, seem to be greatest from nitrate applied as ammonium nitrate and a re-evaluation of the effect of fertiliser composition is now being undertaken (Garwood pers. comm.).

A vast number of studies have been carried out to examine the effects of fertiliser rates upon crop yield and it is clear that, over a wide range of conditions, yields increase markedly with increased applications. Until about 1950, however, few studies demonstrated the full nature of the crop response curves. Since then, it has become clear that yields characteristically rise to a peak as fertiliser applications increase, then decline. Various attempts have been made to fit a mathematical function to these responses and the FAO (1966), for example, have found that three equations fit the data over a wide range of conditions. One, the Misterlich equation, is exponential in form and only applies to conditions up to optimum yields:

$$y = y_0 + d\,(1 - 10^{-kx}) \qquad\qquad [5.1]$$

Where y is the yield with fertiliser
  y_0 is the yield without fertiliser
  d is the limiting response
  k is a constant (assumed to be constant for each major group of fertilisers)
  x is the rate of fertiliser application.

The other two equations suggested by the FAO (1966) are the quadratic equation:

$$y = a + bx + cx^2 \qquad\qquad\qquad\qquad\qquad\qquad [5.2]$$

and the square root equation:

$$y = a + b \sqrt{(x + cx)} \qquad\qquad\qquad\qquad\qquad [5.3]$$

Where a, b and c are constants and y and x are as before. These give generally parabolic curves which fit the data over a wider range of fertiliser applications.

The parabolic relationship between crop yield and fertiliser application is not surprising; in simple terms it reflects the fact that either too little or too much of a particular nutrient suppresses yields. It should also be possible, in theory, to make use of these equations to find the optimum rate of fertiliser application. A range of factors, however, complicate the calculations. First is the fact that the constants in the equations vary from one crop to another; the experimental data on one crop cannot be used to predict responses of other crops. Secondly, environmental variables such as weather, soil type and water availability all influence the relationships. Thirdly, management factors play an important part and variables such as sowing date, previous fertiliser practice and cropping sequence may all affect crop responses. Finally, the relationships are also sensitive to nutrient interactions, so that responses depend upon the degree of limitation imposed by other nutrients (Fig. 5.3).

As a consequence of these considerations, fertiliser recommendations, such as those outlined in Table 5.8, can only be regarded as general estimates. Local factors will undoubtedly modify the requirements of individual crops to individual fields. Morrison *et al.* (1980), for example, found that although the mean response curve to nitrogen for grass swards

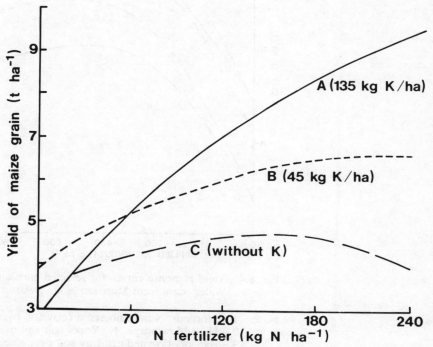

Fig. 5.3   Effects of application of potassium fertilisers on the response of maize to nitrogen fertiliser (from G. W. Cooke 1975).

Table 5.8 Fertiliser recommendations for selected crops

Crop	Fertiliser recommendation (kg ha^{-1})		
	N	P	K
Winter wheat (after cereals)	150–200	50–100	50–100
(after roots)	50–100	40–60	0–60
Spring barley (after cereals)	125–150	50–100	50–100
(after roots)	40–75	40–60	0–60
Sugar beet	125–150	65–125	150–200
Potatoes	120–200	150–220	150–250
Oil-seed rape	120–200	60–80	60–80
Maize	100–150	60–90	60–90
Mown grass	200–400	80–100	60–200
Grazed grass	150–350	50–60	60–100

Source: After Eddowes 1976 and G. W. Cooke 1975

from a wide range of environments approximated to an inverse quadratic form, considerable variation occurred between sites (Fig. 5.4). More general examples are provided by G. W. Cooke (1974, 1975).

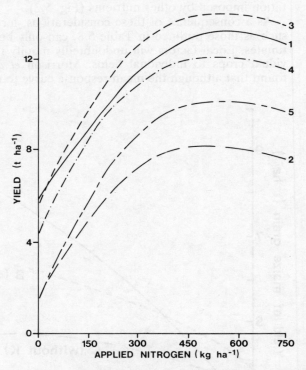

Fig. 5.4   Yield response curves for selected grassland sites in England and Wales (data from Morrison *et al*. 1980).

Notes: 1 – Morpeth, Northumberland (clay soil over till);
2 – High Mowthorpe, N. Yorks (silt soil over chalk);
3 – Bangor, Gwynedd (loamy soil over glacial drifts);
4 – Rocester, Staffordshire (loamy soil);
5 – Cannington, Somerset (loamy-silty soil over sandstone).

5.4.2 Timing of fertiliser applications

The nutrient demands of most plants are not constant throughout the year, but show marked peaks associated with specific growth periods. In addition, fertilisers applied to the soil are rarely static; they may be redistributed through the soil after application and significant proportions may be fixed within the soil or lost through leaching and volatilisation. For these reasons, the timing of fertiliser applications is critical. Fertilisers must be available in the right place at the time they are needed by plants. This means that the time of application must be selected to take account of the nutrient demand characteristics of the crop, where the fertiliser is placed within or on the soil, the mobility of the fertiliser and the prevailing weather and soil conditions.

In an environment which does not limit growth, most plants show a generally sinusoidal growth curve over time (Fig. 5.5), most rapid growth occurring during the early to middle part of the growing season. During this growth period, the nutrient content of the plant characteristically increases until, during senescence, there may be a slight decline in nutrient contents as parts of the plant start to die and decay. The rate of nutrient uptake, however, is often highest early in the growth period, reaching a maximum during the tillering and flowering stages and declining during the ripening phase as nutrients are translocated from the leaves and stalk to the grain (Fig. 5.5). The detailed pattern of nutrient uptake varies, however, from crop to crop and nutrient to nutrient.

A further factor that influences the temporal pattern of nutrient demands is the rate at which nutrients are released from reserves within

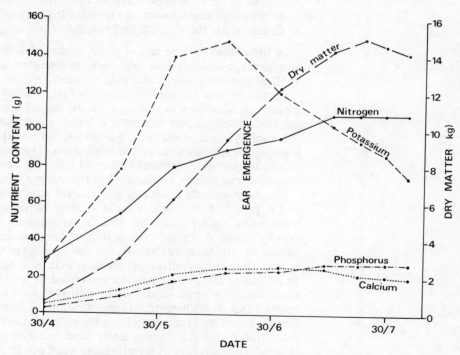

Fig. 5.5 Changes in dry matter weight and nutrient content of wheat during growth (data from Russell 1973).

compounds such as iron and aluminium phosphates, or by exchange from adsorption sites on the colloids. Plants draw most of their nutrients from the soil solution by mass flow of water into the roots. The rate of uptake from the soil solution may be very rapid and continued growth is only possible if the nutrients in the soil solution are replenished by release from the labile pool. This appreciation has led to the concepts of *nutrient intensity* and *nutrient capacity*. The nutrient intensity refers to the concentration of nutrients in the soil solution; it is a measure of the immediately available nutrient content of the soil. The capacity factor relates to the rate at which these nutrients are released from the labile pool; it is a measure of the ability of the soil to sustain the supply of available nutrients in the medium to long term. Studies by Mattingly *et al*. (1963), for example, showed that phosphate uptake during the first two weeks of rye-grass growth was closely correlated with the phosphate intensity ($r = 0.96$–0.98), but after 4 months was related to the phosphate content in the labile pool ($r = 0.98$). Similarly, Mattingly (1965) found that growth of ryegrass after about six weeks was dependent upon the phosphate intensity factor, but after twelve to fifteen weeks was more closely related to the labile pool of phosphate.

The provision of adequate supplies of available nutrients therefore involves consideration not only of the time when plants need their nutrients but also of the ways in which the nutrients are made available. Moreover, in many cases, it is necessary also to take into account the potential losses of fertiliser nutrients through leaching and volatilisation. Thus, the general principles of fertiliser timing include:

1. Allowing sufficient time for the fertiliser compounds to dissolve;
2. Allowing time for movement of the nutrients to plant roots;
3. Minimising waste losses of nutrients from the soil;
4. Catering for the fixation and release of nutrients within the soil.

In the context of the last of these factors, it is important to distinguish between applications that are made to stimulate short-term growth and applications designed to maintain general nutrient contents of the soil.

Fertilisers can be applied either before, during or after sowing. The major consideration in selecting the timing of application is the type of nutrient. Nitrogen, for example, is highly soluble and thus highly mobile. It is rapidly made available to the plants, but it is also quickly lost through leaching, denitrification, and volatilisation. The main constraint upon the timing of nitrogen applications (apart from the obvious constraints of labour and machinery availability) is the danger of waste nutrient loss.

We have already discussed the general structure of the nitrogen cycle in Chapter 1. Plants obtain most of their nitrogen either in the form of ammonium (NH_4) or nitrate (NO_3). Ammonium, however, tends to be rapidly oxidised through tbe action of bacteria and algae to nitrite and thence to nitrate (nitrification), the end product of which is highly soluble. It is in the form of nitrate, therefore, that most losses of nitrogen occur. In addition, the nitrate may be converted to oxides of nitrogen (e.g. nitrous oxide or nitric oxide) or to nitrogen gas by processes of denitrification and volatilisation. Denitrification is essentially an anaerobic process (see Ch. 4) resulting from the microbial extraction of oxygen from the nitrates, and it is consequently most common under waterlogged conditions. The duration of waterlogging need not be long, however, and the process is also encouraged by the presence of readily decomposable organic compounds such as cellulose (Bremner and Shaw 1958). As a result, losses of nitrogen are often greatest after autumn ploughing has incorporated crop residues; the warm, moist soils favour rapid decomposition and denitrification.

In Britain, the danger of nitrogen losses during winter deter autumn applications of fertiliser, and most nitrogen is applied in spring. In the case of spring cereals application is commonly by combine drilling – that is to say, the fertiliser is applied by the same implement that sows the seed – while for winter-sown cereals, application is made either as a side-dressing or as a top-dressing (see section 5.4.3). Early spring application is clearly favourable to some extent, for this ensures rapid growth at the start of the season and may extend the growing season by several weeks. There are, however, factors which militate against applying fertiliser in March or early April; in particular, heavy spring rains may still cause considerable losses, while in wet years the land may not be accessible to machinery. Dressings later in the spring (late April–May) avoid these problems and also give protection against lodging of cereals. In the United States the advantages of spring sowing are less marked and the trend is towards pre-sowing applications of nitrogen, mainly in the autumn (Tisdale and Nelson 1975).

Nitrogen applications to grassland involve rather different problems. An early spring (February or March) application encourages growth and provides grass for an 'early bite' so that animals may be released on to the land sooner (assuming soil moisture conditions are suitable). Subsequent top dressings are often given before each grass cut or each grazing period to stimulate growth. Where the grass sward contains nitrogen-fixing plants such as clover, however, the effects are more complex. Applying nitrogen early in the season ensures greater uptake by the grass and gives higher initial yields, but more of the nitrogen is retained in the crop and removed during harvesting or grazing. The result is that the clover is suppressed and the long-term residual effects of the nitrogen are reduced. With grass-clover swards, therefore, it is usually more profitable to rely upon the clover to give early growth, then apply nitrogen in the middle and end of the growing season to maintain later yields.

Because of the economic and environmental significance of nitrate losses from fertilisers, considerable attention has recently been given to the processes of nitrate leaching, and numerous attempts have been made to model the dynamics of the system. The main pedological factors influencing nitrate movement are drainage conditions, water-holding capacity and pore structure. Leaching tends to be greatest in soils with low moisture content at field capacity (i.e. coarse, sandy soils) because these require relatively small rainfall inputs to displace the pre-existing soil water, and with it the nitrate in solution. As Burns (1977) shows, the rate of leaching also increases with increasing pore size. Water moves rapidly through the larger pore spaces, as it does so displacing the nitrate ahead of it, and carrying it into the subsoil. This results in dispersion of the nitrate throughout the soil (e.g. Wild 1972). Conversely, water movement through finer pores is slower and nitrate displacement is much less rapid while dispersion is reduced; as a result, in soils dominated by a fine pore system, the nitrate is moved slowly downwards as a distinct leaching front (Fig. 5.6).

These factors have to be taken into account in any attempt to model leaching of nitrates. Many of the models that have been proposed are too complex, and require too much data, to have practical value, but Burns (1974, 1977) has developed a relatively simple model that gives good pre-

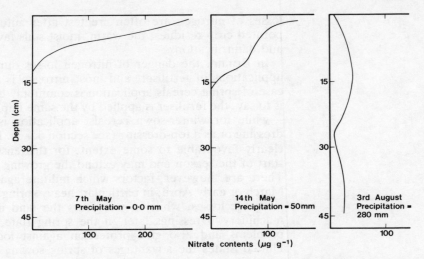

Fig. 5.6   Nitrate leaching profiles in a sandy loam soil supplied with nitrogen
fertiliser (after Burns 1977).

dictions of leaching losses over a wide range of conditions. The general
model can be expressed as:

$$f_h = \left(\frac{100P}{100P + V_m} \right) x \qquad\qquad [5.4]$$

where f_h is the fraction of nitrate below depth h (cm) in the soil
V_m is the moisture content of the soil at field capacity (%)
P is the amount of drainage water that has passed through the soil
(cm)
x is a constant depending on the initial distribution of nitrate in
the soil ($x = h$ for nitrate applied at the surface; $x = h - w/2$ for
nitrate applied at w cm depth; $x = h/2$ for uniformly distributed
nitrate).

The model was tested by monitoring chloride and nitrate leaching on
soils ranging from clay to sand, and a close correlation was found between
predicted and observed amounts of nitrate displacement (Fig. 5.7). More
recently, Burns (1980a, b, c) has extended the model to include consideration of crop characteristics, such as rooting depth. Long and Huck
(1980) have shown that actively growing roots of maize may intercept
downward percolating nitrogen, and undoubtedly root characteristics
have a significant influence on leaching losses. Many other factors are also
likely to be relevant including, in particular, soil structural characteristics,
soil layering and rainfall intensities. Nevertheless, although models such
as this are a considerable simplification of reality, they are clearly valuable
in helping define optimal rates, timing and location of fertiliser application, and may be used to minimise leaching losses.

In contrast to nitrogen, phosphorus is almost immobile in the soil, even
in its 'water-soluble' form. This is because it is actively fixed by chemical
reactions in the soil. A wide range of reaction products may be formed,
depending upon the nature of the fertiliser, the composition of the soil
and moisture conditions. An example is illustrated in Fig. 5.8. As this
shows, the monocalcium phosphate granule is gradually saturated as water
vapour moves into the granule under the influence of the osmotic poten-

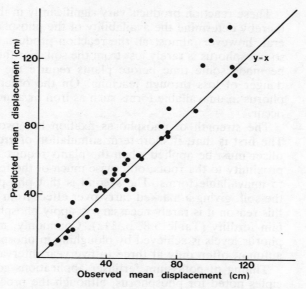

Fig. 5.7   Predicted and observed mean displacement depths for nitrate
fertilisers in soils (from Burns 1977).

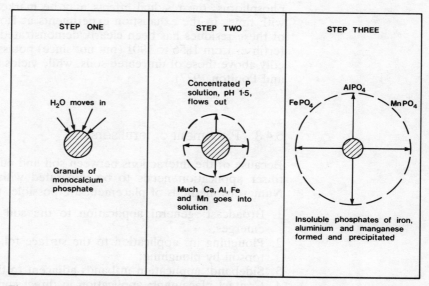

Fig. 5.8   Diffusion of phosphate into the soil from a granule of monocalcium
phosphate (from Tisdale and Nelson 1975).

tial. This produces a saturated solution of monocalcium phosphate ($CaH_2PO_4.2H_2O$) and dicalcium phosphate dihydrate ($CaHPO_4.2H_2O$) which moves out of the granule, leaving behind a residue of dicalcium phosphate ($CaHPO_4$). The mobile, saturated solution is extremely acid (pH 1.5) and in acid soils it reacts with iron, aluminium and, occasionally, manganese to form phosphates of these elements. In calcareous soils, the phosphate front is precipitated as dicalcium phosphate around the particles of calcium carbonate.

These reaction products vary significantly in their solubilities, and they thereby determine the availability of the phosphate to the plants. In general, however, almost all the reaction products are resistant to leaching so phosphorus is rarely lost from the soil. For this reason applications can be made some time before plants require the nutrient without serious danger of loss through leaching. On the other hand, fixation of phosphorus in unavailable forms such as iron or aluminimum phosphates may occur.

The strength of phosphorus fixation has two important implications. The first is that, if short-term stimulation of growth is required, the fertiliser must be applied when the plants require the nutrient and in close proximity to the roots; otherwise much of the fertiliser may be converted to unavailable forms. The second is that phosphorus residues persist in the soil, giving a marked carry-over effect from one year to another. For this reason it is rarely necessary to apply phosphorus every year to maintain fertility (Table 5.8, p. 112). Commonly, maintenance of soil phosphorus levels is achieved by ploughing in phosphate fertilisers during the autumn, often only at three to five year intervals.

Timing of potassium fertiliser applications generally follows the principles noted for phosphorus, although the processes involved within the soil are different. Potassium is a cation and is consequently strongly adsorbed to exchange sites on the soil colloids; in this way it is protected from leaching and reserves are maintained for subsequent crops. As with phosphorus, the residual effects may be marked, although they decline with time. In the exhaustion experiments at Rothamsted the persistence of these residues has been clearly demonstrated. In plots treated with K fertiliser from 1856 to 1901 (but not since) potassium levels are still markedly above those of untreated soils, while yields are also higher (Johnston and Poulton 1977).

5.4.3   Placement of fertilisers

Because of the interactions between soil and nutrients, the timing of fertiliser application needs to be integrated with appropriate placement. Numerous methods of placement are possible, the most common being:

1. Broadcast: general application to the soil surface before the crop emerges;
2. Ploughing in: application to the surface followed by mixing into the topsoil by ploughing;
3. Sideband: application in bands adjacent to the seed;
4. Contact placement: application in direct contact with the seed (combine drilling);
5. Side-dressing: placement in narrow rows at the surface after crop emergence;
6. Top-dressing: general application to the crop after emergence.

The selection of an appropriate placement method must take account not only of the crop type, but also of the nature of the fertiliser, soil conditions and weather.

Application close to or in contact with the seed or roots ensures rapid uptake during early growth but it may have a number of detrimental effects. High concentrations of fertiliser close to the seed may cause scorching and reduce germination; urea is particularly dangerous in this respect, while potatoes, sugar beet and many other row crops (e.g. peas, beans, swedes) are especially susceptible to damage. High fertiliser concentrations may also encourage reduced root growth which, in dry areas, can cause problems of moisture stress later in the year. Moreover, soluble nutrients (e.g. nitrogen) may be lost by leaching before plants exploit the fertiliser.

Surface applications (e.g. broadcasting, top-dressing, side-dressing) are only viable if the fertiliser is able to move through the soil into the rooting zone and are therefore most appropriate for nitrogen. Both phosphorus and potassium are fixed by the soil before they can reach the roots, although small quantities may be absorbed through the plant leaves in the case of top-dressed fertiliser.

Sideband drilling is widely used in the United States but, in the case of relatively insoluble fertilisers such as phosphorus and potassium, the distance between the seed and the fertiliser must not be too great if fixation is to be minimised. In Britain, contact placement, by direct drilling, is widely used to give early growth, while ploughing in is used for maintenance dressings of phosphates and potassium.

5.5 CONCLUSIONS

It is apparent that fertilisers play a vital role in most intensive agricultural systems; it is also clear that the variety of fertiliser practices available to, and employed by, the farmer are extensive. They depend upon the crops being grown, the soil conditions on the farm, general climatic and shorter-term weather conditions and the availability of fertilisers and of manures. In addition, as with all management procedures, infrastructural and cultural factors are also important. The availability of labour, particularly at key periods such as the spring and late summer – early autumn, imposes a major constraint upon the timing of fertiliser applications. Moreover, as farms become more capital and less labour intensive, year-round applications are encouraged which are made whenever soil conditions are suitable and labour and machinery free. This is particularly the case in areas of specialisation, where large areas of land require specific treatments during short periods of the year.

The importance of fertilisers to increased agricultural productivity can hardly be under-estimated. Perhaps more than any other single factor, improvements in crop nutrition are responsible for the rise in yields over recent decades. It is becoming apparent, however, that fertilisers affect more than agricultural systems. Losses of fertiliser to streams and lakes are contributing to eutrophication of surface waters (e.g. Green 1973), while there is evidence for marked increases in nitrate concentrations in groundwaters. (e.g. Cooke 1976; Wellings and Bell 1980; see Ch. 9). Moreover, the present dependence on fertiliser to maintain yields is at least in part a response to the relative cheapness of inorganic fertilisers. This is a situation that is beginning to change as energy costs rise. In the future, there may well be reasons to reassess the policy of fertiliser usage in intensive agricultural systems.

6 PEST CONTROL AND CROP PROTECTION

6.1 INTRODUCTION

An agricultural pest may be defined as any organism that interferes with the production of an agricultural crop. As such, pests obviously include parasitic organisms such as aphids and slugs, and vertebrates such as foxes and rabbits. In addition, however, they comprise both diseases and weeds.

The effect of these pests upon production can be considerable. In 1970, as much as 40 per cent of the total crop in Latin America was damaged; in Africa during that year losses amounted to about 30 per cent of the total harvest (Fletcher 1974). In temperate and developed areas of the world losses are much less, though, even so, often in the order of 20 per cent or more (Table 6.1). It has not always been so, however. Weeds and crop diseases may have been as important as fertility problems in limiting yields during the Middle Ages. Ernle (1961), for example, refers to Marshall's 1789 survey of Gloucestershire which showed 'beans hidden among mustard growing wild as a weed; peas choked by poppies and corn marigolds, every stem of barley fettered with convolvulus; wheat pining in thickets of couch and thistle'. Ernle concludes: 'It is not surprising that the yield of wheat was anything from 18 bushels an acre down to 12 or 8 bushels.' Since then, however, and particularly during the last twenty to thirty years, the investment in methods of crop protection has increased considerably and protection methods – especially the use of chemical pesticides – now represent a major input in most farming systems (Table 6.2). In this chapter, agricultural pests will be considered from the point of view of their impact on farming productivity. The ecological aspects of modern pesticide use will be more fully considered in chapters 10 and 13.

Table 6.1 Estimated crop losses from pests and diseases in England

Crop	% Crop loss			Actual yield
	Mildew	Rust	Others	(t ha^{-1})
Barley	34.5	7.6	3.8	3.54
Wheat	20.0	9.5	4.7	3.82

Source: From Crowdy 1967

Table 6.2 Costs of crop protection in different farming systems in Britain

Crop	1967		1975	
	Cost (£ ha^{-1})	% of total cost	Cost (£ ha^{-1})	% of total cost
Winter wheat	1.90	2.4	7.20	3.1
Early potatoes	6.50	1.8	12.50	1.7
Maincrop potatoes	8.00	2.1	40.00	4.4
Sugar beet	11.50*	5.2*	42.00	10.4

Source: Data from Eddowes 1976
* 1968 data

6.2 THE NEED FOR CROP PROTECTION

6.2.1 Weeds

Weeds present one of the most widespread and intractable of agricultural problems, for they not only compete for light but also for water and plant nutrients. Moreover, they make harvesting more costly and, if incorporated into the harvested crop, reduce its quality. In addition, weeds often provide habitats for other pests and diseases.

A large range of weeds occur in intensive farming systems of temperate areas. Arable systems are particularly affected by wild oats (*Avena* spp.) and black-grass (*Alopecurus myosuroides*). Griffiths (1972) estimated that wild oats were a serious problem on about 500,000 ha of arable land in Britain, while a further 500,000 ha were lightly infested. Black-grass occurred on some 300,000 ha of arable land, of which about a third was seriously affected. Both of these weeds may reduce yields by 20–30 per cent when infestations are heavy. Wild oats, in particular, seem to be an increasing problem. They spread in a variety of ways – by birds, by the sowing of contaminated seeds, and by dispersal in straw bales, grain loads, farmyard manure and machinery – and they are rapidly infesting previously unaffected areas (Elliott 1972).

Both arable and grasslands are also affected by a variety of perennial grass weeds. Amongst the most widespread are couch-grass (*Agropyron repens*), black-bent (*Agrostis gigantea*) and creeping bent-grass (*Agrostis stolonifera*). Invasion of weeds into pastureland often occurs when damage by over-grazing, moles or excessive winter kill of the sward leaves areas of exposed soil. To a great extent grazing controls the spread of many weeds, but where grazing pressure is less intensive, weeds such as dock (*Rumex* spp.), thistles (*Cardium* spp. and *Cirsium* spp.) and, especially in upland areas, bracken (*Pteridium* spp.) and rushes (*Juncus* spp.) present serious problems. Tivy (1973), referring to the 'bracken problem' in Scotland, comments: 'the problem of bracken infestation of upland pasture has become much more acute in approximately the last hundred years, and has stimulated a flood of research on the utilization and control of the fern' (see also Ch. 12).

6.2.2 Animal pests

Insect pests are probably the main cause of crop loss throughout the tropics and subtropics, but are of relatively minor significance in many temperate areas. Together with slugs, nematodes and birds, insects cause localised damage in some years, largely in response to specific weather conditions. During 1980, for example, prolonged hot, dry conditions throughout much of central USA resulted in widespread damage of maize and other cereal crops by insect pests. Extensive infestation by aphids similarly followed the drought of 1975–76 in western Europe, while there have been many incidents of widespread eelworm damage in cereal crops in East Anglia during recent decades, apparently due to the occurrence of mild winters which have allowed the eelworm larvae to survive in greater numbers from one season to another.

Insect pests fall into two broad categories: those that feed off the plants mainly during the larval stage and those that attack the crop throughout their life. Among the former are the caterpillars of butterflies and moths (e.g. *Pieris* spp., *Agrotis* spp.), wireworms (*Agriotes* spp.), weevils (*Sitona* spp.) and fly pests such as the cabbage root fly (*Erioischia brassicae*) and wheat bulb fly (*Leptohylemyia coarctata*). One of the most serious larval pests is the frit fly (*Oscinella frit*) which attacks pasture grass. According to Clements (1980) this may cause losses of up to 30 per cent in ryegrass yields in Britain. The main pests which feed on plants in their adult stages as well are aphids such as the grass aphid (*Metopolophium fescue*) and grain aphid (*Macrosiphium avenae*), which attack wheat, and blackfly (*Aphis fabae*) which attacks sugar beet.

The damage done by insect pests varies. Some, such as the caterpillars, eat the leaves and thereby reduce the photosynthetic efficiency of the plant. Others, including the wheat bulb fly in Europe, and, in North America, the wheat-stem sawfly (*Cephus cinctus*) and the European corn-borer (*Pyrausta nubilalis*), bore into the stems and damage the xylem cells, diminishing the rate at which water and nutrients are transmitted to the leaves. As a result, yields are reduced, cereals may fail to develop ears and, when damage is severe, the stems may break. Yet other pests, such as the corn earworm which is estimated to destroy about 2 per cent of the maize yield annually in the United States, attack the ear or fruit of the plant. Root-eating insects like the eelworm (*Heterodera* spp.), wireworm (*Agriotes* spp.) and leatherjacket (*Tipula* spp.) attack the roots or tubers of crops such as potatoes, sugar beet and cereals. This inhibits water and nutrient uptake and reduces yields considerably. In addition to reducing the yield, the effect of all these pests is to diminish the quality of the crop.

A further detrimental consequence of damage by pests is the encouragement they give to diseases. Particularly in the case of sucking insects, such as the aphids, damaged areas may become susceptible to diseases including moulds and viruses. Even more importantly, the migrating insects often carry with them virus diseases which are transmitted to the host plants. Potato leaf-roll, sugar beet yellow, and barley yellow dwarf virus disease are common examples.

Non-insect pests are locally important, although their significance is often over-estimated, largely because the animals involved are visible and well known. Rabbits were once a major problem, although since the introduction of myxomatosis into Britain their effect has been reduced. However, in some areas they are now again considered to be a serious pest. Otters, badgers, and, with more justification perhaps, squirrels and deer have also been cited as pests in some areas, while many birds, including pigeons, pheasants and rooks may consume measurable quantities of cereal seed. Many of these aspects are considered in more detail in Chapter 10.

The assessment of damage by both insect and other pests remains a problem, however, for in many cases the close interactions between the various organisms mean that they have indirect as well as direct effects.

6.2.3   Diseases

Far more important than animal pests in most agricultural systems are plant diseases. Jenkins (1973) estimated that, in 1973 alone, leaf diseases caused a £70 m. loss of cereal yields. The fungus, take-all (*Ophiobulus graminis*), may cause total loss of cereal yields at the height of its cycle, while King (1973) calculated that powdery mildew (*Erysiphe graminis*) may cause losses of 8–14 per cent in barley yields. Similarly, Eddowes (1976) quotes a survey by Campbell which found a 10 per cent reduction in yields between 1969 and 1973 in the west Midlands of England due to barley mildew.

Crop diseases are transmitted in two main ways: either through the soil or through the air. Soil-borne diseases generally require the presence of protective habitats such as weeds, stubble or other crop residues if they are to survive the winter. Two of the most common soil-borne diseases in intensive cereal systems are take-all and eyespot. Take-all affects both wheat and barley, especially on calcareous soils. Populations show a marked cyclical pattern, building up to a maximum every four or five years in wheat, or five to six years in barley, then declining (Fig. 6.1). Eyespot is caused by the fungus *Cercosporella herpotrichoides* which commonly overwinters on stubble, then releases myriads of spores during the spring. These attack the seedlings, damaging the tillers and weakening the stems. As the ears of the corn fill, the stems may break (Eddowes 1976).

Air-borne diseases are transmitted either by wind or by organisms such as aphids. Powdery mildew, rusts and smuts are common examples in

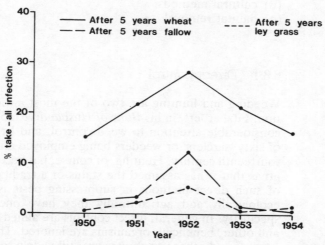

Fig. 6.1   Levels of take-all infection in continuous wheat grown after different crops (data from Hood and Proctor 1961).

cereal crops. Yellow rust (*Puccinia striformis*) attacks wheat, causing bright yellow pustules on the leaves and clusters of spores on the grain, which ultimately shrivels to give greatly reduced yields. Stem rust (*Puccinia graminis*) is the most destructive disease of wheat in the United States, and is apparently favoured by warm, moist weather (Martin and Leonard 1967). Glume blotch (*Septoria nodorum*) is extensive on wheat in western Britain. According to Gair *et al*. (1972) it is the major limitation on the growth of winter wheat in the wetter parts of the country. Virus yellow is one of the most serious diseases of sugar beet in Britain, and is transmitted by the green aphid (*Myzus persicae*).

Once a crop is infected by a pathogen, of course, there is a frequent danger of the perpetuation of the disease in the seeds of the rootstock of the infected plant. Potato blight (*Phytophthora infestans*) is a typical example, and is probably the most common cause of disease in potatoes. Infected tubers may be left in the soil, stored in clamps, or disposed of in dumps; in all cases the disease may then spread to the new crop. A survey in East Anglia, for example, showed that over 50 per cent of the early outbreaks of blight were derived from the rotting dumps of discarded potatoes (Boyd, quoted in Eddowes, 1976).

6.3   METHODS OF CROP PROTECTION

The wide variety of weeds, pests and diseases threatening agricultural crops demands a corresponding range of crop protection methods. These include five main techniques, each of which may be used in an attempt to suppress, deter, eradicate or prevent infestation:

(a) direct control (e.g. weeding);
(b) chemical control;
(c) biological control;
(d) cultural methods;
(e) habitat removal.

6.3.1   Direct control

Weeding and hunting are two of the most ancient forms of crop protection. Fitzherbert, in his *Boke of Husbandrye* (1523), for example, devoted considerable attention to weed control, and Ernle (1961) quotes the use of sixty 'sarclers' or weeders being employed on one Suffolk manor in the fourteenth century. Hunting, of course, has become so established in some areas that it has assumed the status of a traditional sport. The efficiency of such direct methods in suppressing pests is often controversial, but ecologically and agriculturally they have one major advantage: their specificity. In general, direct controls are aimed at clearly indentified pests and other, beneficial, organisms are ignored. Thus, foxes, deer, squirrels, pigeons and other vertebrates are still widely controlled by hunting, while hand-weeding is employed in horticulture and, to a lesser extent, in cereal farming (e.g. the hand-rogueing of wild oats).

The main disadvantages of direct methods of crop protection are that they are time-consuming and, in some cases, costly; Eddowes (1976) quotes a rate of 2.5 man hours/ha for hand-rogueing of wild oats. On the other hand, hunting costs can often be offset by hiring out sporting rights to the land.

6.3.2 Chemical control

Prior to the nineteenth and early twentieth centuries a variety of chemical poisons were developed for use in agriculture. These included sulphur, copper sulphate and lime, lead arsenate, and hydrocyanic acid (Martin 1964). The main stimulus for research in the development of pesticides came, however, during the Second World War, and as a result a number of important compounds emerged in widespread use during the 1950s, including DDT, BHC, 2,4-D and MCPA. In the years since then a large range of pesticides has been developed, many of them highly specific and toxic; they include compounds to attack insects (insecticides), weeds (herbicides), fungi (fungicides) and worms (nematicides).

Insecticides

The composition of these pesticides varies according to their purpose. Initially, many insecticides were based upon chlorinated hydrocarbon compounds. These were highly persistent and had a relatively low toxicity. Amongst their disadvantages were their tendency to accumulate in the soil, runoff waters and biota (Edwards 1969a), and the fact that organisms often developed resistance to them. More recently, organophosphate and carbamate pesticides have been introduced (e.g. D-D, aldrin, lindane and phorate). Most of these are less persistent, are more highly toxic and are more specific.

Organophosphate insecticides operate by damaging the nervous systems of the target organisms. Some, such as phorate, are systemic in that they are absorbed through the roots or leaves of the plants and are then ingested by insects feeding off them (Table 6.3). On the whole, the systemic

Table 6.3 Commonly used insecticides

Pesticide	Main uses
DDT	Leatherjacket; bean weevil; cutworm
Aldrin	Leatherjacket
Dieldrin	
Heptachlor	
BHC	Flea beetle; bean weevil; wireworm; leatherjacket
Phorate	Aphids; cabbage root fly; frit fly; eelworms
Malathion	Aphids
Primicarb	Aphids; grain weevil, saw-toothed grain beetle
Diazinon	Cabbage root fly; carrot fly
Chlorfenvinphos	Cabbage root fly; carrot fly; frit fly; wheat bulb fly
Methiocarb	Slugs
Aldicarb	Eelworms
HCH	Wireworms; leatherjacket; flea beetle

insecticides do not kill predatory organisms unless the organisms come into direct contact with the compound on the plant leaves. Systemic insecticides are generally used either as plant sprays or as granules or solutions which are applied to the soil. Non-systemic insecticides attack the pests directly and are normally applied as sprays.

Herbicides

Herbicides include compounds which are highly selective to particular groups of plants, and also more general pathogens that attack all green plants (Table 6.4). Most of the latter are residual herbicides, which are applied to the soil and thence are taken up by the plants. The selective herbicides, on the other hand, are commonly applied as foliar sprays. They operate either on contact, or by absorption through the leaves, from where they may be translocated through the plant, often to the roots and shoots. They act through their effects upon phytological processes or cell division and elongation. Thus, 2,4-D, which is widely used to control broad-leaved weeds in grass or cereal crops, inhibits cell division and elongation and thereby prevents root extension. Roots lose some of their ability to take up nutrients and water, photosynthesis is reduced, and the efficiency with which the phloem tissue carries food material through the plant is diminished (Harrison and Slife, quoted in Eddowes 1976).

Table 6.4   Commonly used herbicides

Type of herbicide	Examples	Mode of action	Main uses
Phenoxy acids	MCPA; 2,4-D; mecoprop	Growth regulator	Selective herbicide in cereals
Carbomates	propham; barban	Growth inhibitor	Pre-emergence control in sugar beet
Heterocyclic nitrogens	simazine; atrazine	Photosynthesis inhibitor	Total control of herbaceous weeds (high concentration); selective control of annual weeds in maize, field beans
Phenylureas	linuron; diuron	Photosynthesis inhibitor	Pre- and post-emergence control of annual weeds
Acedamides	propachlor	Photosynthesis inhibitor	Pre-emergence control of grasses in vegetables
Trichloroacetic acids	TCA	Growth inhibitor	Control of couch-grass and wild oats
Nitriles	ioxynil	Growth inhibitor	Post-emergence control of cleavers, mayweeds in cereals
Bipyridiliums	paraquat; diquat	Photosynthesis inhibitor	General weed control
Benzoic acid	dicamba	Growth regulator	Broad-leaf weed control in cereals (mixed with other herbicides)

Fungicides

Fungicides are used both to treat seed prior to planting and as foliar sprays (Table 6.5). Cereal seeds, for example, are often treated with organomercury or carboxin to control a variety of fungal diseases including smuts and bunts. More generally, a wide range of fungicides are given as foliar sprays to prevent diseases such as mildew and rust in barley and wheat and blight in potatoes. Most of these operate on contact, so leaves forming after spraying have little protection. Thus Brooks (1970) found that the degree of control of mildew depended on a variety of factors, including the inherent activity of the fungicide, the rate of application, the persistence of the compound and, most critically, the timing of the application. In this context, systemic fungicides, such as benomyl and ethirimol, which are absorbed by the plant and transmitted through the leaves, are more effective than the non-systemic compounds (e.g. drazoxolon).

Table 6.5 Commonly used fungicides

Fungicide	Main uses
Ethirimol	Cereal mildews; septoria
Benomyl	Eyespot; leaf and glume blotch in cereals
Captafol	Mildew; rhynchosporium; septoria in cereals
Organomercury	Bunt; covered smuts; leaf stripe in cereals
Carboxin	Loose smuts in cereals
Tridemorph	Cereal mildews
Carbendazim	Eyespot in cereals
Copper salts	Potato blight
Butafume	Gangrene; skin-spot in potatoes

Advantages and disadvantages

The use of chemical compounds to control pests, diseases and weeds has increased considerably in recent years, and they undoubtedly have a large number of advantages. Among these is their selectivity and their ease of application; they can be used quickly in the face of a specific threat to the crop. In the past they have also been relatively cheap, but with the increase in oil prices costs have risen considerably. Moreover, it is undeniable that some chemical pesticides have adverse environmental effects. Some of the effects on the soil will be discussed later in this chapter; more general ecological consequences are considered in Chapter 10.

6.3.3 Biological control

Biological controls are used mainly to limit pests and weeds. In general, control is achieved by use of predatory species that specifically attack the target organism. A critical factor in the effect of predatory control is the density-dependence of the organisms involved: the activity of the predator is assumed to be dependent upon the population density of the prey or host. Population increases in the target organisms theoretically result in a proportionate increase in the activity (or numbers) of the predators; a decline in the prey causes a related reduction in predator activity. In this way, a balance can be maintained between the two groups of organisms.

The population level at which this balance occurs depends to a great extent upon the degree of specificity of the predator. Organisms that are highly selective tend to maintain low populations in their prey and react very rapidly to changes in their numbers. Conversely, organisms with a wider food base are less sensitive to changes in any one pest, and consequently permit the prey to reach higher population densities.

Several other factors also influence the control efficiency of predatory organisms. The reproductive capacity is important, for this determines the ability of the organisms to expand their populations in response to increases in their food supply. The search effectiveness of the predator is also crucial. This affects the extent to which the predatory organisms can locate their prey, particularly when numbers are scarce. Search effectiveness depends upon both the inherent mobility of the organisms (both predator and prey) and the character of the environment. In the case of soil organisms, for example, soil structure and porosity exert a major constraint on mobility; in poorly structured clay soils, particularly, predatory organisms may be severely restricted. Related to this is the question of adaptability and tolerance. The most effective predators tend to be those that can adapt to, or are tolerant of, a wide range of conditions.

Numerous examples of biological control of pests can be cited. The toxic weed *Hypericum perforatum* in California was controlled by the importation of two beetles, *Chrysolina hyperici* and *C. quadrigemina*. Within 10 years the weed had been reduced to 1 per cent of its maximum distribution and in the seven years from 1953 to 1959 a total saving of over $20 m. was made through improved livestock performance (DeBach 1964). Similarly, the green vegetable bug (*Nezara viridula*), which attacked tree, arable and vegetable crops in the Americas, Africa and Australasia during the middle part of this century, was controlled by various predators and parasites. Australia, for example, introduced the egg-parasite *Trissolcus basalis* from Egypt with great success in the coastal areas, but less satisfactorily inland. Subsequent importation of different strains from Pakistan completed control in these inland areas (F. N. Ratcliffe 1965). Similar success has been achieved in New Zealand (Cumber 1964) and many other parts of the southern hemisphere (DeBach 1974).

The main advantages of biological control are the avoidance of environmental damage, the permanence of the effect, its relative cheapness, its wide applicability and the fact that target organisms rarely, if ever, develop resistance. Against this must be set the difficulties of finding suitable predators or parasites, and the still relatively frequent failure of selected species to establish themselves in their new habitat. In addition, there is often a danger of unforeseen side-effects, in that the predators may attack beneficial organisms and may themselves become pests.

6.3.4   Cultural methods

Cultural methods are still probably the most widely used processes of crop protection in many parts of the world. As we saw in Chapter 3, one of the main purposes of tillage is to control weeds, and it also helps to control many soil pests and diseases, for burial of crop residues, removal of weeds and acceleration of organic decomposition destroy habitats within which the pests would otherwise survive the winter.

Under continuous arable cultivation there are also several plant pathogens that build up from year to year within the soil; for example take-all (*Ophiobolus graminis*) and eyespot (*Cercosporella herpotrichoides*). These are controlled by crop rotations; grass or root crops (e.g. potatoes or sugar beet) are often used in rotation with short periods of continuous cereals, thus breaking the cycle of disease. Similarly, arable weeds such as wild oats (*Avena* spp.), which are not easily eradicated by herbicides, may be reduced or eliminated by putting the land down to grass for several years.

In addition, improvement of soil conditions by fertilising, liming and draining may help control many soil-borne diseases. The effect results partly from the elimination of conditions that favour survival of the pathogens; for example, both slugs and brown foot rot (*Fusarium* spp.) – a major problem with sheep in upland Britain – are encouraged by wet soil and can therefore be controlled by improving drainage. It results also from the more indirect effect of increased plant growth and vigour; in this way the field crop is better able to compete with or recover from the consequences of the pest.

In many ways, cultural methods provide the most effective method of pest control, for they often ensure almost complete eradication of the pests and have prolonged residual effects. Many cultural techniques also fit readily into the established cropping system and thus avoid or reduce the need for additional inputs (e.g. of chemical pesticides). Additionally, methods such as crop rotation, for example, may have beneficial side-effects on soil fertility, while the ecological effects (in terms of pollution or damage to surrounding ecosystems) are small. Nevertheless, several factors militate against the use of cultural methods of pest control. Crop rotation, for example, is often labour-demanding and, because it involves giving the land over to relatively unprofitable crops, may be uneconomic. In intensive arable systems, therefore, chemical methods of crop protection are widely preferred, and in systems based on direct drilling, the use of pesticides is essential to control weeds and insect pests (Ch. 3).

6.3.5 Habitat removal

Many of the weeds and pests that attack agricultural crops do so from bases within the semi-natural vegetation scattered throughout the landscape. Thus, bird pests, such as pigeons and rooks, and animal pests such as rabbits and foxes, often breed in woodland and coppice, but feed in farmland. Areas of rough pasture, moorland or heathland are also important refuges for wildlife, and therefore for agricultural pests. Possibly the most significant semi-natural vegetation, however, is provided by hedgerows. A wide variety of animals live within these varied habitats (Pollard *et al.* 1974) and several of them may forage on adjacent agricultural land. In addition, the continuous network of hedgerows gives ready access for these animals to farmland. Weeds and diseases, too, may overwinter in hedgerows, while it is also possible that plants may spread along these natural corridors, and thereby infest agricultural land.

The removal of habitats such as coppices, hedgerows and moor is therefore one of the main ways of protecting agricultural crops from pests, and, in the long term, possibly the most effective means of eradication. The method is not without its problems, however, as will be considered in more detail in Chapter 11. The habitats that are destroyed are important as breeding grounds not only for unwanted organisms such as weeds and animal pests, but also for many beneficial creatures. Bees (*Apis* spp.) for example, which help pollinate plants and, even with self-pollinating crops, seem to contribute to higher yields (MAFF 1970b), ladybirds (*Coccinellidae*), which attack pests such as aphids, and many birds which help control both soil- and air-borne pests all live within these semi-natural refuges. Elimination of the habitats removes these creatures also. In addition, hedgerows and woodland often provide useful windbreaks, giving shelter to livestock (Sturrock and Cathie 1980) and protection both against wind damage to crops and against soil erosion (Radley and Simms 1967; Briggs and France 1982), while their contribution to the visual quality of the landscape is considerable. Particularly on a large scale, therefore, habitat removal may have adverse ecological, aesthetic and even agricultural effects.

6.4   PESTICIDE ACTIVITY

With the expansion in the use of pesticides over the last thirty years has come the appreciation that pesticides may have considerable impact upon the environment, and ultimately upon man himself. The result is that a great deal of attention has been devoted to pesticide activity. The main areas of concern in this context are:

1. Pesticide toxicity;
2. Pesticide specificity;
3. Pesticide persistence.

6.4.1   Toxicity and specificity

Toxicity and specificity are closely related. Toxicity refers to the ability of the pesticide to kill the target organism; specificity relates to the degree of selectivity of the effects. Pesticide toxicity may be quantified in a variety of ways. One of the most common is in terms of the LD_{50} value: the lethal dose necessary to ensure a 50 per cent kill of the target organism. LD_{50} values are affected by a number of factors, including the chemical composition of the pesticide, its method of application, the physiology and activity of the target organism, and environmental conditions at the time of application. On the whole, as has been noted, organophosphate insecticides are more toxic (i.e. they have lower LD_{50} values) than organochlorine compounds. This is partly because they are more volatile and soluble and spread more rapidly through the soil. Similarly, insecticides applied as fumigants tend to be more toxic than solid or liquid forms.

In the case of soil organisms, size and activity are of paramount importance. More active organisms are often more prone to pesticides than are relatively static organisms, for in moving through the soil they assimilate larger quantities of the compound. For this reason, predatory organisms are often more intensely affected than their prey (Fig. 6.2).

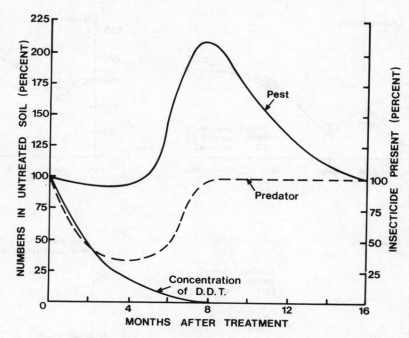

Fig. 6.2 Effects of DDT on predator and prey organisms in the soil (after Edwards 1969b).

Edwards *et al.* (1967) analysed the effects of DDT and aldrin on the soil fauna and found that while neither pesticide affected *Lumbricidae, Enchytraeidae* or *Nematoda*, DDT caused significant reductions in the populations of *Acarina*, while aldrin attacked *Colembolla* and *Pauropoda* (Fig. 6.3). The interdependence amongst soil organisms, demonstrated by Edwards (1969b), means that effects are often indirect.

It is notable that not only insecticides, but also herbicides and fungicides may affect organisms in this way, although in the case of herbicides rates of application are normally too low to be significant. Fungicides, on the other hand, commonly 'serve as partial soil sterilants, killing large segments of the saprophytic population and thereby eliminating the many reactions these micro-organisms catalyse', (Alexander 1969). These reactions include, in particular, processes of nitrification, and it has been shown on a number of occasions that far from excessive concentrations of fungicide may inhibit nitrate formation (e.g. Eno 1957; Koike 1961).

Environmental conditions influence both the toxicity and the specificity of pesticides to a major degree. Weather conditions at the time of application are especially important for they control the rate at which the pesticide is washed into, through and out of the soil, and also influence processes of retention and degradation; weather therefore determines the amount of applied pesticide available to attack actively the target (or other) organisms. Harris (1964, 1966) and Harris and Mazurek (1964) considered the effects of soil conditions upon the toxicity of insecticides in the soil and demonstrated that soil texture, organic matter content and soil moisture content all have a significant effect. LD$_{50}$ values increased curvilinearly with organic matter content in moist soils, but in dry soil the relationship broke down. Muck (organic) and clay soils also reduced the

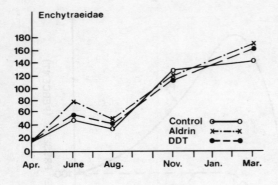

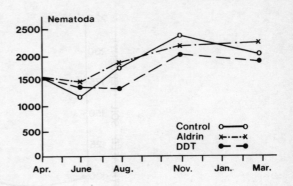

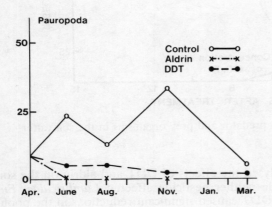

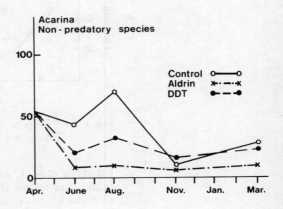

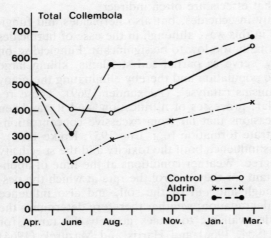

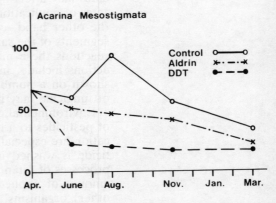

Fig. 6.3   The effects of DDT and aldrin on selected groups of soil organisms in a sandy soil (after Edwards *et al.* 1967).

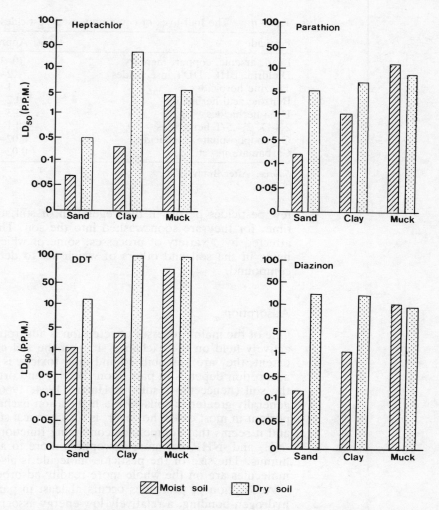

Fig. 6.4   Effects of soil type on the toxicity of four insecticides (after Harris 1966).

toxicity of the pesticides relative to sandy or loamy soils (Fig. 6.4). Increasing soil moisture content reduced LD_{50} values markedly in the case of water-insoluble compounds such as DDT, heptachlor and diazonin, but conversely increased LD_{50} values (i.e. reduced toxicity) in the case of water-soluble insecticides.

6.4.2   Persistence

The persistence of a pesticide in the environment is generally measured in terms of its half-life; the length of time it takes for the quantity of compound to be halved. The half-lives of commonly used pesticides vary considerably (Table 6.6); they may be several years in the case of chemically stable organochlorine insecticides such as DDT and dieldrin, or only a matter of days in the case of many fumigants and herbicide sprays. Very

Table 6.6   The half-lives of commonly used pesticides

Pesticide	Approximate half-life (years)
Lead, arsenic, copper, mercury	10–30
Dieldrin, BHC, DDT insecticides	2–4
Triazine herbicides	1–2
Benzoic acid herbicides	0.2–1.0
Urea herbicides	0.3–0.8
2,4-D, 2,4,5-T herbicides	0.1–0.4
Organophosphate insecticides	0.02–0.2
Carbamate insecticides	0.02–0.1

Source: After Brady 1974

few pesticides persist in the vegetation or soil surface for any length of time, for they are soon washed into the soil. There, however, they are affected by a variety of processes, some of which act to retain the pesticide in the soil and others of which act to decompose or remove the compound.

Adsorption

One of the main processes of retention is adsorption. Many pesticides are actively held on the exchange sites on the soil colloids, where, to some extent, they are immobilised and their toxicity is reduced. The degree of adsorption depends in part upon the clay and organic matter contents of the soil (hence the results of Harris, 1964, 1966, quoted above) and is generally greater in soils with a high cation exchange capacity. More important in most cases, however, is the chemical structure of the pesticide, and it seems that compounds containing functional groups such as -OH, $-NH_2$ and -NHR are more susceptible more to adsorption, especially on humus. The size of the pesticide molecule is also significant, and larger molecules are on the whole more readily adsorbed than smaller ones.

Adsorption of pesticides occurs at least in part through the action of hydrogen-bonding, a relatively low-energy association in which hydrogen atoms share links with other molecules. Protonation also encourages adsorption, however; in this case H^+ ions are added to a group such as an amino acid. In addition, adsorption is facilitated in the case of pesticides which produce cations upon decomposition (e.g. paraquat and diquat), although this is generally highly pH-dependent, adsorption declining as pH decreases.

Leaching

Pesticides which are not adsorbed onto the colloids in the soil are often available for leaching. Water-soluble pesticides are particularly prone to loss in this form, and thus many herbicides may be washed from the soil. Leaching is at a maximum in coarse-grained, permeable soils lacking organic matter. It is also encouraged by rainfall soon after pesticide application, for this allows large quantities of the compound to be removed before decomposition has converted it to more stable forms, and before it can become bound by adsorption or by microbiological assimilation.

Volatilisation

Volatilisation of pesticides involves the decomposition of the compounds to gases such as oxygen, carbon dioxide, chlorine, nitrogen and hydrogen. The process is most important in the case of fumigants, such as HCN and ethyl bromide, but as several studies have shown (e.g. Harris and Mazurek 1966; Harris 1970, 1973) it may also cause significant losses of insecticides such as aldrin, heptachlor, phorate, and thionazin. The factor that distinguishes these volatile pesticides from more stable compounds (e.g. DDT, y-chlordane and dieldrin) is the high vapour-pressure of these substances.

Losses by volatilisation may be considerable, and Spencer and Cliath (1974) reported losses in laboratory experiments of up to 40 per cent when the soil was warm and moist. As this indicates, temperature and moisture content are critical controls on the rate of volatilisation, and losses are much reduced in dry or cool soils. Similarly, Burkhard and Guth (1981) found that the insecticides methidathion, diazonin and izaphos, the herbicide metolachor and the fungicide metalaxyl showed increasing losses by volatilisation with increasing initial concentration, temperature and air flow rate. High clay and organic matter contents also diminish volatilisation, apparently by encouraging adsorption and 'fixing' the pesticides before they can be reduced to gas (McEwen and Stephenson 1979; Burkhard and Guth 1981). Losses by volatilisation do not occur only from the soil, of course. For volatile pesticides applied to the foliage or to the soil surface, volatilisation may represent one of the main pathways of removal, particularly when the compounds are exposed to warm, moist atmospheric conditions.

Chemical decomposition

Chemical breakdown occurs largely by photodecomposition and hydrolysis. Neither are major processes of loss in most cases, and their effect varies considerably from one pesticide to another. Thus, chemical degradation of atrazine (Armstrong and Chesters 1968; Kells *et al.* 1980) and diazinon (Getzin 1968) is greatest under acid conditions, but the herbicides dichlorprop, linuron and propyzamide are more susceptible to decomposition at high pH (Hance 1979). Moreover, in some cases the products of chemical degradation may be as – if not more – toxic than the original compounds; organomercury fungicides, for example, may decompose to the more toxic mercury form through reactions with clays.

Biological decomposition

By far the most important decomposition processes involve soil organisms. Relatively few organisms seem to possess the enzymes capable of destroying pesticides, and not all pesticides are susceptible to microbial attack. The organochlorine compounds, for example, are almost immune to microbial metabolisation and it is largely for this reason that they are so persistent in the soil. In contrast, many of the organophosphates are rapidly decomposed, and it seems that certain groups in their structure, such as -OH, $-NH_2$ and $-NO_2$ are particularly prone to microbial attack. It is also clear from studies of herbicides that quite minor differences in

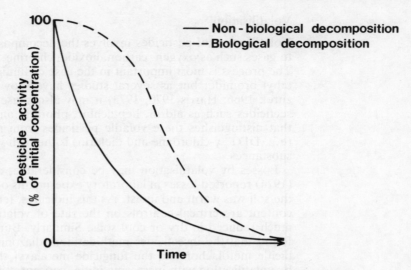

Fig. 6.5   The pattern of decomposition of pesticides in the soil.

chemical structure may markedly alter the resistance of the pesticide to metabolic decomposition. 2,4,5-T, for example, has a long half-life in the soil, while its close relative 2,4-D breaks down much more rapidly. The difference is apparently due to the structure of the aromatic nucleus of the pesticide (Alexander 1969; Loos 1975). Pesticides that are attacked in this way tend to show a logarithmic or sigmoidal disappearance curve (McEwen and Stephenson 1979) (Fig. 6.5).

The rate of microbial decomposition of pesticides is strongly influenced by soil conditions. Adsorption, or rapid removal of the pesticide by leaching or volatilisation, reduces micro-organic attack, whereas factors that increase biological activity tend to increase microbial degradation. Organic matter operates in both ways, encouraging adsorption yet also stimulating microbial activity, so the effects depend upon the balance between these opposing influences. On the other hand, organic decomposition is rapid in warm, moist soil with a moderate pH and reasonably high levels of nutrient availability.

Pesticide persistence in the field

The persistence of pesticides under field conditions depends upon all the processes discussed previously. Each of the processes is affected, however, by environmental and management conditions, so considerable variations in both the persistence and toxicity of pesticides are found in the field. Moreover, many of the factors affecting pesticide decomposition are counteractive, increasing rates of loss by some processes but reducing losses by others. As a consequence, the concept of a specific and constant half-life for individual pesticides is not realistic, and rates of decomposition change greatly over time and from one place to another.

Numerous studies have illustrated the complexity of the interactions which influence pesticide decomposition. Kuwatsuka and Igarishi (1975), for example, found that persistence of the pesticide PCP increased with decreasing soil organic matter content. Meikle *et al.* (1973), however, detected no effect of soil properties on pesticide persistence, whilst Burkhard and Guth (1981) found that persistence of several pesticides increased with increasing organic matter content due to reduced rates of volatilisation. These apparently contradictory results reflect not only differences in the composition and the decomposition pathways of different pesticides, but also differences in soil type and other experimental conditions in the individual studies.

Short-term variations in climatic or atmospheric conditions may cause similar variations in rates of decomposition, for many processes are influenced by factors such as air and soil temperature, rainfall and relative humidity. Thus, Rahman (1978) noted that phytotoxic residues of the fumigant ethofumesate in the soil were greater under cold, dry conditions than under warm, wet conditions. As the data in Table 6.7 also indicate the half-life of the herbicide napropramide is over two times greater in a dry soil at 14 °C than in a wet soil at 28 °C.

Table 6.7 Effects of soil moisture and temperature on the half-life of napropramide

Temperature (°C)	Herbicide concentration (kg ha^{-1} to 5 cm)	Half-life of herbicide (days) at soil moisture contents (%) of		
		10.0	7.5	3.5
28	4.50	54	63	90
14	4.50	102	112	—

Source: After A. Walker 1974

All the factors controlling pesticide decomposition are themselves affected by management practices. The nature of the crop cover, for example, influences microclimatic conditions at the ground surface, and thereby affects processes such as volatilisation and photodecomposition; tillage and drainage affect soil moisture retention and organic activity and consequently influence the rates of leaching and biological decomposition. As a result, the persistence of pesticides in the soil varies with management conditions, though as noted before the effects are not always readily predictable because of the often conflicting implications of management practices. Hance *et al*. (1978a, b) found that atrazine and propyzamide tended to be more persistent in untilled plots than in tilled plots at high rates of application (2.5 kg ha^{-1}), though at lower doses differences were negligible. Similarly, Kells *et al*. (1980) noted that atrazine residues were greater on untilled soil. Liming also affects persistence, and Best *at al*. (1975) found that liming of a silt-loam soil greatly increased the persistence of atrazine, though it had no effect upon prometryn. Indeed, as noted earlier, variations in pH have markedly different effects on the rate of decomposition of different pesticides due to their differing susceptibilities to adsorption, leaching and volatilisation. Variations in crop type and cropping sequence may have comparable effects through their influence on soil conditions such as organic matter content and soil temperature. Thus, Stryk and Bolton (1977) showed that atrazine losses from continuous maize were greater than those from plots under rotation; the implication is that residues will be more persistent under the latter cropping system. Nevertheless, Hance *et al*. (1978b) detected no effect of either plant cover or previous soil treatment on the persistence of the herbicides linuron, simazine and triallate, while A. E. Smith (1980) found that the persistence of 2,4-D was unaffected by pretreatments of herbicides of insecticides.

As all these examples indicate, prediction of pesticide persistence under field conditions is highly complex, and a wide variety of factors are likely to affect the pattern of degradation. Work by Walker (1974, 1976a, 1976b, 1977) and Walker and Bond (1978), however, illustrates what can be achieved and demonstrates the main variable controlling persistence in the field. Using a model incorporating data on soil moisture content, soil temperature and the activation energy of the pesticide, Walker simulated changes in pesticide concentrations over periods of several months. As the example in Fig. 6.6 shows, close agreement was found with actual conditions. Nevertheless, Walker (1974) noted that these simulation methods are valid only when pesticide decomposition follows the laws of

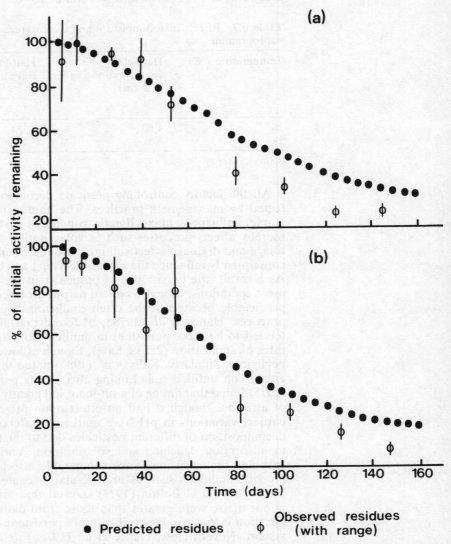

Fig. 6.6 Predicted and observed residues of propyzamide in (a) a sandy loam and (b) a clay-loam soil (from A. Walker 1976b).

first order kinetics: namely, that the rate of degradation is directly proportional to the concentration present. This is true in the case of pesticides that do not degrade by biological processes, but, as we have seen, is less true for those susceptible to rapid microbial attack (Fig. 6.5). As a consequence, degradation of the pesticide does not necessarily imply a proportionate reduction in its activity. Moreover, some pesticides break down into other toxic compounds – for example, aldrin into dieldrin – and in these cases predictions of residues require knowledge of the degradation kinetics of both substances. In other words, the dynamics of pesticide activity in the soil remains a difficult problem.

6.5 EFFECTS OF PESTICIDES ON CROP GROWTH AND SOIL FERTILITY

It is apparent that, by removing the detrimental effects of weeds, insect pests and diseases, the use of pesticides can result in greatly increased yields of agricultural crops. McEwen and Stephenson (1979), for example, quote studies in Illinois which showed that crop losses due to weeds could be as high as 100 per cent and averaged about 81 per cent in unweeded plots. Similar studies in Minnesota showed losses of maize yield of 16–93 per cent, with an overall average loss of 51 per cent. The use of herbicides to control weed infestations in these situations could clearly provide major benefits in terms of increased output.

Nevertheless, the effects of pesticides on crop yields are generally found to vary markedly. These variations in crop response depend upon a wide range of factors, including the nature and extent of the original infestation and the ability of the pesticide to eliminate the pest. As might be expected, yield increases tend to be greatest in situations where the infestations are particularly severe, an effect demonstrated by Baldwin (1979) in a study of chemical control of wild oats and black-grass in England. He showed that in crops with wild oat infestations of 0–10 plants per square metre, yield increases from the use of pesticides averaged only 0.15 t ha^{-1}. At infestations of 30–50 plants per square metre, however, yields increased by about 0.9 t ha^{-1}, and control of infestations of greater than 80 plants per square metre gave increases of 1.8 t ha^{-1}. Similarly, Henderson and Clements (1977), surveying responses of grassland to pesticide treatment in England, noted that increases in dry matter output ranged from 0 to 32 per cent, as a result of suppression of aphids, plant-sucking bugs, stem-boring larvae and, possibly, nematodes. The great variation in crop responses was presumably related in part to the different levels of infestation at different sites and the different degrees of control provided by the pesticides used.

Numerous other factors also influence crop response to pesticides. Responses vary, amongst other things, in relation to the nature of the crop, the soil and climatic conditions during and after application, the timing of the application and the extent to which other factors limit yields. This last factor is particularly important, for it is clear that there is little to be gained from controlling pest problems on crops if other factors continue to limit yields. As a result, the largest responses are often found where pesticide applications are combined with other methods of soil improvement, such as increased fertiliser usage, irrigation or drainage. The significance of soil conditions is illustrated by Boyd (1973). Reviewing two experiments in Scotland, he noted that control of potato blight resulted in average yield increases of 3.0 and 5.3 t ha^{-1} over a 9 to 10-year period. In areas affected by tractor wheelings, however, no improvements in potato yields could be detected. The effects of different types of pesticide and methods of application are shown by Golisch (1977). Taking the yields of winter barley achieved with surface applications of herbicides as a standard, he noted that contact herbicides gave a 4 per cent improvement in yields, whilst systemic herbicides resulted in a relative reduction of 3 per cent.

The effects of pesticides on crop yields are further complicated by the wide range of ways in which individual crops respond to management factors. As we will see in Chapter 8, the final, gross yield of the crop is a function of several different components, including the number of seedlings per unit area, the number of flowering tillers (or other reproductive shoots) per plant and the amount of grain or seed per tiller. All these components of yield are intricately interrelated, with the result that changes in one component tend to cause corresponding changes in others. As a consequence, the effects upon the final yield may be magnified or reduced. Stahlecker (1977) demonstrates the effect of pesticides on seedling emergence, noting that application of lindane at a rate of 0.8 kg ha^{-1} before sowing resulted in a 6 per cent increase in the emergence of sugar beet. The effects upon other components of yield were shown for sorghum by Duncan and Boswell (1981). Methiocarb, applied at a rate of 1.2 kg ha^{-1}, increased average plant height by 8 cm, increased grain yield per plant by 4 g, and raised the final yield from 4.3 to 5.4 t ha^{-1}. These changes were associated with a 39 per cent reduction in the number of plants attacked by birds.

In addition to the quantitiative changes in yield, pesticides also result in changes in the quality of the crop. The elimination of weeds which might otherwise be incorporated in the harvest, and reductions in damage to the crop, improve the purity of the harvested material. Thus, use of pesticides to control scab and nematode attack greatly increases the quality of potato crops (e.g. Eddowes 1976), whilst applications of herbicides have been shown to improve the culinary properties of potato crops (Daniel 1977). In the case of pastureland, improvements in quality are expressed in terms of better grazing conditions for livestock; Baker et al, (1980), for example, found that treatment of rangeland with atrazine and fertilisers increased the crude protein content, yield and condition of range-pasture during the summer, thus providing improved herbage for grazing.

Although the use of pesticides is generally found to increase crop yield and improve crop quality, the effects are not always beneficial. Increasingly it has become apparent in recent years that pesticides may have adverse side-effects on crop growth, resulting in some cases in reduced yield. We have already noted the suggestion that the use of herbicides may cause damage to the roots of cereal crops and thereby increase the susceptibility of the crops to disease (section 6.3.2). Similar root damage may inhibit water and nutrient uptake and reduce yields. The application of pesticides as foliar sprays may also cause leaf scorch, with consequent decreases in the photosynthetic activity and transpiration rate of the plant. Thus, Penny and Jenkyn (1975) reported that spraying of winter wheat with fungicide-nitrogen fertiliser mixtures resulted in significant leaf scorch and loss of yield. The effects were more marked when a herbicide was added to the mixture.

In the long term, doubts also exist about the effects of pesticide usage on soil fertility. Edwards (1969a) noted a number of incidences in which pesticide residues had reached potentially toxic levels in agricultural soils, though problems were confined mainly to situations in which excessive quantities of persistent organochlorine pesticides had been applied over a long period of time. With the change in recent decades to the use of less persistent organophosphate pesticides, contamination of soil has probably become a less serious concern. Fryer *et al*. (1980), for example, reported results from a long-term study in which MCPA, triallate, simazine and linuron were applied to experimental plots. Before the eighth application, the soil nutrient status and pH, and the growth of test crops were assessed. No significant differences were found between the control plots and treated plots, nor were effects noted when the pesticide applications were doubled.

It is clear that the responses of crops and soil fertility to pesticides are extremely complex, and general principles are emerging only slowly. Without doubt, much more research is needed on the effects of pesticides, taking account particularly of the way in which these effects are modified by different environmental and management conditions As ever, such studies are rendered difficult by the large number of interactions which must be allowed for and the need for relatively long-term, replicated experiments. It must also be remembered that the soil-crop relationship is only one part of the whole system in which pesticides take part, and attention also needs to be given to the wider implications of pesticide usage. We will consider some of these questions in Chapters 10 and 13. In this context, however, Way and Cammell (1979) raise the fundamental question of whether increasing levels of pesticide applications need to be made in agriculture, or whether there is not both an environmental and an economic case to limit pesticide usage, albeit with the consequence of achieving somewhat lower yields. Clearly, the need exists to consider pesticide usage in terms of wider cost-benefit relationships in this way if a balance is to be achieved between the agricultural merits and ecological demerits of crop protection.

PART 3

AGRICULTURAL SYSTEMS

7 GRASSLAND AND GRAZING SYSTEMS

7.1 INTRODUCTION

7.1.1 Distribution and character of temperate grasslands

Over 40 per cent of the land surface of the temperate world is devoted to grassland. Relatively little of this is natural climax vegetation. Only in areas where marked moisture deficits occur, or where radiation inputs are too low to support a forest cover, do grasslands form the climax biome. Such areas are confined mainly to the more arid extremes, or the higher altitude and higher latitude fringes of the temperate region. Elsewhere, the grasslands are a product of human activity. Over wide areas, they have developed as a plagioclimax, consequent upon clearance of the original forest and maintained by grazing which prevents the regeneration of scrub or trees; only when these pressures are released, by abandonment of the land or by deliberate exclosure of animals, does the vegetation follow its natural succession back to woodland. Much of the present grassland, however – and certainly the agriculturally more productive part – is a result of more deliberate and intensive management, including ploughing, fertilising and reseeding of the land with a carefully selected range of plant species.

The character of temperate grasslands therefore varies considerably. We will discuss these variations, from an ecological viewpoint in Chapter 11. In terms of their agricultural use, however, Semple (1970) defined five types of grassland: natural pasture or range (though much of this should more correctly be called semi-natural), permanent pasture, rotation pastures or leys, temporary pastures and supplemental or complemental pastures. The first three categories are most significant here.

Natural pastures include most of the prairie rangeland and steppe of temperate America and Eurasia, together with the mountain meadows of the Alps, Andes and Urals and the moorlands of Britain and Ireland. These pastures consist of native grasses, forbs (herbaceous plants other than grasses) and shrubs and are for the most part relatively extensively managed. Throughout much of the steppe-land, grasses belonging to the genera *Stipa* predominate. The prairies comprise several different types of grassland. The true prairie is now almost extinct but was originally dominated by grasses such as *Agropyron smithii, Andropogon scoparius* and *Bouteloua curtipendula*, together with common stipa grass (*Stipa spartea*). The mixed prairie is composed of both medium height and dwarf grasses; the former include *Stipa comata, Agropyron smithii* and *Koeleria cristata*, while the latter consist mainly of buffalo grass (*Buchloe dactyloides*) and grama grass (*Bouteloua gracillis*). Overgrazing and cultivation have often modified these prairie grasslands, so that in northern areas of the Great Plains, for example, the short-grass species predominate, while cactus (*Opuntia* spp.) and sagebrush (*Artemisia tridentata*) have also become established.

In Europe, natural pastures are less common; the nearest equivalents are the Alpine meadows and the moorlands and heath lands. In Britain these latter are characterised by grasses such as purple moor grass (*Molinia caerulea*), mat grass (*Nardus stricta*), wavy hair grass (*Deschampia flexuosa*) and the bent-grasses (*Agrostis* spp.), together with shrubby plants such as heather (*Erica* spp.) and ling (*Calluna vulgaris*).

Permanent pastures are widespread in Europe, USA and New Zealand. They consist of artificially grown grass, normally established on land that has been ploughed, but maintained for long periods between reseeding by use of fertilisers and controlled grazing. The grass sward varies regionally, but cultivated species such as the fescues (*Festuca* spp.), meadow grasses (*Poa* spp.), cocksfoot (*Dactylis glomerata*), perennial ryegrass (*Lolium perenne*) and the bents (*Agrostis* spp.) are common. One of the difficulties in categorising permanent pasture, however, is the fact that management practices vary considerably, and a gradual transition occurs from permanent pasture to long-term ley pasture (which is resown at intervals of several years). In general, ley pastures contain more nutritious grasses in greater quantities; in particular ryegrass (*Lolium* spp.), cocksfoot, timothy (*Phleum pratense*) and meadow grasses (*Poa* spp.) together with legumes such as clover (*Trifolium* spp.).

7.1.2 History of grassland systems in Britain

As has been noted, most of the grasslands of the temperate world are a product of human activity. Until the seventeenth century, however, agricultural management of the grasslands was relatively limited; most of the grazing lands were permanent pastures that lay outside the enclosed farmland and received little attention. The practice of sowing or reseeding pasture was only widely adopted with the development of what has been called 'up and down' or 'convertible husbandry'. Under this system, the land was held in 'closes' which could be converted into 'up' (i.e. arable land) or 'down' (pasture). It was a practice which had been recommended by Fitzherbert (1523), but which was only widely adopted in Britain during the period 1590 to 1640. It led to the conversion of both arable and permanent grasslands to a simple and somewhat haphazard ley pasture system.

During the seventeenth century, the introduction of new fodder crops and the development of more rigorous rotational farming systems led to considerable improvements in grassland fertility. Clover, sainfoin, coleseed, rape and lucerne were all found extensively by 1700, leading to marked increases in yields, better animal nutrition and greater livestock productivity. At the same time, a decline in grain prices led to a widespread shift into grassland farming throughout much of Europe, although not in England; here, although some areas were turned over to grass, grain production increased as a result of technological innovations and the further development of mixed farming. A further impetus for grassland farming came in the nineteenth century. Repeal of the Corn Laws and improved international transport facilities led to an even more marked slump in the grain market and, eventually, a fundamental shift into grassland farming. Within a few decades, the acreage of permanent pasture rose by a quarter of a million hectares in the grazing areas of England and by over half a million hectares in the arable areas. Low grain prices ensured cheap fodder, while prices of animal products remained high; in much of Britain it was the heyday of grassland farming.

Although the First World War led to some recovery in arable areas, the pattern was almost repeated in the 1920s as grain prices once more slumped. In Britain, between 1920 and 1935 almost two million hectares of land were taken out of arable production, and by 1939 the area of land used for cereals in Britain was less than that in 1914. As ever, the decline of arable farming implied a rise in the area of grassland, and in 1938 almost 15 million ha of grassland were to be found in Britain, representing approximately 80 per cent of the total cropped area. Once more, however, history repeated itself, and the Second World War restored the fortunes of cereal farming and led to a marked decrease in the area of permanent pasture. Since then, there have been fluctuations in the proportions of permanent and temporary pasture, largely in response to changing concepts of management and the increased attention given to ley pasture, but the total area of grassland has remained relatively constant at about 12.5 million ha (Eddowes 1976; see also Lazenby and Down 1982).

7.1.3 Agricultural use of grasslands

Most of the grasslands of the temperate world are used primarily for animal grazing. The type of livestock grazed and the management system under which they are maintained vary considerably. On the poorer, climatically or edaphically marginal lands, goats and sheep often dominate, and stocking densities are low – frequently no more than 0.1–0.2 animals/ha (Table 7.1). Where conditions are better and biomass production greater, beef cattle are commonly reared, as for example over many of the prairie grasslands of the Americas and Eurasia. Stocking densities here obviously vary considerably from area to area but generally average about 0.5–1.0 animals/ha.

By comparison, permanent and ley pastures are intensively used. In the Netherlands, permanent pastures carry up to 3–4 cattle/ha, while in New Zealand stocking densities of 5–6 animals/ha are not uncommon. Beef cattle and dairy cattle tend to predominate on permanent and ley pastures, and profitability in these intensive grazing systems may be comparable with that of relatively high input arable systems. As G. W. Cooke (1979) has noted, grass yields in these systems may be close to their physiological optimum.

Grassland systems comprise three main components: the vegetation, the soil and the livestock. Each of these components interacts closely with the others (Fig. 7.1). Soil fertility, for example, partially controls grass growth and herbage production, which in turn affects animal behaviour and development. In its own turn animal behaviour influences the sward and the soil.

Table 7.1 Typical stocking rates on pastureland

Type of grassland	Location	Stocking rate (LU ha^{-1})	Source
Prairie	Californian chapparal	0.02–0.4	Duckham and Masefield 1970
Poorly drained hill pasture	Co. Kerry, Republic of Ireland	0.2–0.4	Kerry County Committee of Agriculture 1972
High moorland	Scottish highlands	0.1	Department of Agriculture for Scotland 1952
Poorly drained lowland pasture	Co. Kerry, Republic of Ireland	0.8–1.1	Kerry County Committee of Agriculture 1972
Well drained lowland pasture	Co. Kerry, Republic of Ireland	1.2–1.5	Kerry County Committee of Agriculture 1972
Lowland, permanent pasture	Vale of Glamorgan, Wales	1.0–1.2	Crampton 1972
Lowland, permanent pasture (high fertiliser inputs)	Netherlands	2.9–3.1	de Boer et al. 1979
Lowland, permanent pasture (high fertiliser inputs)	Netherlands	2.0–2.5	Duckham and Masefield 1970

Note: LU (livestock units): 1 LU ha^{-1} is equivalent to 1 mature dairy cow per hectare (see, for example, Kerry County Committee of Agriculture 1972, Appendix B).

The aim of the farmer is to control all three components – by using fertiliser, by selecting and sowing the grass seed, by harvesting the crop, by controlling the livestock. He thereby regulates the inputs to and the outputs from the system, as well as its internal structure. It should also be remembered that he is affected by the operation of the system, not always directly, but often indirectly through the economic implications of grassland productivity.

7.2 MANAGEMENT OF GRAZING SYSTEMS

Management techniques vary considerably, partly in response to local conditions, partly in accord with developments in the scientific knowledge of agricultural systems, but also in relation to the personal preferences and experience of the farmer. For these reasons it is difficult to distinguish general systems of management that are widely used. Nevertheless, certain basic techniques are employed; in general they are aimed at:

1. Encouraging sward growth (e.g. burning, fertiliser application);
2. Controlling animal behaviour (e.g. grazing regimes);
3. Harvesting the products (e.g. mowing, ensiling).

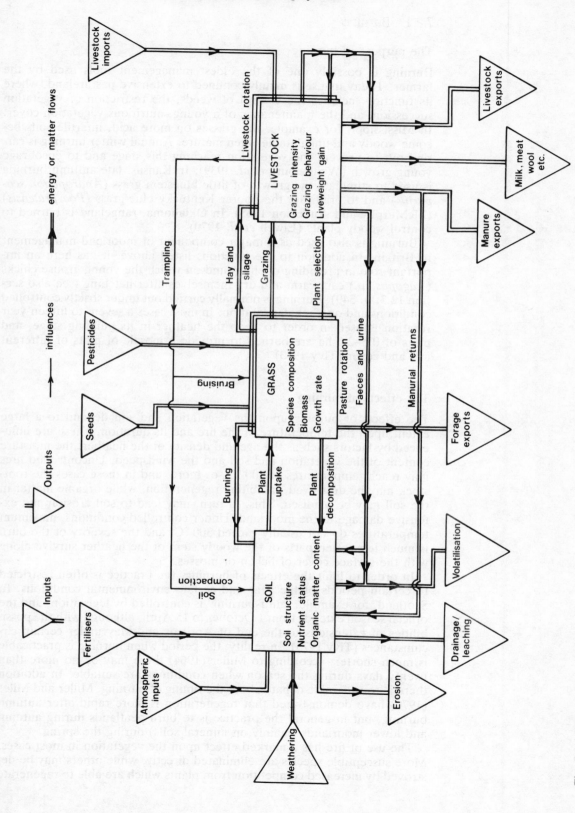

Fig. 7.1 The grassland system.

7.2.1   Burning

The purpose of burning

Burning is possibly one of the oldest management tools used by the farmer. Today its use is mainly confined to extensive pastureland, where its function includes the control of weeds, the restriction of vegetation succession, and the maintenance of a young, nutritious vegetation cover. In Mississippi, for example, the grasses on more acid, infertile soils become woody and unpalatable when mature. Annual winter burning is carried out to prevent the vegetation reaching this stage and to encourage young growth (Wahlenburg *et al*. 1939). In Kansas, late autumn burning is used to stimulate the growth of little bluestem grass (*Andropogon scoparius*) and to eradicate the poorer Kentucky blue-grass (*Poa pratensis*) (McMurphy and Anderson 1965). In Oklahoma, rangeland is burned to control woody plants (Elwell *et al*. 1970).

Burning is also used as a major component of moorland management in Britain. In addition to the functions listed above, it has here an important role in providing open ground on which the young grouse chicks (*Lagopus* sp.) can warm and dry themselves after hatching (see also section 12.3, p. 349). Burning is normally carried out under strictly controlled conditions and on a regular rotation. In most cases a seven- to fifteen-year rotation is used in order to keep the heather in its building stage, and units of 0.5–2.0 ha are burned to provide a mosaic of plots of different age and cover (Tivy 1973).

The effects of burning

The effects of burning upon the vegetation and soil depend to a large extent upon the temperature of the fire and its duration. These are influenced by factors such as the age and density of the heather, the moisture content of the vegetation and soil and the windspeed. Uncontrolled fires may reach temperatures of 900 °C, or more, and in these cases the rootstock may be destroyed, inhibiting regeneration, while organic matter in the soil may be oxidised. This, in turn, may lead to soil erosion and extensive damage to the moorland. Under controlled conditions, maximum temperatures do not usually exceed 300 °C, and the severity of the burn is much less; often parts of the woody stem of the heather survive along with the surface cover of lichen or mosses.

In order to limit the effects of burning, the practice is often restricted to certain periods of the year and certain environmental conditions. In Scotland, for example, 'muir-burning' is controlled by legislation, and the official season extends from 1 October to 15 April, although with the possibilities of extensions to the end of April or mid-May under certain circumstances (Tivy 1973). In reality, the period when burning is practicable is much shorter; according to Miller (1964) there may be no more than twenty days during the season when conditions are suitable. In addition there is considerable debate about the timing of burning. Miller and Miles (1970) have demonstrated that regeneration is more rapid after autumn burning, but in general the practice is to burn peatlands during autumn and lower moorlands (mainly on mineral soils) during the spring.

The use of fire has a marked effect upon the vegetation in most cases. More susceptible species are eliminated directly, while others may be destroyed by increased competition from plants which are able to regenerate more effectively. In Britain, it is believed that the decline of the common juniper (*Juniperus communis*) in heathlands has been due to regular burning (Gimingham 1972). Nevertheless, the effects are not always as marked as might be expected; Powell *et al.* (1979), for example, found very little effect on species composition through regular burning of rangeland in Oklahoma. More subtle and indirect effects upon the system are also instigated, however, through the influence of burning upon the nutrient cycle. Burning represents an important cause of nutrient loss and several authors have suggested that it might lead to a progressive decline in soil fertility and productivity (e.g. Elliott 1953; Chapman 1967).

Losses of nutrients as a result of fire occur in a number of ways. Considerable quantities of material may be lost directly to the atmosphere, either in gaseous form during burning or as particulate matter carried away in the smoke. The amounts involved depend very much upon the temperature of the fire (Table 7.2), but in general nitrogen losses are particularly severe. Allen (1964) recorded losses of almost 80 per cent of the total nitrogen content of the vegetation during burning, mainly in the form of gas. Similarly, over 50 per cent of the carbon and sulphur was lost. Some, at least, of the fraction carried into the atmosphere may be returned to the system downwind, either by dry fall out or condensation, so these losses may not be absolute. Nonetheless, Evans and Allen (1971) found that only between 2 and 56 per cent of the particulate matter was recovered, and in most cases almost all the nutrients in the smoke were lost. Comparable results have also been found by Smith and Bowes (1974) in studies of field burning in southern Ontario.

Table 7.2   The effects of temperature on nutrient losses during heather burning

Nutrient	Losses (% of original content, corrected for dry weight)	
	Normal burn 550–650 °C	Severe burn 800–825 °C
Potassium	1.4	4.9
Calcium	0.1	2.4
Magnesium	0.4	2.1
Carbon	60.5	67.5
Nitrogen	67.8	76.1
Phosphorus	0.8	3.5
Sulphur	50.2	56.3

Source: From Allen 1964

The burned residue of the vegetation is returned to the soil as ash. Further losses of nutrients may then occur, for the nutrients in the ash tend to be highly soluble so that they are readily leached from the surface. Potassium, in particular, is lost in large quantities through leaching of the ash, but significant amounts of calcium and magnesium may also be removed. Some of the nutrients may, of course, be held by adsorption lower down within the soil profile, although much seems to depend upon the soil type and climate. Elliott (1953), for example, found marked losses through leaching, but in experiments by Allen (1964) and Allen *et al.* (1969) most of the nutrients in ash derived from *Calluna* burning were retained in the surface organic layers of the soil.

Because of the numerous factors that influence nutrient losses during burning, it is difficult to quote general quantities; the data in Table 7.2

are probably characteristic only of upland heath conditions on organic soils. What is clear, however, is that the losses may not be rapidly replenished by inputs from rainfall or weathering, and in the absence of fertiliser inputs a gradual decline in the nutrient content of the system may occur. Chapman (1967) and Kenworthy (1964) both indicate that phosphorus losses are likely to exceed inputs under regular moorland burning and the same may be true of nitrogen (Table 7.3); for these nutrients, at least, the type of decay curve illustrated in Fig. 7.2 may well apply.

Table 7.3   Nutrient balance for burned heather moorland in southern England

Nutrient	Loss by single burn (kg ha^{-1})	Input by rainfall (kg ha^{-1} y^{-1})	Difference (kg ha^{-1} y^{-1})	Years to achieve balance
Na	1.5	25.4	+23.9	0.06
K	8.3	1.2	− 7.1	6.9
Ca	12.5	4.7	− 7.8	2.7
Mg	4.0	5.6	+ 1.6	0.7
P	2.2	0.01	− 2.19	220.0
N	173.1	5.2	−167.9	33.3

Source: Data from Chapman 1967

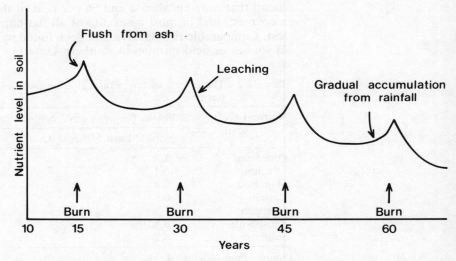

Fig. 7.2   Theoretical pattern of change of soil nutrient contents over time on a burned heather moorland.

7.2.2   Fertilisers

General principles of fertiliser management have been considered in Chapter 5. It is clear, however, that in recent decades the use of fertilisers in grassland systems has increased markedly. Grass has been found to respond well to large inputs of fertiliser nitrogen, even on relatively fertile soils, and in intensive systems nitrogen losses in the harvest, in drainage waters and through volatilisation need to be replenished by fertilisers.

Innumerable studies have examined the response of grassland yields to fertiliser nitrogen, and many of the classical experiments are reviewed by Cooke (1974).

In these early studies, fertiliser application rates were generally relatively low and maximum responses were rarely obtained. In more recent experiments, higher rates of nitrogen application have been used, and the full character of the yield response curve has become apparent (Fig. 7.3). The response to fertilisers, however, varies markedly with both management and environmental conditions. Richards (1977), for example, noted that maximum yields for grazed grasslands were achieved at nitrogen application rates of about 300 kg ha^{-1}, but in ungrazed swards Morrison *et al.* (1980) showed that yields continued to increase up to application rates of 540–678 kg ha^{-1}. The difference is apparently due to the adverse effects of trampling and grass pulling on swards receiving high doses of nitrogen, and the availability of nitrogen in animal faeces (see section 7.3).

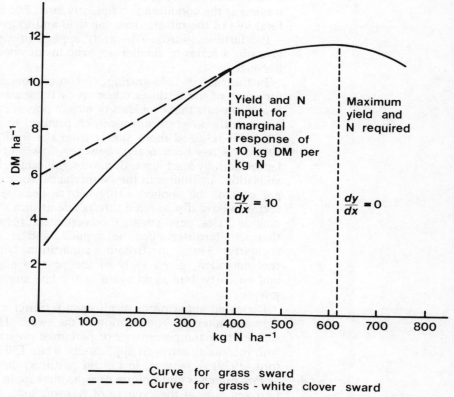

Fig. 7.3 Effects of nitrogen fertiliser on the dry matter yield of grass and grass-white clover swards (from T. E. Williams 1980).

Morrison *et al.* (1980) also analysed results from a range of experiments in Britain and computed a response index (Y_{10}) for each site: this was the yield at the point at which an incremental increase in nitrogen application produced an increase in yield of only 10 kg ha^{-1}. Values of Y_{10} were found to vary with environmental conditions, and the following relationship was established between the yield response index and the available water capacity (W) and amount of summer rainfall (R_s):

$$Y_{10} = -0.3186 + 0.02996 \, W + 0.0392 \, R_s - 0.00004379 \, R_s^2 \qquad [7.1]$$

Thus, responses to fertiliser applications continue up to higher yields in areas where the soil available water capacity is high and show a positive, curvilinear relationship with summer rainfall.

Fertiliser applications result in significant variations in processes of nutrient cycling, though the effects depend upon the timing of the applications and the management system. While the grass crop is actively growing, most of the applied nitrogen is taken up by the sward and leaching is slight; significant losses of N, in fact, only occur when heavy rains occur immediately after application. When the crop is harvested, however, the nitrogen is removed, with the result that little is available to subsequent growth. Moreover, if the nitrogen-rich forage is fed to cattle there is a serious danger of nitrate poisoning; the nitrate is reduced to nitrite in the rumen and as a result methemoglobin is produced in the bloodstream (Kemp *et al.* 1978; Geurink *et al.* 1979). Without immediate treatment the condition is frequently fatal. For these reasons – to prevent total loss of the nitrate from the field and to avoid the danger of toxicity – the fertiliser nitrogen is rarely applied in a single, large dose, but is given in a series of smaller applications of about 80–100 kg before each harvest.

In the case of field-grazing, of course, losses of nitrogen are far less, for much of the fertiliser taken up by the grass is returned in the faeces. It also appears that problems of nitrate poisoning are much reduced, even when cattle do graze nitrogen-rich pasture, mainly it seems because the nitrate is released more slowly from the fresh material (Geurink *et al.* 1979). As has been noted, however, losses of nitrogen occur from the faeces both by leaching and, more significantly, by denitrification and volatilisation. In studies in the Netherlands, for example, losses of up to 80 per cent of the applied nitrogen have been reported, and most of this seems to have disappeared through denitrification and volatilisation (Penning de Vries, pers. comm.). Nevertheless, recycling of the nitrate means that less fertiliser need be applied, and it can also be applied less regularly. Thus, in Britain, applications of 100–200 kg are often recommended, given early in the year to ensure rapid spring growth and an early 'bite', and again in the late summer to encourage autumn growth.

Repeated applications of nitrogen fertiliser not only stimulate growth; they also alter the composition of the sward. Hopkins and Green (1979) showed that the percentage of perennial ryegrass in the sward increases with increasing nitrogen application, while Elliott *et al.* (1974) and Sandford (1979) show that, in upland pastures, more nutritious species such as *Poa pratensis* and *Lolium perenne* increase in abundance with increased nitrogen rate, at the expense of *Agrostis* spp., *Nardus stricta* and *Festuca ovina*. The effect varies according to soil type (Davies 1979).

It is apparent that nitrogen applications can result in significant increases in grass yield, and these may be translated to higher output of animal products either by allowing more rapid liveweight gain or milk yield, or by permitting higher grazing intensities. Nonetheless it is notable that profitability does not always increase with higher rates of N application, for other overheads (such as costs of animal housing, labour, machinery) also rise. Thus, Stewart and Haycock (1980) have shown that high N rates may reduce the profitability of beef-rearing systems in Northern Ireland. In these systems clover is found to provide a much more economic source of nitrogen.

Fertilisers other than nitrogen are rarely needed in such large quantities to maintain yields, and heavy doses of phosphate and potassium need normally be applied only on freshly sown or reclaimed pasture. Subsequently, especially in grazed pasture, the residual effects of the recycled nutrients are sufficient to maintain growth for several years. In ungrazed grassland, of course, where cycling of nutrients is short-circuited, and where large amounts of P and K are removed in the harvested crop, regular replenishment is necessary. Again these fertilisers may alter sward composition. Hopkins and Green (1979), for example, argue that experimental studies demonstrate that phosphate applications favour the more productive species such as *Poa trivialis* and *Dactylis glomerata*, and suppress *Agrostis tenuis* and possibly *Festuca rubra*. Potassium fertilisers similarly encourage ryegrass growth at the expense of *F. rubra* and *Agrostis* spp.

A more common deficiency in many pastures is lime. Especially in upland grassland, where leaching is active and the parent materials are generally nutrient deficient, the soils are often acid, with a pH between 3 and 4.5. In these circumstances, nutrients are scarce, not only because of rapid leaching but because of the lack of microbial activity and the slow rate of organic matter decomposition and nutrient cycling. Application of lime raises the soil pH, encourages soil organisms, speeds up decomposition and aids the release of nutrients. As a consequence, yields are increased and sward changes take place. (e.g. Williams 1974).

The effects of soil pH are well exemplified by studies at Great House Experimental Husbandry Farm in Rossendale, Lancashire. On acid soils (pH 4.8), pastures sown with a perennial ryegrass-timothy-white clover mixture experienced marked declines in the percentages of sown species when unlimed. On plots receiving lime, however, the soil pH was raised and the sown species accounted for over 70 per cent of the sward even after thirteen years (Cromack *et al.* 1970). Similarly, results from the GRI-ADAS National Farm Study (reported in Hopkins and Green 1979) show that perennial ryegrass contents of pasture swards are closely correlated with soil pH while *Agrostis* spp. and *Holcus lanatus* predominate under more acid conditions. Nonetheless, the relationships between soil pH and sward composition are more complex than data such as these imply. Van den Bergh (1979), for example, noted that *Dactylis glomerata* responded to pH mainly during the period before the first harvest due to poor tillering under acid conditions. Following cutting, growth was similar on soils ranging from pH 4.2 to pH 6.7. He also showed that adjustments in botanical composition to changes in fertiliser practice may take up to sixty years.

7.2.3   Grazing management

One of the main ways by which the farmer controls the operation of the grassland system is through his choice of grazing system. It has often been said that there are as many different grazing systems as there are farmers, and certainly it is often difficult to distinguish clear systems which are strictly adhered to. Rather, there are several different methods of grazing management, each of which may be more or less modified to take account of local conditions, the type of livestock being reared, and, above all, the preference (or whim) of the farmer. Five general grazing systems can be identified (Fig. 7.4):

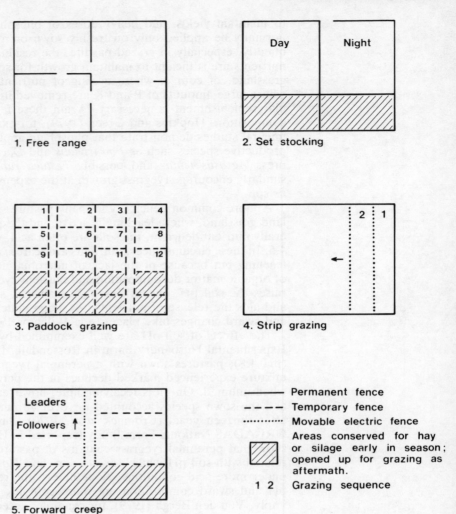

Fig. 7.4   Grazing systems.

1. Free-range;
2. Set-stocking (continuous grazing);
3. Rotational (paddock) grazing;
4. Strip-grazing;
5. Zero-grazing.

Many intermediate forms may also be encountered, however, including dual systems such as the forward creep method used widely with sheep (MAFF 1978) (See Fig. 7.4).

Free-range systems

Without doubt, free-range systems are the most widely used throughout much of the world. They are also the most variable. Essentially, the animals are allowed to roam and graze unrestricted over the grassland, although at various times of the year the livestock may be moved to on new pasture or be excluded from certain areas. It has the advantage of allowing the livestock to adjust their feeding habits to their own requirements, and it is clearly a low-cost system, minimising the need for supervision or fencing. On the other hand, the very lack of control means that it is difficult to keep a check on the progress of grazing; lack of food, disease or other problems may go unseen. In addition, the system allows animals to graze the pasture selectively, so that marked variations in growth and, ultimately, sward composition may occur (see section 7.4). It is a system, therefore, characteristic of extensive grasslands, where animals need access to large areas of land and where inputs need to be minimised.

Set-stocking

In recent years, many grassland farms have adopted methods of set-stocking. This involves the use of relatively large fields which are grazed continuously for long periods but under much greater control than with free-range systems. Typically in England the system might comprise three fields, one used for daytime grazing, one used for night and one at weekends. Outlying fields are initially used for grass conservation but, as the grass growth declines, these are opened up to give animals access to the aftermath. Thus, the size of the set-fields is adjusted to maintain a balance between herbage demand and supply.

Compared to other grazing systems, it is relatively rigid and, for that reason, easy to employ. Moreover, because the grazing intensity at any time is low, the risk of soil damage may be reduced. Its labour requirements are also low, and this has certainly been a major factor in the adoption of the system. In addition, because the animals graze the same pasture for long periods of time, milk yields tend to remain more constant than under rotational grazing. There are, though, inevitable drawbacks. The very inflexibility of the system may be a disadvantage, for the presence of large fields means that it is difficult to restrict grazing should this be necessary. By the same token, selective grazing may occur leading to deterioration in the sward, while the need to gather the animals from a wide area can be a problem, particularly on dark, winter mornings! Above all, it is often difficult to judge the amount of herbage on offer to the animals, since the sward is never allowed to grow to any height and it must be assessed over a large area.

Paddock grazing

Paddock grazing, by contrast, involves the careful partitioning of the pasture into a number of small paddocks which are grazed in strict rotation. Animals often spend one day (in the most stringent systems) to several days in each paddock. Early in the season, when grass growth is rapid, relatively few paddocks are used and the rotation may be in the order of ten to twenty days, the additional paddocks being used for hay or silage. Later, as grass growth rates decline, more paddocks may be brought into the rotation, which may be extended to thirty or more days.

There are several advantages of this system, not least the fact that it is extremely flexible. Extra paddocks may be brought into or taken out of the rotation according to the prevailing balance between herbage supply and demand, while it is less labour-intensive than strip-grazing. On the other hand, it requires careful management, in particular because grazing intensities in each paddock are high and considerable damage can be done to the soil if it is grazed when wet. Additionally, of course, the initial capital investment in fencing is high, and the system is constrained in many cases by topography and farm layout since it requires easy access to all parts of the farm.

Strip-grazing

Strip-grazing is broadly similar in design to paddock grazing, save that the pasture is divided on a day-to-day basis, using temporary fences, in order to ensure a careful balance between herbage demand and supply. Typically, a small strip of land is fenced off with an electric fence and the animals allowed to graze the pasture for a day or so. As the herbage is grazed, the fence is moved to open up new areas of grass. Over time, therefore, the animals have access to a progressively larger area of the field, although grazing tends to be concentrated in the newly opened strip. The main advantage of the system is clearly the fact that the quantity of herbage available to the animal can be carefully adjusted to allow for factors such as changing demand or differences in growth. The major disadvantages, apart from the heavy load it places upon the labour-force, are that it is often difficult to organise the system so that the whole field is grazed before the grass starts to mature to such an extent that its nutritional value starts to decline; and that because animals have access to different parts of the field for different lengths of time, the problem of uneven grazing and trampling pressure is exacerbated. To avoid this latter problem, the pasture in some strip-grazing systems (e.g. in the Netherlands) is also bounded by a 'back fence' which restricts access to previously grazed land and allows these areas to recover.

Zero-grazing

A recent development has been the introduction of what are called zero-grazing systems. As ever there are many variations to the same theme, but in general they involve stall-feeding animals with forage cut from the field. The grass may be supplied in a variety of ways: as fresh grass, mown daily with a forage-harvester; as wilted (partially dried) grass; or as compressed, dried grass-cake.

The clearest advantage of this system is that it minimises waste involved in field-grazing by the animals; there are no losses through sward rejection in the field, through trampling or through fouling. As a result, pasture growth is characteristically uniform. Conversely, other difficulties are encountered. Housing of the animals involves considerable outlay on buildings, while the cost of regular cutting and transport of the forage is also substantial. Even more critically, in many cases, the disposal of animal wastes presents difficulties. The wastes cannot always be returned directly to the land without risk of damaging the sward by scorching, yet storage is often costly. On the other hand the manure represents an important source of nutrients that will be lost if the wastes are disposed of elsewhere; moreover, increased environmental control is eliminating the opportunity of local disposal in streams or open tips. Recent research in the Netherlands has been examining the possibility of subsurface application of the manure through the growing sward. At present, however, waste management is a major constraint upon adoption of zero-grazing.

Comparisons of grazing systems

There have been a number of comparative studies of different grazing systems (see, for example, Journet and Demarquilly 1979) although no clear pattern of their relative merits has emerged. To a great extent, this is not surprising. In comparing grazing systems we are comparing totally different farming practices, in which almost all the variables – including many of the environmental conditions – vary. Additionally, as we have seen, the grazing systems themselves show a wide range of forms, and each 'system' comprises a considerable spectrum of management techniques. Consequently, it is almost impossible to control the myriad of variables in anything like a rigorous fashion.

Hodgson (1974) has reviewed some experiments and quotes, amongst others, the results of Hughes and Redford (1952) who found that beef production was slightly higher under strip-grazing than under rotational grazing. Similarly, Alder and Chambers (1958) concluded that rates of calf growth were greater under set-stocking than under strip-grazing. A more extensive comparison was made by McMeekan and Walshe (1963) who found that rotational grazing systems had a slight advantage over other methods, but that the increased yield was generally no more than 5–10 per cent, rather than the 30–50 per cent claimed by many earlier workers. Cooper (1961) reviewed the debate on set-stocking and rotational grazing systems, recalling the influence of early experience in New Zealand on the attitudes of agriculturalists during the 1950s. Initially it was estimated that rotational (paddock) grazing had a yield advantage of some 50 per cent over set-stocking. Subsequently, results from Australia failed to substantiate this, but 'it was felt, if not said, that Australians were not New Zealanders . . . so matters of grassland management were outside their province' (Cooper 1961). Cooper also cites later work which reveals that the yield advantage of rotational grazing is often small, but on balance he favours the system because it allows higher grazing intensities to be maintained and it also restricts disease problems; with lambs and young cattle this may be serious. He also argues that rotational systems prevent selective grazing of the sward. More recently, Ernst *et al.* (1980) have reviewed experiments comparing rotational and continuous grazing systems. They argue that previous studies indicate a small advantage to rotational grazing, with increased milk yield per head of about 1.5 per cent under dairying systems and an increased liveweight per hectare of about 6 per cent under beef systems. Nevertheless, in practice this is offset by the significantly lower labour requirement (Table 7.4) and the greater resilience to weed invasion of continuous grazing.

It is apparent, therefore, that in terms of yield there is often little to choose between the various systems, so long as they are managed

Table 7.4   Labour requirements for rotational grazing and set-stocking

| | Labour input (man hours/ha) | | | |
	N fertilising	Fodder conservation[a]	Pasture topping[b]	Total
Rotational	11.1	5.4	2.7	19.2
Set-stocking	3.5	4.2	0.9	8.6

Source: After Ernst *et al.* 1980
Notes: [a] on 60% of total area
[b] on 40% of total area

efficiently. The choice tends to be made on the basis of other considerations, such as the ease of application, the availability of labour, farm layout and stocking rates.

7.2.4   Harvesting

Harvesting in grazing and grassland systems may involve a wide range of
practices. In some cases, the grass itself is harvested, either to supply fresh
forage for stall-fed animals (e.g. in zero-grazing systems) or for conservation purposes (i.e. storage as hay or silage to supply fodder during the
winter). In addition, the grazing animal itself may be removed. All these
practices involve the removal of dry matter and nutrients from the system
and consequently have major effects upon the structure of the system and
on the processes of nutrient cycling.

As we will see later, grass cutting has a significant influence on the pattern of sward growth for it removes the flowering head and causes reproductive growth to cease (see section 7.4.1). Factors such as the height,
frequency and timing of cutting therefore determine the character of the
growth curve and the ultimate yield of the crop. In general, increasing the
frequency of cutting reduces the total yield as shown in Table 7.5. Optimum yields are normally obtained when a first cut is taken soon after
emergence and two or three subsequent cuts are taken at intervals of approximately eight weeks (Holmes 1980). Cutting height is also critical.
Yields of perennial ryegrass, cocksfoot and timothy appear to be maximised with a cutting height of about 25 mm (e.g. Blood 1963; Reid 1966),
although Spedding and Diekmahns (1972) recommend a minimum cutting
height for conservation purposes of 50 mm.

Table 7.5   The effects of cutting frequency on the yield and digestibility of
           grass

Cutting interval (weeks)	Annual DM yield ($kg\ ha^{-1}$)	Dry matter digestibility (%)	Digestible DM yield ($kg\ ha^{-1}$)
4	12,690	73.8	9,350
6	13,360	71.1	9,530
8	14,050	67.0	9,420

Source: From Murdoch 1980

Grass conservation is a complex topic and only the basic principles need
concern us here (for details, see, for example, Watson and Nash 1960;
Murdoch 1980). Traditionally, hay-making has been the main method of
conservation and in 1960 accounted for about 55 per cent of the total
production of conserved grass in Britain. By 1978, however, its contribution had fallen to about 25 per cent (MAFF 1980a). One of the reasons
for this decline is that conservation as hay requires drying of the grass to
a moisture content of 25 per cent or less in order to suppress fungal or
bacterial decomposition. When drying is carried out in the field, this process is highly dependent upon weather conditions. It has been estimated,
for example, that there are on average only 2–2.5 periods of 3 or more
days of fine weather between May and August in southern England each

year, and such conditions are rarer in northern Britain. For this reason, use of machine-drying methods has increased in recent years, so that the crop can be gathered in a partially wilted state.

Silage-making practices vary considerably. In Britain, it is common to make silage from fresh-cut grass; it is treated with a preservative such as formic acid to prevent microbial putrefaction and stored in open silos. In much of continental Europe, in contrast, silage is produced from wilted grass which is stored, often untreated, in air-tight silos. The relative advantages of these two methods are slight; silage made from wilted grass tends to reduce the acidity of the fermented material, and as a result intake by cattle may be greater (Alder *et al.* 1969), but against this must be set the fact that digestibility is often less than with unwilted silage, while the need for field-wilting of the crop makes the process susceptible to weather conditions (Wilkins and Wilson 1974). It should also be noted that the quality of both hay and silage is affected by sward composition. Weeds, of course, reduce the nutritional value of the material, but clover, too, may cause problems in conservation. Although clover has the advantage of maintaining its nutritional value throughout the growing season, it has the disadvantages of taking longer to dry than grasses, and being subject to a clostridia-dominated fermentation which results in a butyrate-type silage.

Whatever methods of cutting and conservation are used, harvesting of the sward involves a major loss of nutrients. Because nutrients tend to be concentrated in the upper parts of the plant, disproportionate amounts of nutrients may be harvested. Moreover, frequent harvesting of relatively short, young grass probably removes greater quantities of nutrients than less frequent harvests, although the yield may be lower. Data on nutrient exports through cutting are surprisingly few, though G. W. Cooke (1975) quotes values of about 300 kg of nitrogen, 300 kg of potassium and 60 kg of phosphorous for a high-yielding grassland (12–14 tonnes per hectare per year). Theoretically, not all this is lost; some may be returned to the soil while the grass is drying on the surface; more importantly, large amounts should be returned in the manure from stall-fed animals. In addition, significant losses occur during storage. Most of this waste occurs through fermentation, but nutrients are also lost by removal in effluents (Table 7.6) and these may represent a major source of pollution for streams and lakes (see Ch. 9).

Table 7.6   Losses from silage during storage

	Wilted silage	Unwilted silage
Dry matter (%)	30.2	26.4
Source of loss (%)		
Spoilage	0.0	0.0
Effluent	1.6	7.4
Gaseous	8.8	10.2
Total loss	10.4	17.6

Source: After Gordon *et al.* 1959

On grazed pasture the losses are much more restricted. Animals retain relatively little of their nutrient intake within their bodies and the majority is returned to the land in urine and dung. The direct losses in milk, wool or carcases are therefore small (Table 7.7) and are normally amply replenished by inputs in rainfall and fertilisers.

Table 7.7   Nutrient losses in animal products, dung and urine in grazing
systems

	Loss (% of total intake)				
	P	Ca	K	Mg	Na
Urine	0	5	82	11	56
Dung	61	75	10	80	30
Milk	30	8	5	4	7
Carcase	9	12	3	5	7

Source: Data from Davies *et al.* 1962

7.3   THE ANIMAL COMPONENT

The effects of the animals in grassland systems depend upon the species
and breed, but in general animals:

1. Harvest the growing sward;
2. Return plant residues to the soil;
3. Alter the soil environment by trampling.

Each of these activities is spatially inequable, so that over time the animal
tends to produce inhomogeneity in the system. Moreover, through a
range of positive feedback effects, these influences are often self-
reinforcing.

7.3.1   Herbage demands

A mature cow consumes something in the order of 10–30 kg of dry matter
(DM) per day, a mature sheep probably 1–2 kg DM day^{-1}. The requirements of the animal vary, however, throughout the season in response to
its physiological condition, to environmental factors and to the nutritional
quality of the herbage. It is one of the fundamental problems of most
grazing systems that the requirements of the animal do not coincide with
the productivity of the sward (Fig. 7.5), which is why grass has to be conserved (or other sources of forage have to be used) to supplement the diet
during certain periods of the year.

The main factor determining the food demand of the animal is its liveweight. In general, as the animal gets older its liveweight and therefore
its food demand increases, although it ultimately does so at a decreasing
rate. Additionally, food requirements increase markedly during preg-
nancy and lactation. Another important factor is the activity of the animal; energy expenditure during grazing, for example, increases the
nutritional demands of the animal. Thus, animals which are stall-fed for
all or part of the year generally have lower requirements than grazing
animals.

7.3.2   Selectivity of grazing

The ability of the grass sward to satisfy the demands of the animal is not

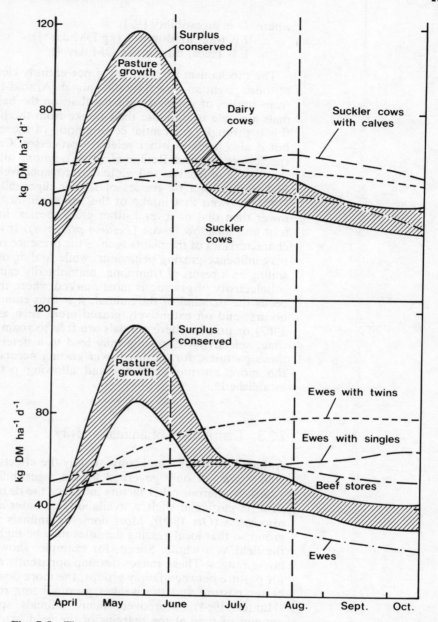

Fig. 7.5 The seasonal pattern of grass growth and herbage needs of grazing animals (from Holmes 1980).

constant. It varies from one species of grass to another, and it varies from one time of year to another. It also varies between different parts of the same plant. To a great extent this depends upon the digestibility of the plant material: the proportion of the total intake that is retained by the animal is:

$$D = \frac{I - F}{I} \times 100 \qquad [7.2]$$

where D is digestibility (%)

I is total food intake (kg DM day^{-1})

F is faecal output (kg DM day^{-1}).

The mechanism of selection is not entirely clear, but involves the use of touch, sight and smell by the animal (Arnold 1970). It is also apparent from studies of voluntary grazing that, on the basis of this selection, animals are able to optimise their intake from the herbage on offer. In part this is through preferential consumption of the more digestible material, but it also involves other selective processes. Comparisons for different species are few, but Reid *et al.* (1966), among others, have demonstrated that intake of lucerne and white clover respectively may be 20–40 per cent higher than that of grasses of similar digestibility whilst Minson *et al.* (1964) showed that intake of the S48 strain of timothy was significantly lower than that of several other grass species, including ryegrass, cocksfoot and meadow fescue (*Festuca pratensis*). It is also clear that specific characteristics of the plants such as the presence of external hairs or spikes may influence grazing behaviour, while fouling of the sward by faeces, or soiling as a result of trampling, undoubtedly causes rejection.

Selectivity of grazing is most marked where the supply of herbage exceeds the demand of the animal; it is thus common where undergrazing occurs, and on extensively grazed areas such as moorlands (e.g. Eadie 1967) or prairie where animals are free to roam over a large area. Over time, selectivity of grazing may lead to a deterioration in the sward of these pastures, for localised over-grazing occurs, reducing the vigour of the more nutritious grasses and allowing poorer species to become established.

7.3.3   Distribution of animal activity

Animal behaviour – and consequently the effects of animals on the grassland – are not only governed by the digestibility or palatability of the sward. Numerous other factors influence the distribution of animals in the field, in particular shelter, availability of water and their instinctive gregariousness (Orr 1980). Most domestic animals wander as a loosely knit group so that local grazing densities may be high, whatever the mean for the field as a whole. Sheep, for example, show marked preferences for home ranges. These ranges develop apparently as a result of competition for pasture between family groups, the more dominant groups gaining the better pasture and the weaker groups being restricted to poorer areas (Hunter 1964). Moreover, many animals spend a disproportionate amount of time at the margins of the field, partly it seems to gain shelter. Similarly, trees or buildings may be used for shelter and localised zones of intensive exploitation are commonly found around them. In addition, where water troughs or feed-lots are provided, there is a natural concentration of activity.

The overall effects of this behaviour are to give very uneven intensities of exploitation across a field. The data in Fig. 7.6A provide an example; it shows the distribution of fourteen heifers in a field in County Kerry, western Ireland, monitored at intervals over a period of two weeks. It is notable that the animals favoured the downwind edge of the field – and, of course, the area around the water trough – despite the presence of a high sheltering wall on the upwind boundaries.

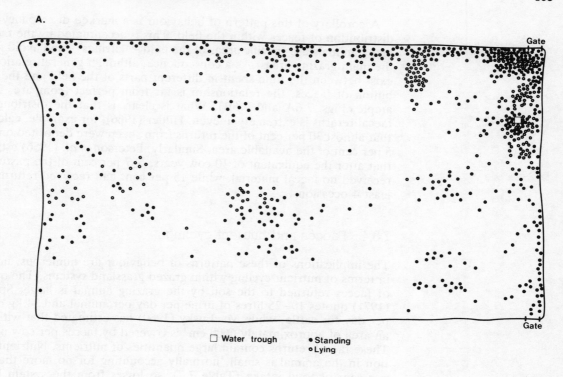

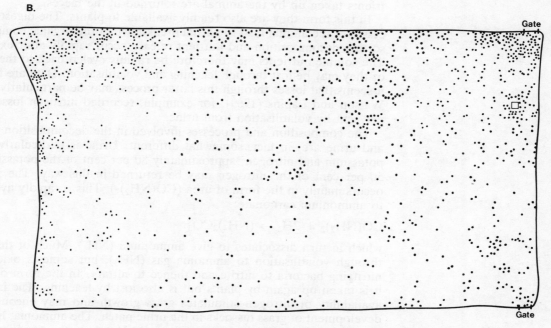

Fig. 7.6   Spatial pattern of field usage by grazing animals in a 4 ha paddock in County Kerry, Republic of Ireland (a) distribution of 14 heifers on 57 sampling occasions over a 2-week period (b) distribution of faeces after 14 days' grazing (survey by D. J. Briggs).

A corollary of this pattern of behaviour is a marked discrepancy in the distribution of feaces within the field. This is accentuated by the fact that many animals use specific areas of the field – 'latrines' – and avoid fouling the main grazing areas. As a consequence, although general relationships exist between the time spent in different parts of the field and the distribution of faeces, the relationship is far from perfect (compare, for example, Figs. 7.6A and 7.6B). What is clear, is that the distribution of faecal returns is extremely uneven. Hilder (1966), for example, calculated that almost 30 per cent of the returns from sheep were deposited on about 5 per cent of the available area. Similarly, Peterson *et al.* (1956) estimated that after the equivalent of 10 cow years, 6–7 per cent of the pasture had received no faecal material, while 15 per cent had received returns on at least 4 occasions.

7.3.4 Faeces and nutrient cycling

The implications of these patterns of behaviour are numerous, not least in terms of nutrient cycling within grazed grassland systems. The quantity of faeces returned to the soil by the grazing animal is large; Spedding (1971) quotes 10–25 litres of urine per day per animal and 34 kg of dung per day for cattle, while MacLusky (1960) has estimated that with cattle an area of approximately 6600 cm^2 is covered by faeces per cow per day. These faecal returns contain large quantities of nutrients. Nutrient retention in the animal is small, normally accounting for no more than 5–10 per cent of total intake (Table 7.7), so losses from the system by harvesting of the animal are relatively limited. The great majority of the nutrients taken up by the animal are returned in the faeces.

In this form they are also readily available to plants. The digestive processes of the animal lead to partial breakdown of the organic compounds, so that decomposition and nutrient release occur rapidly after excretion. The soluble nutrients may therefore be rapidly cycled back to the plants, or they may be lost through leaching and volatilisation. Data are few, but it seems that losses through this latter process may be particularly severe; Watson and Lapins (1969), for example, recorded nitrogen losses of 84 per cent by volatilisation from urine.

The composition and processes involved in the decomposition of dung and urine are, in fact, somewhat different. Urine is particularly rich in potassium and nitrogen; approximately 80 per cent of the potassium and 50 per cent of the nitrogen may be returned in this way. The nitrogen occurs mainly in the form of urea ($CO(NH_2)_2$). This is rapidly hydrolized to ammonium carbonate:

$$CO(NH_2)_2 + 2H_2 \rightarrow (NH_4)_2CO_3 \qquad\qquad [7.3]$$

which in turn dissociates to give ammonium (NH_4). Much of this is lost through volatilisation to ammonia gas (NH_3), but some is oxidised by nitrifying bacteria to nitrite and thence to nitrate. In the form of nitrate it is taken up again by plants but is also lost by leaching. The increased availability of nitrogen stimulates grass growth and may encourage the development of grass tussocks in the urine patch. The ammonia, however, also ties up hydrogen ions and raises soil pH; where the rise in pH is excessive, scorching of the vegetation may occur.

Losses of potassium through volatilisation are small, but leaching may result in considerable loss; Davies and Hogg (1960) reported as much as

30 per cent. On the whole, however, the strong absorptive tendency of the potassium helps retention and in most cases the vast majority is recycled to the plants.

Nutrients returned in the faeces are, by comparison, less readily available and less mobile, though still far more so than the nutrients returned in the undigested plant material. Phosphorus, calcium and magnesium are all abundant (Table 7.7, p. 162), and the latter may be responsible for marked changes in soil pH beneath the dung patch; Davies *et al.* (1962) recorded an increase from 6.15 to 6.7 in one study. Decomposition of the faeces is dependent upon the action of soil organisms, particularly beetles, earthworms, nematodes and fungi and the availability of these determines the rate of breakdown and nutrient release. In Australia, where the dung beetle is naturally absent, decomposition is excessively slow with the result that nutrient deficiencies severely limit grass growth. With the artificial introduction of suitable species of beetle, however, more rapid cycling of nutrients has been achieved and pasture productivity greatly increased.

The uneven distribution of faeces results in a similarly irregular distribution of nutrient returns to the soil (Gillingham and During 1973). As has been noted, animals tend to avoid fouling the grazed areas of the field, so an internal cycle tends to develop involving an export of nutrients from the grazed areas and an import to the camp and 'latrine' areas. This redistribution of nutrients may be marked and persistent; Briggs (1978) found that exchangeable potassium contents in heavily used parts of a pasture field were six to eighteen times those in the lightly used areas. Data presented in Figure 7.7 show the concentration of phosphorus and potassium in the intensively fouled marginal areas of a field near Portland, Dorset, England.

The pattern of faecal returns to the soil has a major effect upon grazing behaviour. Animals tend to reject fouled grass, and in some cases the dung patches and surrounding areas may be ungrazed for several months. Norman and Green (1958), for example, noted that fouled areas were avoided by cattle for up to eighteen months after fouling. The degree of rejection depends upon the grazing intensity, however, and as intensities increase the extent and persistence of rejection declines. Yiakoumettis and Holmes (1972) recorded a decrease in the rejected area from 42 to 29 per cent as the stocking rate was increased from 6.4 to 10.3 cattle per hectare. Where grazing intensities are low, rejection exacerbates the selective effects of grazing and intensifies the tendency for herbage deterioration. Similarly, where urine scorch occurs, weed species such as couchgrass (*Agropyron repens*) may become established.

7.3.5 Trampling and poaching

As the grazing animal moves around the field it tends to exert a considerable pressure on the surface which modifies both the grass sward and the soil. The average weight of a mature sheep is about 60–80 kg; that of a mature cow about 500 kg. Given total hoof areas of 80–100 and 250–350 cm^2 respectively, this indicates that the animals may exert a force of up to 1600 g cm^2 (Table 7.8). In the case of younger animals, the body weight is clearly low, but because the ratio between weight and hoof area remains fairly constant, the pressure exerted on the soil is much the same.

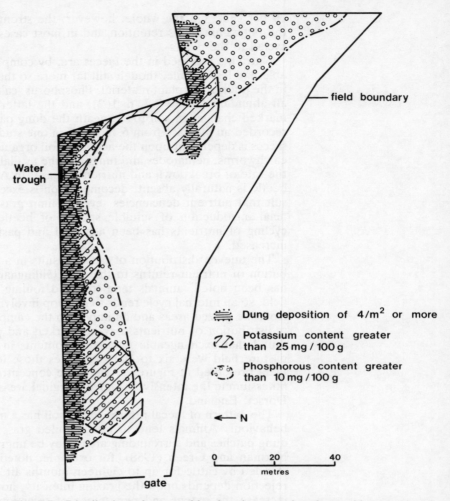

field boundary

Water
trough

Dung deposition of 4/m² or more

Potassium content greater
than 25 mg / 100 g

Phosphorous content greater
than 10 mg / 100 g

N

0   20   40
            metres

gate

Fig. 7.7 Distribution of extractable phosphorus and exchangeable potassium in
          a field in Dorset, England, 1 week after a 6-week grazing period by
          30 Friesian cattle (from Salmon 1980).

Table 7.8 Hoof pressures exerted by grazing animals

Animal	Total hoof area (cm²)	Animal weight (kg)	Pressure (g cm⁻²)
Sheep (Kerry Hill)	92.4	74.3–87.5	800–950
	79.8	59.0–73.5	740–920
Cattle (South Devon)	350.0	500–560	1430–1600
Cattle (Jersey)	250.0	320–365	1280–1460

Source: From Spedding 1971

When moving, an animal's weight is concentrated on one or two hooves,
so the pressure exerted increases markedly. Moreover, the forces exerted
comprise both a vertical (normal) component and a horizontal (shear)
component (Fig. 7.8), increasing the effect on the soil and sward.

The effects of this pressure upon the sward are varied. To some extent

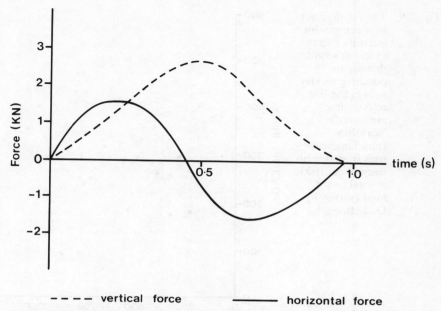

Fig. 7.8   Distribution of vertical and horizontal forces by the hoof of a cow
whilst walking (after Scholefield, pers. comm.).

treading may stimulate growth, apparently by encouraging the development of basal tillers. As trampling pressure increases, however, tearing
and bruising of the plants occurs, reducing growth of the more sensitive
species and causing both a decline in yield and a change in species composition. A survey of seventeen sites on uniform clay soils in South Yorkshire showed an almost linear decrease in the sward cover as stocking rate
increased (Fig. 7.9) and undoubtedly weed invasion tends to occur in the
areas of bare soil. Hodgson and Ollerenshaw (1969) showed that an increase in stocking density from 71 to 200 sheep per hectare resulted in
an increase in tiller damage of the sward from 7.5 to 62 per cent. Several
studies have demonstrated the effects upon yield. Edmond (1958), for
example, showed an almost linear reduction in yield with increasing treading intensity (Fig. 7.10), while Campbell (1966) found a 15 per cent reduction in growth after a single day's grazing by cattle at 150–300 animals
per hectare. Edmond (1963) also indicated the importance of soil conditions in this context, for trampling of a wet soil by 45 sheep per hectare
reduced herbage yields by 90 per cent; on a dry soil the same stocking
density gave a reduction of 53 per cent.

   As this implies, the effects upon the sward are related to the effects of
trampling upon the soil. From the principles discussed in relation to tillage
in Chapter 3 it is clear that the ability of the soil to withstand the applied
pressure of the animal depends upon its shear strength. If the applied
pressure exceeds the shear strength of the soil, then failure occurs and the
animal hoof penetrates the soil.

   In general, failure may be of two types. Plastic failure occurs when the
soil shears internally. It is normally accompanied by marked structural
rearrangement and deformation, but relatively little compaction, so that
bulk densities of the soil do not increase markedly. Plastic failure generally takes place when the soil is relatively wet – beyond its lower plastic

Fig. 7.9 The relationship
between grazing
intensity and the
extent of sward
damage by
poaching on clay
soils in the Isle
of Axholme
area, south
Yorkshire
(poaching scale
runs from 0 = no
damage, to 1000
= total sward
loss) (survey by
D. J. Briggs).

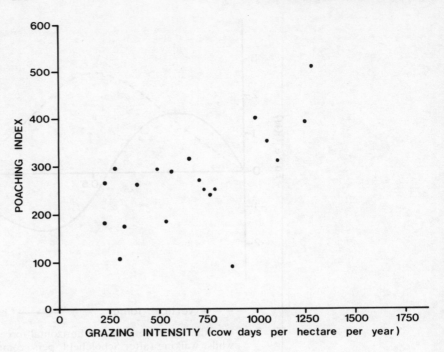

Fig. 7.10 Effects of
different
stocking rates
on the dry
weight of
ryegrass and
clover,
measured after
twenty-eight
days of grazing
(after Edmond
1958).

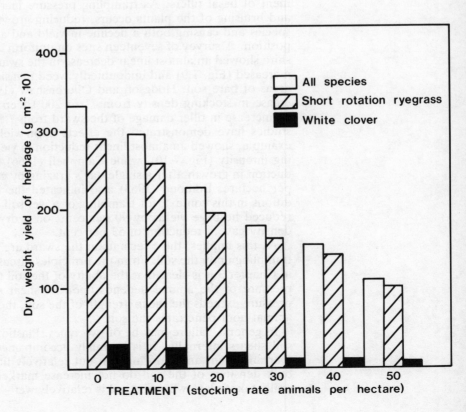

limit – so that the clod shear strength is low while the bulk shear strength is high. Conversely, under drier conditions, compressive failure occurs; shearing takes place not along distinct failure planes but randomly and minutely by movement of individual particles or aggregates relative to each other. The result is a closure of the pores, particularly between the aggregates, and an increase in bulk density. Gradwell (1968) recorded an increase in the bulk density of a volcanic soil from 0.79 to 0.88 g cm^{-3} in response to a change in stocking density from 9 to 53 sheep per hectare; characteristically, this involved a reduction in macropore volume from 13 to 7.1 per cent.

The effects of compaction and deformation by trampling are wide-ranging. Closure of the pore spaces inhibits root development and reduces plant growth; it increases moisture retention and causes a reduction in infiltration capacity (e.g. Gifford and Hawkins 1978, 1979); it destroys soil

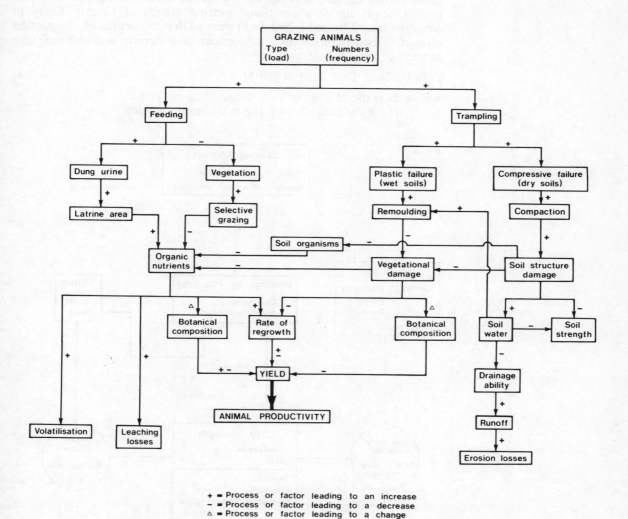

Fig. 7.11 A model of the effects of trampling and grazing on soil, herbage and animals (from Salmon 1980).

organisms and thereby interferes with organic matter decomposition and nutrient cycling (Fig. 7.11). Ultimately it may lead to changes in sward composition (e.g. Charles 1979) and soil erosion.

The extent of soil damage by trampling is a function of two main factors (Fig. 7.12): the animal and the environmental conditions. The animal factor includes variables such as the static hoof weight (itself a product of the species, breed, age and physiological state of the animal), the animal behaviour and the grazing intensity. These determine the magnitude, distribution and frequency of trampling pressures. The environmental conditions control the resistance of the soil to damage: its shear strength. This depends in general upon soil type, but more particularly upon a wide range of edaphic, vegetational and climatic conditions (Fig. 7.11). In a general survey of susceptibility of soils to damage by trampling in England and Wales, Patto *et al.* (1978), took account of rainfall, soil texture, organic matter content and slope. A more detailed investigation of structural changes under a rotational grazing system in County Kerry in western Ireland (unpublished data) showed that the degree of compaction related to two main factors: antecedent bulk density and the moisture content of the soil:

$$D = 0.574 - 1.67\ D_o + 0.660\ M \tag{7.4}$$

where D is the change in bulk density (kg m^{-3})

D_o is the bulk density (kg m^{-3}) before grazing

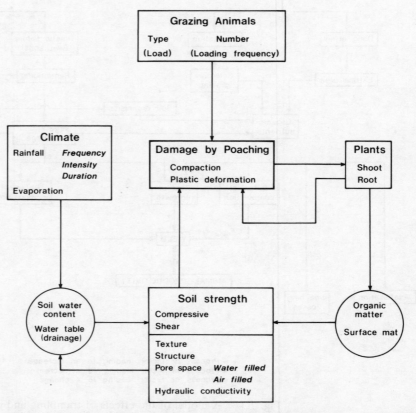

Fig. 7.12   Factors affecting soil poaching by livestock (from Patto *et al.* 1978).

M the moisture content (%);
the coefficient of determination (r^2) was 0.889.

It is apparent that as the soil is deformed by trampling so its resistance to further damage tends to decline. This occurs in part because compaction encourages moisture retention and thereby reduces the soil shear strength; it occurs also because damage destroys the vegetation cover which otherwise provides protection against trampling. For these reasons, repeated trampling tends to result in a cumulative increase in bulk density. On the other hand, given time, the soil is able to recover to some extent from the effects of trampling so that the net effect depends upon the relative rates of damage and recovery; this, in turn, is a function of the length of time between each trampling event. Studies of a paddock grazing system near Barnsley in south Yorkshire, for example, showed that recovery of structural conditions may be almost complete when grazing occurs at relatively long intervals (in this case about fifty days), but may be only partial when the interval is shorter.

Recovery of the soil structure occurs through both physical and biological processes. Desiccation through drying, the action of freezing and thawing, and pressures exerted by ice crystals all modify the soil structure and act to re-create the soil pores. More important in many cases, however, is the activity of burrowing earthworms and of root extension; where these are restricted, as happens for example when the soil is badly damaged, recovery is extremely slow. Thus, Gradwell (1968) found that damage caused during winter grazing failed to be repaired during the summer rest period; a condition also noted by Briggs (1978).

It is apparent that one of the main factors contributing to soil damage by poaching is over-stocking of the land; in very general terms the land can be considered to have a carrying capacity, beyond which poaching is likely to be severe and, through the processes of positive feedback involved, self-reinforcing. In reality, however, the concept of carrying capacity in this context is difficult to define, for, as we have seen, the grazing animal displays a characteristic pattern of behaviour; it is the irregularity of trampling pressure resulting from this behaviour that often causes poaching problems. Thus, where trampling intensities are higher, around feed troughs and water troughs, in gateways and in areas used for shelter, poaching tends to occur even at low overall stocking rates. Moreover, in many of these areas the high inputs of urine raise the soil moisture level and reduce the resistance of the surface to trampling. Although localised trampling of this sort may not cause great losses of yield – particularly when it is appreciated that animals, by preference, are not grazing in these areas – the problems of hygiene, animal cleanliness and health may be considerable.

7.4 THE VEGETATION COMPONENT

As Fig. 7.1 (p. 149) indicates, the vegetation forms a central component in grassland systems. It receives inputs of energy from solar radiation, water and nutrients from the soil. It provides energy and nutrients to the livestock and, through the return of plant residues and the penetration of roots, has a marked effect on soil conditions. It also acts as a buffer between the grazing animals and the soil, absorbing some of the physical pressures exerted by the animals and protecting the soil from physical damage. The sward is, likewise, influenced by animal behaviour including the effects of trampling, selective defoliation and the return of faeces to the surface. All these processes affect the rate of growth and competitiveness of the grass plants, and thereby influence sward composition. These, in turn, control the productivity of the grassland system.

7.4.1   Grass growth

Grasses range in habit from annual (i.e. completing the life cycle in a single year) to perennial (requiring several years to complete their cycle). The habit seems to be largely genetically controlled, but within perennials the longevity of the plant is also affected by environmental and management factors which influence its ability to compete with neighbouring plants. Ultimately, too, it seems that the residues which collect at the surface of the soil may provide habitats for pathogens which attack the plant, reducing its competitive ability if not actually killing it. In addition, as the roots extend they fill the available growth space in the soil, remove the available nutrients and eventually result in a nutrient deficiency within the rooting zone. For these reasons, plants are often more short-lived in the field than might be expected from theoretical considerations.

Within a single growing season, grasses show a fairly consistent growth pattern. Initiation of growth is largely determined by temperature and in most cases growth does not commence until soil temperatures reach about 6 °C. Thereafter growth tends to be rapid and the daily dry matter-production of the sward increases to a peak within six to eight weeks (Fig. 7.13). During this period the sward is growing by two processes; leaves are extending from existing, parent tillers, while new daughter tillers are forming at the base and growing laterally. Thus, this phase of vegetative growth, as it is known, sees an increase in both the mean height of the sward and in its density. In time, the new tillers develop adventitious roots which grow into the ground so that the daughter tillers become independent of their parents.

Throughout this phase, the flowering stems of the plants remain short and close to the ground, but during late spring and early summer (May and June in the northern hemisphere) vegetative growth gives way to reproductive growth, and the flowering stems then extend and ultimately produce a flower. Following inflorescence, the seeds within the flower head are shed and the tiller starts to die. No further extension of the leaves occurs, and the only new growth is through the development of tillers from the dying basal stem. At this stage, therefore, the net production rate of the sward starts to decline as death of the tillers (senescence) begins to balance the effects of renewed vegetative growth. As the season progresses, senescence increases, for stresses upon the plants become greater due to lower radiation inputs and reduced ambient temperature, the plants cease to be able to compete so successfully, and new tillers fail to mature.

If allowed to proceed unhindered, the sward therefore tends to show a relatively simple growth pattern (a in Fig. 7.13), and such a cycle has in fact been illustrated by, for example, Alberda and Sibma (1968) work-

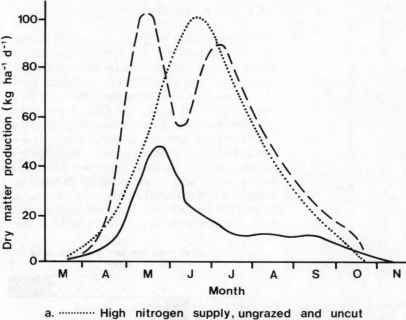

Fig. 7.13 Typical patterns of dry matter production of irrigated pasture under different management regimes.

ing in the Netherlands. Research in Britain, however, has shown somewhat more complex, bimodal growth patterns, not so much, it seems, because of differences in plant behaviour as because of differences in experimental design and sward management. Studies such as those by Anslow and Green (1967) have involved cutting the sward at regular intervals to measure its rate of dry matter production. Cutting, like grazing, removes the developing flower head and truncates the phase of reproductive growth; it causes leaf growth within the reproductive tiller to cease, and, as after inflorescence, further growth occurs only through new tiller development. Thus the sward enters a new phase of vegetative growth, but one which is slower because of the effect on the reproductive tillers. New tiller development initially involves little vertical growth in the sward, so the harvestable yield declines markedly until the new tillers start to grow upwards. At this stage a second, but lower, peak in the net production rate is achieved, after which generally increasing senescence leads to a gradual reduction in growth towards the end of the season (b or c in Fig. 7.13). In addition, it is likely that the early summer decline in growth found in the British studies is due to the presence of moisture stress which occurs despite irrigation of the experimental plots.

The differences between the growth patterns found by Alberda and Sibma (1968) and Anslow and Green (1967) illustrate the difficulties of experimental studies of sward growth in grazing systems. They also show the effect of harvesting upon sward growth, and demonstrate a factor of considerable agricultural importance. Early harvests of the sward, which truncate the period of inflorescence, may result in as much dry matter as all the rest of the harvests during the period of regrowth put together.

7.4.2   Root growth

The growth cycle of the above-ground parts of the sward is mirrored by developments in the root system. The adventitious roots produced by the tillers extend and branch in the soil during growth, although root production is somewhat slower than shoot growth, so that over the season the shoot–root ratio increases. Most of the roots are concentrated in the upper 10–20 cm of the soil (Fig. 7.14), but a minority may reach depths of several metres. Like the above-ground parts of the plant, the roots die off during the growing season, mainly, it seems, in association with tiller death following inflorescence or harvesting (Fig. 7.15). In addition, root senescence may increase when soil moisture stresses develop in the upper layers of the soil, while physical damage by trampling also destroys the rootstock. Much of the return of organic matter, however, takes place during the periods of active root growth; material is sloughed off from the roots, contributing to the food supply of the rich rhizosphere population.

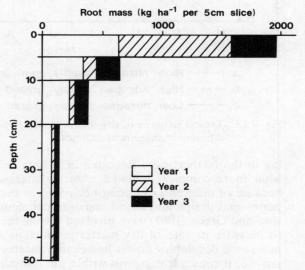

Fig. 7.14   Distribution with depth of roots in a grassland soil (all samples in June) (data from Hoogerkamp 1974).

7.4.3   The lean years

Following sowing, grass swards often show a marked pattern of growth, characterised by several years of high yield, then a period of depressed growth before stabilising at an intermediate–high yield. The period of depressed growth has been identified by a variety of names, including the lean years, the hungry years and the turf-bound phase. It has been cited as one of the main disadvantages of resowing permanent pasture, and one of the main reasons for frequent ploughing and re-establishment of leys; clearly yields can be maintained at a high level by ploughing up or resowing the pasture before the lean years arrive.

   In recent years, it has become clear that the problem of this depressed growth can be avoided by a variety of mechanisms: by careful selection of the grass species, by appropriate fertiliser policy and by strict control of grazing. Nevertheless, the cause of the lean years is far from clear.

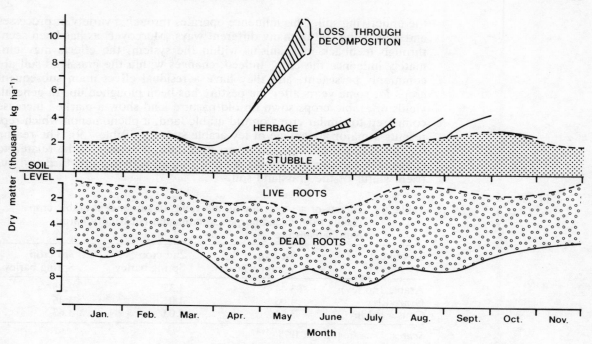

Fig. 7.15 The pattern of herbage production and root growth and decomposition during the year in pasture soils (from Jewiss 1972).

Amongst the explanations put forward have been nutrient deficiencies, structural deterioration, microbial factors, CO_2 toxicity, accumulation of organic residues, physiological characteristics of the grass species and toxic excretions from the sward (Bates 1948; Ahlgren 1952; Klapp 1971). More recently the problem has been reviewed by Hoogerkamp (1979) and it seems clear that some combination of these factors may be involved. He notes the association of reduced yield with changes in sward composition, poorer species such as *Agrostis* spp. and *Festuca* spp. replacing the high-yielding sown species. In part, at least, this could be related to lack of nitrogen resulting from the accumulation of organic matter in the rooting zone; the increased supply of organic material stimulates the microbial populations which tie up much of the soil nitrogen in an unavailable form (Garwood *et al.* 1977). Suitable applications of N fertiliser maintain nitrogen contents and help to prevent invasion by lower-yielding species. Similarly, it is apparent that other causes of species change, such as trampling or inherent weakness of the sown species, may encourage invasion by poorer grasses, with a consequent reduction in yield. What is perhaps less clear is why the sward eventually recovers from the lean years even when corrective management practices are not adopted.

7.5 THE SOIL COMPONENT

The growth, death and decay of plant material and the activities of the grazing animals in the grassland system have an important influence upon the underlying soil. This influence operates through a variety of processes and affects the soil in many different ways. Moreover, as has been seen, through feedback mechanisms within the system, the effects may ultimately influence the crop. Indeed, changes within the grassland soil are commonly persistent, and they have a residual effect upon subsequent crops for some years after the pasture has been ploughed up. In general, yields of arable crops sown on old pasture land show a marked increase compared to similar crops on old arable land, a phenomenon which represents a major advantage of ley-arable systems (Table 7.9). The reason for this residual effect is, however, uncertain; it appears to be related to changes in the organic matter content, faunal activity, structure and nutrient status of the grassland soil.

Table 7.9   Effects of a 1-year ley on the yields of subsequent arable crops: results from Gleadthorpe Experimental Husbandry Farm, Nottinghamshire

Treatment	1st crop Winter wheat	2nd crop Spring barley	3rd crop Spring barley
Arable (no ley)	3.44	3.35	2.75
Grazed ley	4.05	4.01	3.46
Conserved ley	3.81	4.03	3.63

Source: Data from Eagle 1975

7.5.1   Organic matter and faunal activity

It has been recognised for many years that arable land, sown to grass, experiences a relatively rapid and marked increase in soil organic matter contents (e.g. Clement and Williams 1964; Hoogerkamp 1973; Johnston 1973, Garwood *et al*. 1977) (Fig. 7.16). The exact causes of this change are still something of a mystery. In part it maybe due to the greater mass of roots developed in grass as opposed to arable crops; nevertheless, evidence to support this contention is inconclusive, and data presented by, for example, Troughton (1961, 1963) indicate that the biomass of cereal roots is often little less than that of grass. Alternatively, the differences may reflect the lack of removal of the crop during harvesting. On grazed pasture, at least, removal is almost negligible compared to the losses experienced during harvesting of cereals. On the other hand, a similar though reduced effect is found even in mown pastures (Table 7.10). It has also been argued that reduced rates of oxidation, due to lack of tillage and aeration of grassland soils, may favour organic matter accumulation, particularly in the form of humus. As ever, it seems likely that all these factors are involved to a greater or lesser degree.

Whatever the reasons, it is clear that the organic carbon contents often rise sharply following the establishment of the grass crop, particularly in the upper 2–3 cm (Fig. 7.17). Ultimately, the content seems to stabilise at about 3–4 per cent, some 2 to 3 times the levels commonly found in old arable soils. It is also apparent that this effect varies considerably according to the nature of the sward and the management regime. Clement and Williams (1964) compared numerous species under both mowing and grazing regimes and demonstrated that certain varieties of ryegrass and timothy resulted in significantly greater increases in organic carbon contents than the less nutritious species such as *Nardus stricta* and *Agrostis*

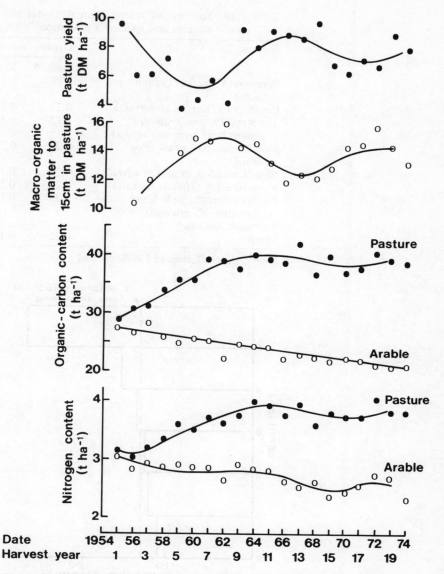

Fig. 7.16 Changes in pasture yield and soil conditions in arable land and pasture (after Garwood *et al*. 1977).

tenuis. They also demonstrated that organic carbon contents were generally higher under a grazed sward than under a mown sward (Table 7.10).

Changes in organic activity occur in association with these changes in organic carbon content. Most soil organisms show a marked increase in numbers under grass as compared with arable crops, but possibly the most notable and important effect is on earthworm numbers. Russell (1973), summarising research, quotes average earthworm numbers of up to $1\frac{1}{4}$ million per hectare in pasture, weighing some 650–1100 kg ha^{-1}. This contrasts with average quantities in arable land of approximately 100 kg ha^{-1}. This vast increase in earthworm populations presumably

Table 7.10 Soil organic matter contents under different grassland management
systems and species composition

Species	Organic matter content (%)	
	Grazed	Mown (with aftermath grazed)
Perennial ryegrass (*Lolium perenne*)	0.19	0.16
Cocksfoot (*Dactylis glomerata*)	0.21	0.11
Timothy (*Phleum pratense*)	0.22	0.10
Common bent (*Agrostis tenuis*)	0.19	0.16
Smooth meadow grass (*Poa pratensis*)	0.21	0.20
Rough meadow grass (*P. trivialis*)	0.27	0.19
Yorkshire fog (*Holcus lanatus*)	0.24	0.18
Ryegrass/timothy/red clover (*L. perenne, P. pratensis, Trifolium pratense*)	0.25	0.12
Mean	0.22	0.15

Source: After Clement and Williams 1964

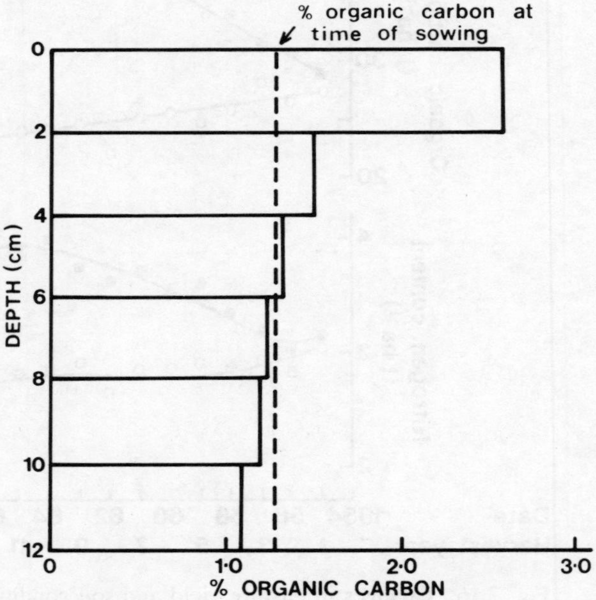

Fig. 7.17 Organic carbon contents with depth in old arable soils put down to
three and a half years of grass (after Clement and Williams 1964).

reflects a number of factors: the lack of death by tillage, the improved food
supply and possibly increased nutrient availability. The numbers vary,
however, according to the grassland management regime and sward composition. Sears and Evans (1953), for example, found a close correlation
between grazing density and earthworm weights, while Heath (1962)
noted that grazing increased both the numbers and the total weight of
earthworms (though not their average weight). Poaching, however, may
greatly reduce earthworm numbers (Briggs 1978). It should also be noted
that changes in earthworm populations involve variations in species as
well as numbers. Evans and Guild (1947) concluded that the small *Allo-*

lobophora chloritica is most abundant in arable soils and, when such land is put down to pasture, their numbers increase rapidly. In time, however, the larger species, *A. nocturna*, becomes dominant. Data on deeper-burrowing species such as *Lumbricus terrestris* and *Octolasium eyaneum* are less reliable, due to the difficulty of ensuring accurate sampling.

The activity of earthworms in the grassland soil has a number of implications. The burrowing and mixing improves aeration, especially in the upper 15–20 cm of soil. Their channels also act as major route-ways for water movement (Ehlers 1975), so they increase infiltration and facilitate drainage. Moreover, earthworms play an important part in organic matter decomposition, by macerating the larger plant residues and making them available for attack by micro-organisms. In addition, they help to concentrate nutrients in the rooting zone of the crop. The nutrients are excreted in the body slime of the worm and in the cast, with the result that both the walls of their burrows and the casts are enriched with nutrients (Table 7.11). Barley and Jennings (1959) found that 6.4 per cent of the non-available N ingested by the worms was excreted in soluble compounds, so it appears that at least part of this enrichment process involves conversion of unavailable to available form. The fact that plants tend to exploit worm channels for rooting also means that the enrichment occurs in areas immediately accessible to plants.

Table 7.11 Nutrient contents of earthworm casts

	Available nutrients (p.p.m. dry soil)				
	Ca	Mg	K	P	NO_3-N
Cast	2790	492	358	67	21.9
Topsoil	1990	162	32	9	4.7

Source: Data from Russell 1973

7.5.2 Soil structure

The effect of even short periods of grass upon soil structure has been vividly summarised by Russell (1973): 'The visual effects of a three-year grass or lucerne ley on the soil condition can be striking. After the ley has been ploughed out, the land may be in magnificent tilth that allows a seedbed to be made over a very wide moisture range, whereas old arable soil may either be too sticky and plastic or so hard that seedbed preparation is impossible.' These differences reflect the ability of the grass or lucerne to improve the structural stability of the soil, a phenomenon noted in numerous studies (e.g. Clement and Williams 1958, 1959; Low 1954, 1972, 1973; Burke *et al*. 1964; Russell 1971a; Johnston 1972) (Fig. 7.18).

The factors involved in this improvement are far from perfectly understood. It is apparent that the effect is confined mainly to the upper few centimetres of soil (Fig. 7.19), although despite this the improved stability persists after ploughing and mixing of the soil. It is also apparent that the effect varies considerably, and on some soils the influence of grass is almost negligible; in general, it is most marked on soils with an inherently good structure and under mown, as opposed to grazed, grassland (for treading by animals damages the structure). Similarly, the effect is dependent upon the composition of the sward, and Clarke *et al*. (1967), in

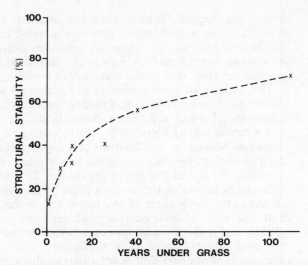

Fig. 7.18 Changes in the structural stability (per cent of soil particles < 2 mm diameter in stable aggregates > 2 mm diameter) in an old arable soil put down to grass (data from Low 1972).

Australia, found that ryegrass was more effective in improving structural stability than phalaris grass. Comparable studies in Britain are less clearcut in their conclusions; Pringle and Coutts (1956) noted that timothy was better than ryegrass, but several studies have revealed little difference between species (e.g. Clement and Williams 1958).

In so far as differences do exist between grass species, they seem to reflect variations in rooting characteristics. The fine roots of some grasses grow through the soil aggregates and help bind them together; the greater the quantity of such roots and the greater their tensile strength, the more effective they are. In addition, fungal hyphae associated with the roots are likely to play an important part in maintaining structure by enmeshing the soil crumbs. The increased organic matter content of the soil may also be significant, for the high exchange capacity may encourage cation bonding, while the humic acids and polysaccharides produced during decomposition improve the water stability of the aggregates. Above all, however, the improvements in structural stability seem to depend upon the increased activity of earthworms and other soil organisms. These stabilise aggregates through the production of gums and slimes and by intimate mixing of the organic material into the mineral soil.

The significance of these structural improvements for crop yields is a matter of some debate. Undoubtedly the increased stability of the aggregates aids drainage, facilitates tillage and reduces erosion, but the direct effects on productivity are difficult to determine, largely because structural improvements are so closely related to other changes in the soil. In the long term, however, it seems clear that the greater flexibility of cropping and tillage which structural improvements offer are major advantages. This, again, is a factor in favour of using ley-arable systems on soils which are inherently unstable or difficult to cultivate.

Not all the structural effects of grassland farming are beneficial. As we have seen, poaching associated with grazing of the sward may cause severe compaction. Compared to arable soils, pastureland may have a relatively high potential for self-regeneration due to the abundant and active

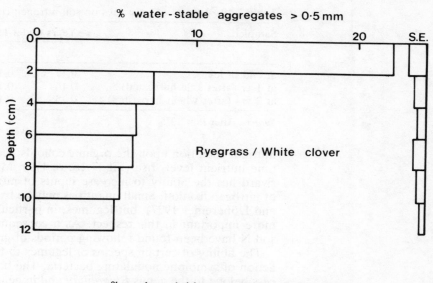

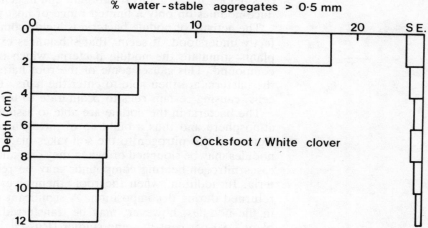

Fig. 7.19   Changes in structural stability with depth of old arable soils after three years' ley grass (after Clement and Williams 1958).

soil fauna, but when compaction results in the exclusion of soil organisms (particularly earthworms), then the capacity for structural recovery is greatly reduced. Moreover, the positive feedback mechanisms within the system (Fig. 7.11, p. 171) mean that compaction, once initiated, encourages further compaction. In these instances, the effects of poaching may be persistent and compaction in one grazing season may carry over to the next, leading ultimately to cumulative deterioration of the soil structure.

7.5.3   Soil nutrients

Probably the most significant of all the residual effects of the grass crop upon the soil is its tendency to increase nutrient availability. To a small degree this is due to the improved nutrient cycling and greater facility for

Table 7.12   Effects of grass leys on soil nitrogen content

Sampled	Years under ley			
	3	2	1	0
at end of ley	0.177	0.165	0.155	0.143
at 1 yr (after kale harvested)	0.164	0.158	0.150	0.139
at 2 yrs (after wheat harvested)	0.158	0.152	0.146	0.138

Source: After Low 1975

nutrient retention upon the organic colloids; leaching is reduced and over time nutrient levels rise (Fig. 7.16, p. 179). More importantly, the grass sward has the ability to increase inputs of nitrogen through the process of nitrogen fixation. Small quantities may be fixed by some grasses (Neyra and Döbereiner 1977), but legumes, in particular clover and lucerne, are more important in this respect. As a consequence, marked increases in soil N have been found following periods of grass cropping (Table 7.12).

The ability of certain species of legumes to fix nitrogen arises from the action of symbiotic nodulating bacteria. The bacteria involved in the process belong to the genus *Rhizobium* and in general individual species produce nodules on only a limited range of host plants.

The activity of nodule bacteria is highly complex and still far from perfectly understood. It seems that substances excreted by the roots of the plants stimulate the nodule bacteria, which in turn produce an irritant compound. This causes some of the root hairs on the plant to curl, and the bacteria are then able to enter the hairs. The bacteria infect the root cells, causing certain cells to proliferate to form a nodule.

The bacteria in the nodule are able to assimilate nitrogen from the soil atmosphere and thus a build-up of nitrogen occurs in the root systems. Return of this nitrogen to the soil takes place in a number of ways. The nodules may be sloughed off when they die, and it also seems that in some cases nitrogen-bearing compounds may be released by the nodule bacteria. In addition, when the roots themselves die, the nitrogen may be returned during decomposition. A significant proportion of the nitrogen in the nodules, however, may be transferred to the upper parts of the plant – 90 per cent in some studies (Russell 1973) – and this is only released through leaching from the leaves and by plant decay after death. This latter process may provide a significant residual source for subsequent crops.

However the process operates, the overall effect can be considerable. At a maximum, as much as 200 kg of nitrogen may be added to each hectare during a single year by clover (Cooke 1974), but average rates of fixation probably amount to about 60 kg ha^{-1} yr^{-1}. Even at this rate, Cooke (1974) estimates that the 7–8 million ha of grassland in Britain may provide as much as 475,000 tonnes of nitrogen each year.

Again it is important to stress that the effects of grass crops upon the soil depend to a great extent upon the management regime. Root nodulation is itself affected by lime and fertiliser application (Nutman and Hearne 1980) while, as we noted earlier, grazing often results in a considerable redistribution of nutrients in the field, so that the spatial variability of chemical properties in grasslands is generally greater than that in arable soils (Beckett and Webster 1971). Moreover, in the long term, persistent discrepancies may develop in nutrient contents in different parts of the pasture. Hilder (1966), for example, recorded markedly higher phosphorus contents in the field margins of grazed grassland, even after fertiliser applications. The degree of persistence of such patterns is also indicated by Ralph (1982), who detected higher phosphate concentrations adjacent to disused medieval field boundaries at Holne Moor, on Dartmoor.

7.6 NUTRIENT CYCLING IN GRASSLAND SYSTEMS

Some indication of the interactions between the various components of grassland systems, and the effects on them of different management procedures, can be seen by considering nutrient cycles at the farm scale. At this level, the main inputs of nutrients are from fertilisers, atmospheric deposition, nitrogen fixation, bedrock weathering and inputs of seeds livestock and foodstuffs. Outputs are by removal of the grass crop and animal products (e.g. milk, wool, carcases), by leaching, volatilisation and erosion.

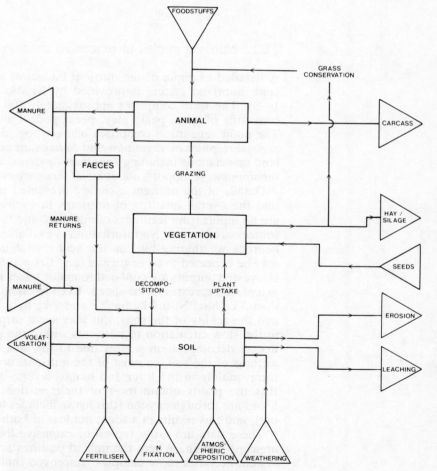

Fig. 7.20 The nutrient cycle in grassland systems.

Within the system, transfers of nutrients occur from the soil to the vegetation and hence to the livestock, and from both these latter back to the soil. In addition, various other flows and stores can be seen, for example in farmyard manure, or in grass used for conservation purposes as hay or silage (Fig. 7.20).

The magnitude of these various transfers, stores, inputs and outputs varies markedly according to the nature of the grassland system. In general, the amount of nutrients in circulation is small in extensive grazing systems,. and increases with grazing intensity. Unfortunately, it is rare for all the components of the system to have been measured in detail; in particular, inputs from bedrock weathering have rarely been assessed while data on the internal transfers within the soil are often incomplete. For this reason the examples presented here show only the main pathways of nutrient cycling within the system. It should also be emphasised that many of the components in these cycles have not been measured directly, but have been estimated by difference. Due to the likelihood of cumulative errors in the analysis such estimates are not always reliable. The data, most of which are taken from Frissel (1978), nevertheless show the overall magnitude of the cycles and give some indication of general trends in nutrient budgets.

7.6.1   Nutrient cycles in extensive grazing systems

A detailed example of the nutrient budget of an extensively grazed, upland moorland system is provided by Newbould and Floate (in Frissel 1978). The farm comprises approximately 500 ha of Pennine moorland, consisting of peat, peaty gley, peaty podzol and brown calcareous soils. The main vegetation comprises blanket bog, dominated by *Calluna vulgaris, Eriophorum vaginitum* and *Sphagnum acutifolium*, and wet moorland associations including *Juncetum squarrosi subalpinum* and *Nardetum subalpinum*. It supports about 8,500 sheep which winter on lower pastures.

Details of the nutrient cycle are presented in Fig. 7.21. It is apparent that the overall quantity of nutrients in circulation is very small. There are no inputs from fertilisers of manures, and the inputs from atmospheric sources are limited. No information is available on the nutrient supply by bedrock weathering but, on the acid crystalline rocks of this area, they may be expected to be negligible (e.g. Crisp 1966). Losses, too, are small. Harvested outputs are confined to minor losses in the wool and, to a lesser extent, the carcases of the sheep. Even leaching losses amount to no more than 3 kg ha^{-1} N, 0.4 kg ha^{-1} P and 9 kg ha^{-1} K. Given the high rainfall and the acidity of the soils, this may seem surprising, but the reason lies in the slow circulation of nutrients within the soil-plant system. Organic matter decomposition is slow due to the lack of soil organisms and the wetness of the soil, so most of the nutrients are stored in an unavailable form, mainly in the litter and humus layers; it is also from these layers that the plants obtain most of their nutrient requirements. The main losses are through erosion (this figure includes losses during heather burning), and this results in a slight net loss of both phosphorus and nitrogen.

These data are fairly typical of extensive livestock farms in moorland areas of Britain. Where the upland pastures are improved, however, the system becomes more complex. Improved (hill pasture) and unimproved (moorland) areas are managed as interrelated subsystems (Table 7.13),

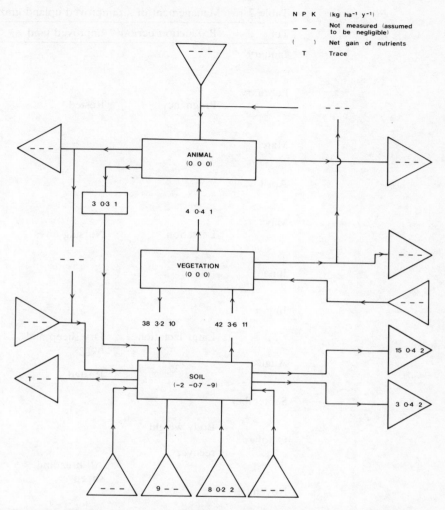

Fig. 7.21   The nutrient cycle of an unimproved upland grazing system: Pennine moorlands England (data from Newbould and Floate, in Frissel 1978).

so that transfers occur from one part of the farm to another (Fig. 7.22). Interestingly, in data for such a farm in Scotland, Newbould and Floate (in Frissel 1978) show that improvement leads to a 60–80 per cent increase in nutrient turnover and to a 100 per cent increase in production.

7.6.2   Nutrient cycles in intensive livestock systems

Some of the most intensive livestock systems in the temperate world occur in the Netherlands, where there is a tradition of high input grassland farming. The data in Fig. 7.23 illustrate the nutrient cycle for such a farm: an intensive dairy farm on clay soils. The grazing density is 4 cows ha^{-1} and the milk production 18,000 litres ha^{-1}, with meat production of 768 kg ha^{-1}.

Table 7.13   Management of an improved upland grazing system (see Fig. 7.22)

Time	Production period	Improved land	Unimproved land
January			
February	Pregnancy	Rested	Breeding sheep storm feeding on hay and beet pulp
March			
April			
May	Lactation	Nursing ewes	Dry sheep and hoggs
June			
July			
	Later lactation	Dry sheep and hoggs	Nursing ewes to weaning
August			
		Rested	All breeding sheep and stock ewe lambs
September			
October	Body weight recovery		
		All breeding sheep	Ewe hoggs
November			
December	Tupping		
	Close of tupping	Rested	All breeding sheep

Source: After Newbould and Floate, in Frissel 1978

In marked contrast to the previous two examples, this farm shows excessively high rates of nutrient cycling. Nitrogen inputs from fertilisers amount to 400 kg ha^{-1} (the national average for the Netherlands), and the massive subsidisation of the livestock by imported foodstuffs (concentrates) represents a major input of all nutrients. No nitrogen fixation occurs, for the swards in such systems contain little or no clover. Losses of nitrogen by volatilisation and denitrification are high, accounting for almost 60 per cent of the applied nitrogen fertiliser. Leaching losses, on the other hand, are low, presumably due in part to the clay soil but more importantly to the ability of the pasture to take up the applied nitrogen before it can be leached. Output of milk is extremely high, so nutrient losses in farm produce are considerable. Nevertheless, the farm as a

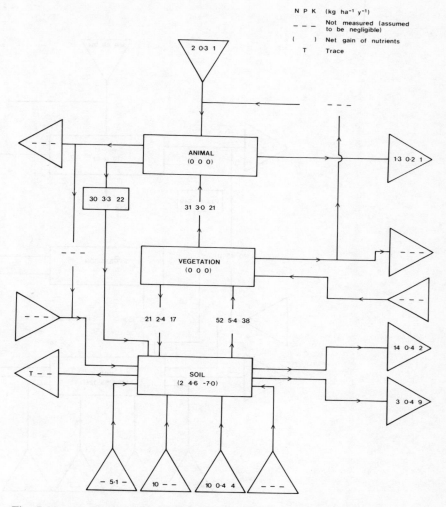

Fig. 7.22   The nutrient cycle of an improved upland grazing system: northern England (data from Newbould and Floate, in Frissel 1978).

whole shows a marked nutrient surplus, with an annual accumulation of 352 kg ha^{-1} N, 65 kg ha^{-1} P and 124 kg ha^{-1} K. Most of this surplus must be accumulating in the soil.

The fate of this nutrient surplus is not clear. Assuming that the data are substantially correct, it is clear that a marked build-up of all nutrients must be occurring. While some of this accumulation may be taking place in the unavailable (labile) pool of nutrients in the soil, it is nonetheless apparent that continuation of this trend would ultimately cause problems of toxicity. Studies from elsewhere in the Netherlands, however, have indicated that no major accumulation of nitrogen is occurring in the soil, and it seems likely that the date presented here underestimate some of the losses; in particular by volatilisation and denitrification (both of which are difficult to measure). It is also clear that the potential for accumulation of nutrients in runoff waters, with consequent problems of eutrophication, is considerable. Without doubt, these large nutrient imbalances

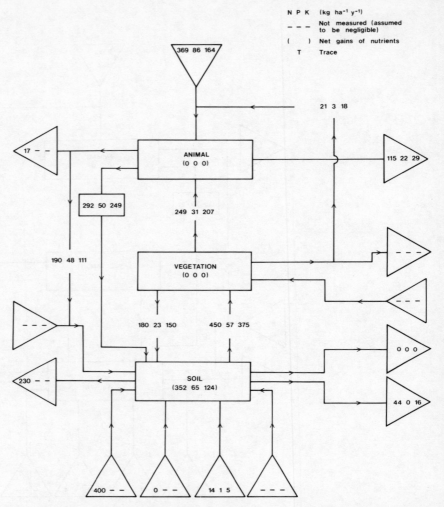

Fig. 7.23   The nutrient cycle of an intensive grazing system: Dutch dairy farm on clay soil (data from Henkens, in Frissel 1978).

are worthy of further study. They also indicate some of the possible ecological implications of intensive farming systems (see Ch. 12).

7.7 PRODUCTIVITY OF GRASSLAND AND GRAZING SYSTEMS

The productivity of grazing systems is a function of three related sets of factors (Fig. 7.24):

1. Those controlling primary production (i.e. grass growth);
2. Those cantrolling sward utilisation (i.e. the ability of the grazing animals to graze the sward as it grows);

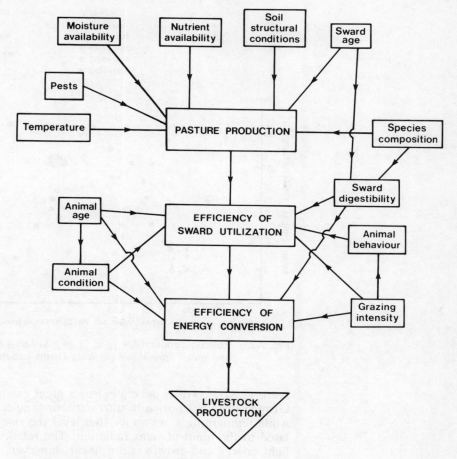

Fig. 7.24   Factors affecting the productivity of grassland systems.

3. Those controlling secondary production (i.e. the conversion of the digested herbage into animal products).

In any situation, any of these may limit overall productivity.

At a general level, it is clear that the primary productivity of grasslands depends upon environmental conditions, particularly climate. Data for fifty-two natural and semi-natural grasslands are presented in Figure 7.25. The data do not provide a representative sample of temperate grasslands, and are heavily weighted towards semi-arid areas, but they show that primary productivity increases with increasing precipitation and, to a lesser extent, temperature.

Similar general climatic relationships have been found for more intensively managed grasslands. Smith (1967) recorded a close correlation between the average effective transpiration and the percentage of farmland devoted to grass. Bendelow and Hartnup (1977) showed the importance of moisture stress on the pattern of grassland farming in Britain; all-grass farms are found predominantly in areas with an average maximum potential soil moisture deficit of less than 100–135 mm.

The more detailed explanation for these broad climatic relationships relates to the role of energy and moisture in grass growth. As we have

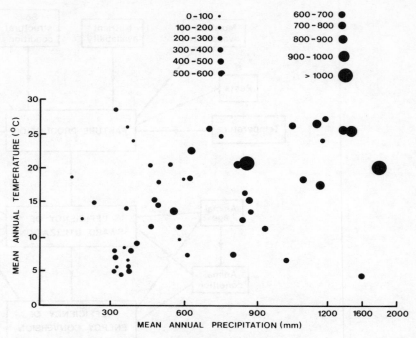

Fig. 7.25   Primary productivity (g m⁻²) of grasslands in relation to precipitation and mean annual temperature (from Lauenroth 1979).

seen, grass growth is dependent to a great extent upon climatic factors. Growth does not normally start until air temperatures exceed 6 °C, and while temperatures are above that level the rate of growth is closely related to the input of solar radiation. The relationship between inputs of light energy and growth is not linear, however, but shows a peak at intermediate energy inputs. The relationship also varies markedly between species. The effect of light energy upon yield is also complex, for, as the sward grows, shading of the lower parts of the plants occurs; thus maximum growth occurs at the point where crop cover is sufficient to absorb all the available incoming light energy without causing significant shading of the actively perspiring lower strata.

Growth is also limited by moisture stress, the degree of limitation increasing as the moisture deficiency increases although, again, not necessarily linearly. Rickard (1973), for example, suggested that growth of perennial ryegrass was unaffected until the moisture content of the soil fell below about two-thirds of the available water capacity. Similar relationships have been quoted or assumed by Bhat *et al.* (1980), Brockington (1978), Garwood and Williams (1967a, b), Fick (1980) and France *et al.* (1981). The effect of moisture stress is apparently to reduce effective transpiration, and, as has been shown, for example, by Penman (1962) and Stiles and Williams (1965), there is an almost linear relationship between yield and effective transpiration over a wide range of conditions. Research by Ben Harrath *et al.* (1977) and Francois and Renard (1979) on a range of grass species has suggested that the relationship between potential evapotranspiration, actual evapotranspiration and relative growth is more complex (Fig. 7.26).

Because of the importance of climatic variables such as radiation inputs,

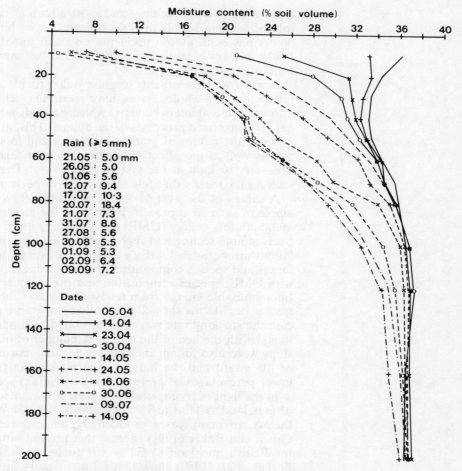

Fig. 7.26   Changes in soil moisture content with depth in a silt soil beneath a grass sward (after Ben Harrath *et al.* 1977).

rainfall and moisture availability, yields of grassland show a marked response to variations in weather. Damage to the crop by adverse weather conditions, such as frost, snow or wind abrasion may reduce yields (Copeman 1979), but more frequently plants experience periods of stress without suffering direct damage. Garwood and Tyson (1979) showed that for the period from 1954–78, moisture stress exerted a major control on grassland yields; 67 per cent of the variation in dry matter yield of a ryegrass-white clover sward could be accounted for by the degree of soil moisture deficit:

$$Y = 12.574 - 0.015\,d_\mathrm{p} - 0.019\,d_\mathrm{c} \qquad [7.5]$$

where Y is the annual yield (t ha^{-1})

d_p is the accumulated moisture deficit during the previous year
d_c the moisture deficit during the current year.

The effect of soil conditions upon grass yields operates in part through their interaction with these climatic factors. The ability of the soil to retain and supply water, for example, determines to a great extent the severity of soil moisture stress. Nevertheless, comparative data for different soil types are generally lacking, and most attention has been devoted to the role of nutrient supply in grass growth. As we have seen, particular emphasis has been placed upon the effect of nitrogen supply (e.g. Cooke 1974), and low N levels are often found to be limiting grassland yield (e.g. Munro *et al*. 1973).

In any case, assessments of grass yield provide only a partial picture of the system as a whole. As we have seen, this yield needs to be converted into utilisable animal products, which depends upon the efficiency of the energy conversion processes of the animal. This, in turn, requires consideration of the nature of grazing behaviour and food utilisation by the animal as well as the effect of the management regime; that is to say, it is necessary to look at the system as a whole. The obvious complexity of such a task means that few attempts have been made.

A survey of the west Midlands by Forbes *et al*. (1977) provided an example of such an approach. It analysed the factors relating to the utilised metabolisable energy (UME) and stocking rate of dairy farms. Using relatively simple scaling and regression techniques, they found significant relationships between UME (a measure of overall production) and nitrogen inputs and species composition of the sward. Stocking rate was correlated with UME, species composition and the quantity of imported foodstuffs (metabolisable energy from feedstuff: FME) (Table 7.14). Similar analyses for beef farms showed generally comparable results, although in this case the stocking rate was also correlated with nitrogen inputs and (negatively) with the age of the sward. A number of other interesting relationships were also found, including a negative correlation between the age of the sward and the index of species composition; older pastures had lower percentages of preferred (high-yielding) species (Table 7.14).

In the light of recent studies, it is clear that increases in the productivity of grassland systems are slowing down (e.g. G. W. Cooke 1979). Nevertheless, in many cases productivity remains below its potential level. Green and Baker (1981) suggest that present output is about 40 GJ ha^{-1} for British grassland (1 GJ = 10^9 Joules = 0.239×10^9 calories), and Forbes *et al*. (1980) show that on many dairy farms the output is as low as 30 GJ ha^{-1}; such levels of production are insufficient to meet the maintenance needs of the grazing animals. It is therefore necessary to consider why this discrepancy between potential and actual yields exists.

The optimum yield of grass crops – as of any crop – is physiologically determined. Inadequacies of climatic or soil conditions tend to limit actual yields, however, so that the average yields obtained in the field are considerably below the physiological maximum. In Britain, even under conditions of optimal moisture availability and nutrient supply, Leafe (1978) has estimated that yields will not exceed about 20 t DM ha^{-1} and this is borne out by the yields achieved in some small experimental plots (Wright 1978). The average yield of grass crops in Britain, however, is only 5–6 t DM ha^{-1}, indicating that in practice soil conditions are rarely optimised. It is therefore argued that better management of the soil and crop would result in higher yields, closer to the maximum possible for existing grass species (e.g. G. W. Cooke 1979).

The questions arise: which conditions limit yields, and what management methods are necessary to remove these limitations? In general, it is accepted that two major constraints are nitrogen availability and water supply. Moisture deficiencies are a common cause of reduced yields and Garwood (1979), analysing data from throughout the UK, found that yields could be increased in many areas by irrigation. The absolute in-

Table 7.14 Relationships between production, management and environmental factors in grassland farming systems in England

(a) 34 dairy farms

UME (GJ ha⁻¹)	1.00					
SR (LU ha⁻¹)	0.68***	1.00				
Nitrogen (kg ha⁻¹)	0.57***	0.31	1.00			
Pref. sp. index	0.41*	0.35*	0.54**	1.00		
Age index	−0.21	−0.30	−0.36*	−0.69***	1.00	
Drainage index	−0.03	0.06	0.01	0.26	−0.32	1.00
FME (GJ LU⁻¹)	0.10	0.68***	−0.03	−0.06	−0.05	0.06
	UME	SR	N	Pref. sp index	Age index	Drainage index

(b) 32 beef farms

UME (GJ ha⁻¹)	1.00					
SR (LU ha⁻¹)	0.39*	1.00				
Nitrogen (kg ha⁻¹)	0.36*	0.57***	1.00			
Pref. sp. index	0.41*	0.61***	0.53**	1.00		
Age index	−0.33	−0.44**	−9.36*	−0.76***	1.00	
Drainage index	0.03	0.14	−0.07	0.39*	0.03	1.00
FME (GJ LU⁻¹)	−0.10	0.60***	0.33	0.21	−0.11	−0.05
	UME	SR	N	Pref. sp index	Age index	Drainage index

Notes:
UME: utilisable metabolisable energy (output)
SR: stocking rate (livestock units per hectare)
Nitrogen: rate of nitrogen fertiliser application
Pref. sp. index: preferred species index (% of preferred species in sward)
Age index: age of pasture (since reseeding)
Drainage index: soil drainage status
FME: metabolisable energy from feedstuff
* significance level: 0.05 ** significance level: 0.01 *** significance level: 0.001

Source: From Forbes *et al.* 1977

crease in yield ($\triangle Y$, kg DM ha⁻¹) was closely correlated with the maximum potential soil moisture deficit (SMD, mm):

$$\triangle Y = 18.57 \text{ SMD} - 670 \qquad (r^2 = 0.64) \qquad [7.6]$$

Yields can also be increased by greater use of nitrogen fertilisers. As we noted in section 7.2.2, grass yields respond to nitrogen application rates up to at least 300 kg ha⁻¹ in grazed swards and up to 540 kg ha⁻¹ or more on ungrazed swards. Down and Lazenby (1981), however, showed that the average application rate for ley grasslands was only about 120 kg ha⁻¹ in Britain in 1979, and for permanent grass no more than 90 kg ha⁻¹. Considerable scope thus exists to increase yields by removing the limitations of nitrogen availability. This has been illustrated, similarly, in eastern Canada by Baier *et al.* (1980). In an analysis of timothy yields from 8 stations for the period 1960 to 1973, they showed that nitrogen application rates alone accounted for 32 per cent of the variation in the yield of the first cut of timothy grass.

In recent years it has become increasingly clear that pests are a further cause of reduced yields on grasslands (e.g. Henderson and Clements 1979; Clements 1980). In some cases, yields may be greatly affected, particularly by virus and fungal diseases, and Carr (1979) has estimated that losses of up to 20 per cent DM can occur due to infestation of ryegrass with crown rust, whilst ryegrass mosaic virus may reduce yields of Italian ryegrass by up to 30 per cent. A'Brook and Heard (1975) have similarly recorded reductions in yield of 30 per cent in plots of Italian and perennial ryegrass infested with ryegrass mosaic virus. Generally, however, the impact of pests and diseases is much less; though intense at a local level, diseases have a relatively small effect on national production.

It is clear that all these factors – nitrogen availability, moisture supply and diseases – may influence yields, and as a consequence marked regional variations in yield potential are seen, reflecting variations in environmental conditions. In Britain, Wilkins et al. (1981) show that total yields of perennial ryegrass at Hurley, Berkshire, on plots receiving 600 kg N ha^{-1}, were 10.7 kg DM ha^{-1}; at Aylesbury Buckinghamshire, the yield was 12.2 kg DM ha^{-1}, and at Wenvoe in south Wales 14.8 kg ha^{-1}. These differences are evidence of the effects of different soil and climatic conditions at the three sites, and they seem to illustrate inherent differences in grassland potential across the country. Using the equation developed by Morrison et al. (1980) (equation [7.1], (p. 153) Wilkins et al. (1981) assessed the potential production (Y^{10}) for grasslands cut at four-weekly intervals, without irrigation, for the whole of Britain (Fig. 7.27). As the results indicate, maximum yields are found in the south-west where rainfall is abundant and potential evapotranspiration is high; minimum yields are found in the north and east where temperatures are low and rainfall often limiting. A rather more complex analysis was carried out for the whole of the European Community by Briggs (1983). On the assumption that yields of mixed species grass swards are correlated with actual evapotranspiration during the growing season, he used a modelling approach to asssess 'biomass potential' from data on soil available water capacity, temperature, rainfall, relative humidity and solar radiation. The results show a similar pattern to those found by Wilkins et al. (1981): namely, potential yields are maximum in the wetter and warmer south and west of the region, and least in the drier or cooler areas.

Although studies such as this indicate the inherent suitability of the environment for grass production, they do not take account of the effects of grazing processes. Under field conditions these are important, and it seems that a major reason for sub-optimal yields in many cases is inadequate utilisation of the grass sward. Grass utilisation – the proportion of the crop eaten by the animals – is often no more than 60 per cent, whereas values of 75–80 per cent are technically feasible. Lazenby and Down (1982) claim that by increasing stocking levels, and thereby improving utilisation of the sward, productivity could be increased by 25 per cent. If nitrogen application rates were also doubled, then mean output from dairy farms would increase to around 80 GJ ha^{-1}.

Clearly the factors involved in controlling productivity of grassland systems are complex, and detailed modelling remains difficult. Attempts have been made, however, by Rose et al. (1972) and France et al. (1981). The latter took account of energy inputs, moisture stress, nitrogen fertiliser applications, nitrogen fixation, leaching, volatilisation, grazing behaviour and energy utilisation to model the productivity of two-year-old wether sheep (castrated rams) grazing a perennial ryegrass sward. While admitting that the data necessary to test the model were lacking, they conclude that the results are of the correct order of magnitude and that the model provides a realistic framework of the system. Clearly, the need

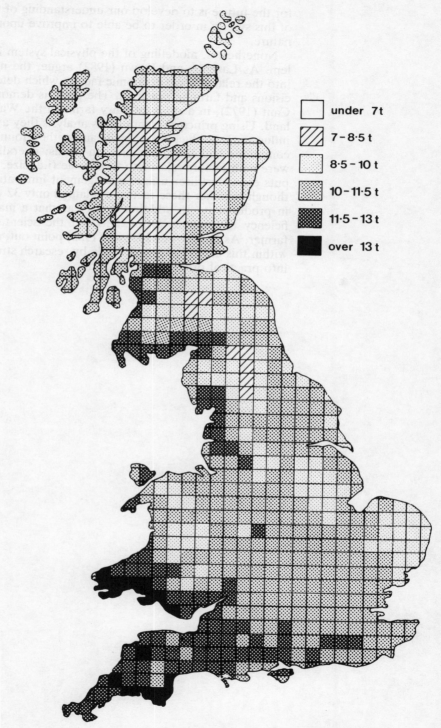

☐	under 7 t
▨	7 – 8·5 t
▦	8·5 – 10 t
▨	10 – 11·5 t
▓	11·5 – 13 t
■	over 13 t

Fig. 7.27   Potential production of grassland in the United Kingdom (t ha^{-1}) assuming cutting at four-weekly intervals and no irrigation (from Wilkins *et al*. 1981).

for the future is to develop our understanding of the various components of this system in order to be able to improve upon general models of this nature.

Nonetheless, modelling of the physical system is only part of the problem. As Lazenby and Down (1982) argue, the need also is for research into the related socio-economic factors which determine management decisions and farming practices. The point is demonstrated by Watson and Cant (1972), in a study of dairy farms in the Waikato area of New Zealand. Using principal components analysis they analysed the main factors influencing productivity. Significantly, they found that soil and climatic conditions had little influence on yields; overall the main relationships were with factors related to farm scale (i.e. size, labour force, capital inputs etc.), fertiliser usage, management innovativeness and topography, though together these factors explained only 32 per cent of the variation in productivity. In addition it is clear that a major constraint is the efficiency of communication between the scientist, the adviser and the farmer. As Lazenby and Down (1982) point out, more attention is needed within this area if advances made by research studies are to be translated into practical progress on the farm.

8 ARABLE SYSTEMS

8.1 INTRODUCTION

8.1.1 Distribution of temperate arable systems

Arable farming represents one of the most important components of agriculture in the temperate world. In 1980 almost 5 million ha (44 per cent) of the agricultural land of Great Britain was devoted to arable cultivation (Lazenby and Down 1982). In the USA in 1971 about 37 per cent was under arable crops and in Canada in the same year about 55 per cent (Grigg 1974). The European Community had almost 35.6 million ha of arable land (38 per cent of the total agricultural land) in 1971. By 1980, with the enlargement of the EEC to 9 members, the area of arable land had risen to 46.2 million ha – 50 per cent of the utilisable agricultural area (CEC 1982).

The distribution of this land is, in general, related to broad bioclimatic conditions. In Britain, for example, 'land under tillage' (mainly arable land) is concentrated in the east and south-east. As Bendelow and Hartnup (1977) have shown, this pattern relates to accumulated temperature and potential soil moisture deficit: arable farming is associated mainly with the warmer, drier areas. This reflects a number of constraints on successful arable farming, including the problems of disease in wet areas and the difficulties of ensuring adequate grain ripening while the soil is still dry enough to allow the passage of combine harvesters in the cooler and moister western and northern areas.

In broad terms, most arable crops have similar environmental requirements, but at a detailed level their tolerances differ so that the distribution of the main arable crops – wheat, barley, oats and root crops such as potatoes and sugar beet – show distinct regional variations. Coppock (1976) mapped the distribution of arable crops in England and Wales in 1970 (Fig. 8.1). Barley and wheat have similar distributions, confined mainly to the east, though barley extends further west and north. Oats are grown mainly in the peripheral areas, beyond the main zones of wheat and barley cropping. Potatoes and sugar beet are most prevalent in eastern and midland counties. These distributions reflect not only climatic factors, but also soil conditions, the pattern of pests and diseases and the influence of historical and economic forces.

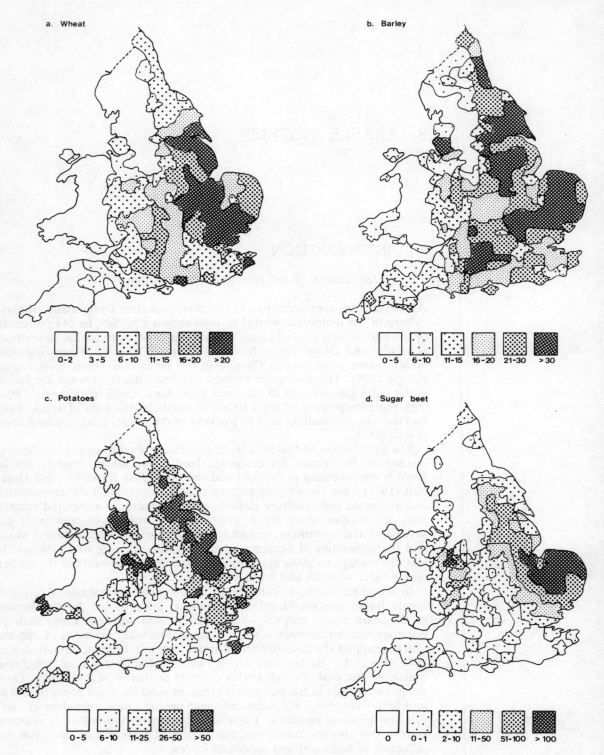

a. Wheat

| 0-2 | 3-5 | 6-10 | 11-15 | 16-20 | >20 |

b. Barley

| 0-5 | 6-10 | 11-15 | 16-20 | 21-30 | >30 |

c. Potatoes

| 0-5 | 6-10 | 11-25 | 26-50 | >50 |

d. Sugar beet

| 0 | 0-1 | 2-10 | 11-50 | 51-100 | >100 |

Fig. 8.1 The distribution of arable crops in England and Wales, 1970 (ha per
1000 ha of agricultural land) (after Coppock 1976).

8.1.2   History of arable systems

The history of arable cultivation in these areas is complex. Cultivation of crops such as wheat and barley can be traced back to prehistoric times when, about 10,000 years BP, wild wheat crossed with a natural goat grass to produce a plumper, more fertile wheat – Emmer wheat. This and barley formed the basis of prehistoric cultivation, and by the first millenium BC much of western Europe was given over to these crops (though not, interestingly, to rye which was then still a weed). In Britain, following the Saxon invasions, the area of cultivated land extended as forests were cleared for settlement and at the same time as wasteland was taken into the open fields. In these, barley, wheat and oats were the dominant crops, their cultivation interrupted only by a year of fallow every three or four years (Ch. 2). Subsequently, with the gradual enclosure of land for grazing, during the sixteenth to nineteenth centuries, and the introduction of root crops by, amongst others, Walpole and Townshend, the open-field system declined, to be replaced in many areas by grassland or rotational farming systems. By the 1900s, however, cereal cropping was again on the ascendent, and there was a move away from mixed farming to all-arable systems (Ernle 1961).

This trend was particularly marked in the USA, where cereal cropping expanded rapidly between 1850 and 1880 (Fig. 8.2), and much of the land taken into cultivation west of the Mississippi was devoted almost entirely to cereals. Grigg (1974), for example, comments that: 'The large mechanised farm, growing wheat almost to the exclusion of other crops, appeared first in California in the 1860s. . . . Many farmers grew wheat continuously for fifteen years or more.' In both Argentina and Australia arable farming developed later, but in both cases intensive cereal production had become established by the early part of the present century. In Canada, development occurred later still, wheat reaching its peak acreage during the 1930s (Fig. 8.3).

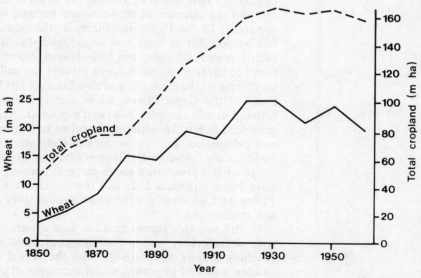

Fig. 8.2   Changes in the area of wheat-land and total crop-land in the USA, 1850–1960 (data from Grigg 1974).

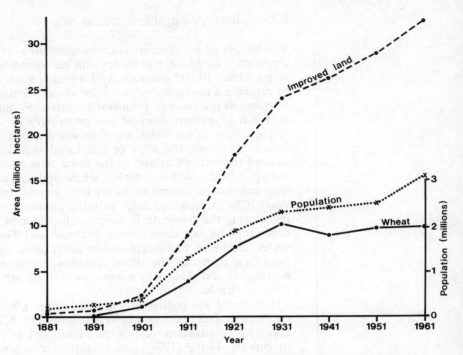

Fig. 8.3 Changes in the area of wheat-land and 'improved land' in Canada, 1881–1960 (data from Grigg 1974).

In the early years of expansion, all-arable farming was often synonymous with monoculture. Wheat, particularly, was grown continuously, largely because throughout much of the late nineteenth and early twentieth centuries wheat prices were high as improved communications, increased per caput income, and the repeal of the Corn Laws in Britain, opened up new markets. During the years of depression, however, prices fell and the dangers of monoculture became suddenly and devastatingly apparent. In the 1930s, for example, the population of the Great Plains fell by 300,000 as land was abandoned due to the combined impact of falling prices, drought and soil erosion. Subsequently, there has been a trend to diversification, both to protect the soil and to broaden the economic base of the farm. Ley-arable farming has been encouraged in wetter areas of the Great Plains, while almost everywhere the principle of rotational arable cropping has been espoused. Acreages of cereal crops have therefore fallen. In addition, in order to combat problems of soil erosion and exhaustion, new techniques of cultivation have been employed: crop residues are conserved, tillage is often reduced to a minimum and a wide range of soil protection methods (e.g. contour ploughing, strip cropping) have been introduced. It is a trend characteristic not only of the Great Plains. In Canada and Australia a similar story can be told; mixed farming has re-emerged.

In Britain, this recent trend is less apparent. The area of arable land rose considerably during the eighteenth and early nineteenth centuries, reaching a peak about 1840. With the repeal of the Corn Laws and the sudden influx of imported wheat, acreages of cereal crops declined steadily. They picked up briefly late in the First World War, then declined again until the 1930s. Since then, arable cropping has expanded, with a further marked increase during the Second World War, and by 1974 almost 31 per cent of the agricultural acreage was devoted to cereal crops, while a further 9 per cent was under roots and vegetable crops. This increase in the arable area has been accompanied by a significant trend towards specialisation, so that more than ever before arable and livestock farming is segregated, not only from farm to farm but also regionally (Fig. 8.4).

During the last half-century, there have also been marked changes in the crops and cropping systems used in arable farming. Traditionally in Britain, the main cereal crops have been wheat, barley, oats and rye, and these have been grown in association with root crops including turnips and swedes. In recent decades, rye and oats have become progressively less important, largely because of their lower yield potential and the decline in demand for 'coarse'-grain products. Turnips and swedes, too, have declined in importance, to be replaced by potatoes, sugar beet, peas, beans and, increasingly, oil-seed rape (Fig. 8.4). Additionally, there has been a marked trend during the 1970s and 1980s towards cultivation of maize, mainly as a fodder crop for associated livestock rearing.

The other main trend in modern arable farming has been the increased dependence upon capital inputs, both of machinery and of chemicals. As we noted in Chapter 2, this reflects a widespread change in farming methods, and it is one that has had considerable significance. Yields have increased, labour requirements have fallen, and the impact of arable farming upon the environment has undoubtedly become more acute. Some indication of the importance of capital inputs in modern farming systems is provided by the data in Table 8.1. The environmental effects of arable systems will be considered in Chapters 10–12.

Table 8.1   Capital costs of crop production in arable farming systems

Input	Winter wheat	Maincrop potatoes	Sugar beet
Seed	27.00	170.00	12.00
Fertiliser	24.00	80.00	66.00
Sprays	7.20	40.00	42.00
Baler twine	3.00	—	—
Marketing board	—	16.00	—
Sacks and ties	—	80.00	—
Transport	—	—	45.00
Rent	40.00	40.00	40.00
Labour and power	47.00	325.00	105.00
Machinery	12.50	110.00	52.00
General overheads	40.00	40.00	40.00
Total	200.70	901.00	402.00

Notes: all costs are £ ha^{-1} at 1975 prices.

Source: Data from Eddowes 1976

8.2   MANAGEMENT OF ARABLE SYSTEMS

The main components of arable systems are the soil and the crop. The farmer manipulates these directly, in order to maintain the system at a high level of production. Manipulations include the regulation of inputs

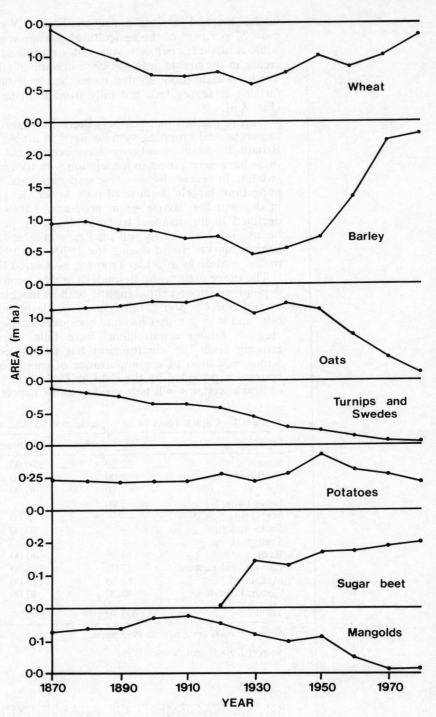

Fig. 8.4   Changes in the acreages of selected crops in Britain (data from
MAFF 1968, Hood 1982).

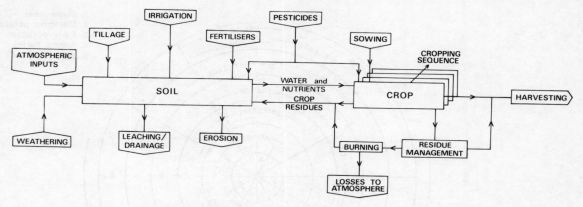

Fig. 8.5 The arable cropping system.

of fertiliser, seeds and pesticides; control of the outputs of the harvested crop and crop residues; and management of the internal structure of the system by, for example, tillage, choice of cropping sequence and decisions upon factors such as sowing date and harvesting date (Fig. 8.5).

Given the variety of crops and cropping systems found in arable farming, it is clearly difficult to make valid generalisations about the management of arable land. Different crops have different management requirements, while each farmer tends to adopt his own regime according to local environmental conditions, cultural and social influences, and personal experience and convenience. The main management operations, however, include seedbed preparation, sowing, fertiliser application, irrigation, pesticide application, harvesting and residue management. These procedures vary from year to year according to decisions regarding cropping sequence, but in any one year they follow a broad pattern, as indicated in Fig. 8.6.

8.2.1 Seedbed preparation

The general principles of tillage have been outlined in Chapter 3. It was noted there that the main purpose of tillage was to prepare a seedbed in which the crop could be sown. Traditionally, it has been argued that an ideal seedbed comprises a loose, finely and evenly granulated surface. This permits the efficient operation of the seed-drill so that an even distribution and depth of seed placement can be achieved, while the seed is at the same time set into a well aerated and reasonably warm environment. During germination and seedling development, good contact between the nascent roots and the soil solids is necessary to enable water and nutrients to be obtained. A loose and friable soil consistence also allows both roots and shoots to force their way easily between the aggregates.

In recent years, this interpretation has increasingly been challenged as it has been realised that adequate yields can be achieved without tillage, and, as we discussed in Chapter 3, methods of direct drilling of cereal crops have been widely adopted in much of the temperate world. The success of direct drilling varies, not least because of differences in the expertise of the farmer, and to date yields on some types of soil have

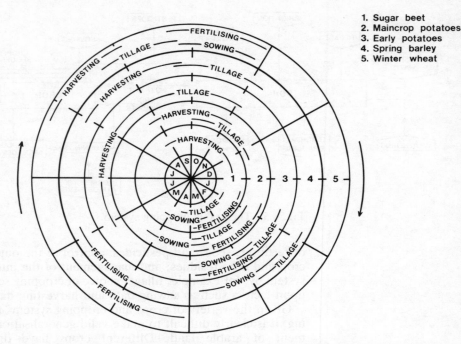

1. Sugar beet
2. Maincrop potatoes
3. Early potatoes
4. Spring barley
5. Winter wheat

Fig. 8.6 The annual sequence of management practices in arable cropping systems in Britain.

failed to match those attained with conventional tillage (see Table 3.5, pp. 68–69). Nevertheless, it is clear that conventional methods of seedbed preparation are not a prerequisite for the growth of arable crops on all soils. Moreover, because seedbed management often requires several operations with high-cost machinery, it accounts for a major proportion of total variable costs in cereal production (Table 8.1). Efficiency of tillage also influences the growth rate and yield of the crop to a marked extent, not only in the short term but, because of its effect on soil structure, in the longer term as well. Indeed, problems of tillage have been identified as a significant cause of reduced yield in many arable systems (see Table 3.1, p. 44).

The range of tillage operations employed in preparing and maintaining a seedbed, as well as their timing, vary according to the crop and soil conditions. With winter-sown crops (e.g. winter wheat) seedbed preparation must be completed by late autumn so there is little scope between harvesting one crop and sowing the next for flexibility in the timing of tillage operations. On light soils, this may present little problem for a fine tilth may be achieved with minimal effort (e.g. in one tillage operation). On heavier soils, however, repeated ploughing and harrowing may be necessary to reduce the aggregates to a fine tilth. As Utomo and Dexter (1981a, b) have shown, tilth mellowing is greatly aided by stresses set up in the soil aggregates by fluctuations in moisture content. Experimental studies have indicated that these fluctuations are accentuated immediately after tillage, so it is beneficial to leave the soil for a few days between tillage operations to maximise the effect of natural mellowing.

In the case of spring-sown crops (e.g. spring barley, potatoes), seedbed preparation need not be completed until March. Nevertheless, primary tillage is often carried out during the preceding autumn to leave the soil open to frost attack throughout the winter; again, this reduces the mechanical effort needed to produce a fine tilth. Secondary tillage operations (e.g. harrowing) are then carried out during the early spring, ideally as soon as evaporation reduces the surface soil moisture content below field capacity. The exact sequence of secondary tillage procedures depends greatly upon specific soil and weather conditions, but often includes one or more passes with a fixed-tine harrow before a final cultivation with a spring-tine harrow or roller.

8.2.2 Sowing

The ultimate yield of the harvested crop is controlled to a great extent by two factors: the total amount of photosynthate each plant has been able to produce during the growth period and the total number of plants per unit area of land. Both these factors depend in turn upon the efficiency and timing of sowing operations.

Sowing date has a particular significance since this limits the length of time available for growth. In general, high yields are favoured by early sowing and L. P. Smith (1972), for example, has shown that yields of spring barley in England and Wales are closely correlated with sowing date. On the other hand, sowing spring cereals too early in the year may lead to crop failure because of rotting of the seeds in the cold and damp soil before germination can take place. Normally, winter cereals are sown between late September and December, while spring cereals are sown from February to early May.

The other main factor controlling the length of the effective growing season is the depth of seed placement. The seeds of cereal crops need to be buried to protect them from birds or other scavengers, from frost, from dehydration and from removal by wind or rain-splash. To some extent, increasing the depth of placement increases this protection, but it may also delay seedling emergence. Soil temperatures at depth are lower than those at the surface during the spring, so the threshold temperature necessary for germination is attained later. Moreover, once the seed has germinated, the shoots must grow further to reach the surface when the seed is placed at depth. Since, during the period prior to emergence, the seedling is still dependent upon food reserves stored in the seed, this may result in plants dying due to exhaustion of their food supply before they emerge. Excessive sowing depths therefore result in patchy emergence, while those seedlings which do emerge suffer from a reduced growing season.

The role of sowing density in influencing the yield of cereals is less clear. As we will see in section 8.3, the final yield of a cereal crop has four components: the number of seedlings per unit area, the number of ears per plant, the number of grains per ear and the average weight of each grain. Clearly as the sowing intensity increases, the number of seedlings per hectare rises also, but compensating changes take place in the development of ears and grain so that there is not necessarily a proportional increase in the total yield (Fig. 8.7). Numerous studies have indicated the complexity of these relationships. They have shown that increasing plant density results in an increase in the number of ears per unit area but a decline in the number of grains per ear (Kirby 1969;

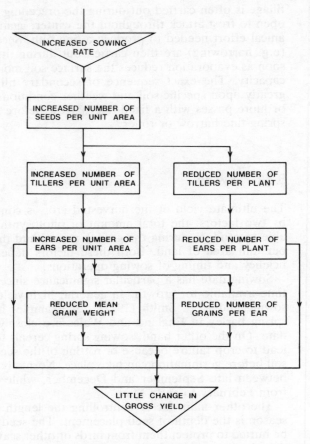

Fig. 8.7 Effects of sowing density on the yield components of a cereal crop.

Cannell 1969; McLaren 1981), and a decrease in the mean weight of the grains (Kirby 1967), presumably due to competition for light, moisture and nutrients. As a consequence, cereal crops often demonstrate an inelastic response to a wide range of sowing densities; yields of spring barley, for example, do not vary much for seed rates between 125 and 180 kg ha^{-1} (other factors being equal), while fairly consistent yields of winter wheat can be achieved with sowing rates between 125 and 180 kg ha^{-1}. A similar tolerance to sowing rate is shown by root crops such as potatoes and sugar beet; increasing the spacing between seeds tends to increase the ultimate size of the individual tubers so that, over a wide range of seed rates (i.e. with a final density of about 60,000 to 100,000 plants per hectare) total yield remains constant. In the past, recommended sowing rates were relatively high, but, in the case of sugar beet, this led to the need for removal of surplus plants by hand. With the increased costs of labour and the development of precision drills, seed rates have been much reduced and today most of the sugar beet in Britain is 'drilled to stand' (i.e. sown at rates which avoid the need for later thinning), normally at an inter-seed spacing of 12–18 cm (Eddowes 1976; Lockhart and Wiseman 1978).

8.2.3 Fertiliser application

The need to apply high levels of fertilisers to arable lands in order to maintain yields arises from the effects of regular harvesting on soil nutrient reserves. The annual demand of cereal crops for nutrients varies in relation to yield, but at modern levels of output of 5–6 tonnes/ha, wheat and barley extract about 100 kg ha^{-1} of nitrogen, 20 kg ha^{-1} of phosphorus and 75 kg ha^{-1} of potassium (G. W. Cooke 1975) (see Table 5.2, p. 103). A large proportion of these nutrient uptakes – 30–90 per cent depending upon the management practice – is removed from the system during harvesting. Moreover, the natural supply of these elements to the soil, by weathering and in atmospheric inputs, is small. Over time, therefore, without replenishment by fertilisers, the soil nutrient reserves are depleted and crop yields decline (see Fig. 5.2, p. 104).

Many studies have been carried out to examine the response of cereals to fertiliser inputs, and it is clear from the exhaustion experiments at Hoosfield and Brooms Barn in England that with adequate supplies of nutrients either from inorganic fertilisers or from farmyard manure, long-term average yields may be increased three- to fourfold compared to those on unfertilised soils (Jenkinson and Johnston 1977). Earlier experiments were reviewed by Cooke (1974). More recently, results from several long-term studies have been reported by Draycott *et al.* (1978), Widdowson and Penny (1979), Widdowson *et al.* (1980) and Rehm *et al.* (1981). In the Woburn Reference Experiment, an arable rotation comprising barley, wheat, potatoes, sugar beet and a grass-clover ley has been in operation since 1877. Reviewing the results from the 1970–74 cycle, Widdowson and Penny (1979) note that application of nitrogen fertilisers increased yields of all crops, especially the ley, while potassium gave marked increases in all but the wheat crop. Phosphorus had little effect, largely because the soil contained adequate contents of phosphorus, derived in part from residues from previous fertiliser applications. Farmyard manure also increased the yields of all crops. In the case of the potatoes, increases were greatest when FYM was applied with NPK fertilisers, but for other crops the effect of farmyard manure was greater when applied alone.

Similar results have been obtained from the parallel Rothamsted Reference Experiment (Widdowson *et al.* 1980). Again application of farmyard manure increased crop yields, though wheat yields declined when both FYM and nitrogen fertiliser were applied. Differing fertiliser treatments, however, had different effects on the nutrient balance of the system. Without farmyard manure, nitrogen applications were never sufficient to balance losses by crops (Table 8.2), though to some extent the deficit may have been made up by nitrogen fixation by the clover in the ley crop. Phosphorus and potassium, on the other hand, had surpluses when NPK fertilisers were applied. Farmyard manure, either alone or in combination with NPK, gave surpluses of all nutrients. The results in Table 8.2 illustrate some of the complexity of the nutrient balances in this system. It is notable that with the N1,PK treatment (i.e. when 352 kg ha^{-1} N, 1402.5 kg ha^{-1} K and 137 kg ha^{-1} P were applied during the five-year rotation), nitrogen deficits were greater than when no fertilisers were applied. Similarly, increasing the rate of nitrogen (from 352 to 703 kg ha^{-1}) led to reduced surpluses of P and K. These effects presumably reflect the fact that increased crop growth, made possible by

Table 8.2   Nutrient balances in the Rothamsted Reference Experiment
(1971–75)

	Fertiliser treatment								
	None			N1,PK			N2,PK		
	N	P	K	N	P	K	N	P	K
Added	0	0	0	352	137	1042	703	137	1042
Removed	269	37	157	625	99	783	745	107	867
Difference	−269	−37	−157	−273	+38	+259	−42	+30	+175
	FYM			FYM + N1,PK			FYM + N2,PK		
Added	760	128	1008	1112	265	2050	1463	265	2050
Removed	564	97	748	743	126	1104	842	132	1208
Difference	+196	+31	+260	+369	+139	+946	+621	+133	+842

Note: all quantities in kg ha^{-1}, rounded to 1 kg.
Key
FYM: applied as two 50 t ha^{-1} applications to all crops each year
N1: nitrogen applied as follows: barley, 57 kg ha^{-1}; ley, 19 kg ha^{-1}; potatoes, 75 kg ha^{-1};
wheat, 75 kg ha^{-1}; kale, 126 kg ha^{-1}
N2: double application above
P: phosphorus applied at rate of 27.4 kg ha^{-1} to all crops each year
K: potassium applied at rate of 208.5 kg ha^{-1} to all crops each year

Source: From Widdowson *et al.* 1980

increasing the availability of the most limiting nutrients, results in additional uptake of other nutrients.

The complexity of plant responses and nutrient interactions in relation to fertiliser management means that it is difficult to define optimal rates of fertiliser application for individual crops. Moreover, crop responses vary according to changes in soil type, weather, cropping sequence, and other management practices (e.g. drainage, irrigation, tillage). In general, however, marked increases in yields of wheat and barley are found for nitrogen inputs of up to 100–125 kg ha^{-1} y^{-1}, and phosphate and potassium applications of up to 50–75 kg ha^{-1} y^{-1} each. For root crops, inputs of up to 200 kg ha^{-1} y^{-1} of nitrogen, 200 kg ha^{-1} y^{-1} potassium and 250 kg ha^{-1} y^{-1} of phosphorus are recommended (see Table 5.8, p. 112).

Methods of application of fertilisers for maximum response vary from one crop to another. With winter-sown cereals, there is little benefit to be gained from applying nitrogen before the spring; the majority is lost by leaching and denitrification before the root system develops. Commonly, nitrogen is applied as a top-dressing (see Ch. 5) during March–June, though experimental results show little difference in crop response to single or split doses, or to variations in timing of applications within this period. With spring cereals, nitrogen is often applied by combine drilling or by contact placement. Additional doses may be given as a top dressing after seedling emergence, especially to replenish losses during heavy rain. Several studies have also indicated that increases in yield can be obtained by giving nitrogen as a foliar application to cereal crops after ear emergence (e.g. Penny and Jenkyn 1975; Polous 1977), though the effect is much diminished when water supplies are limited by summer drought (e.g. Alston 1979). Phosphorus and potassium are generally combine drilled.

8.2.4 'Crop protection

Weeds, animal pests and diseases have always been a major problem in arable farming, and the problem remains, despite the development of chemical pesticides. Until the introduction of MCPA in 1942, weeds in particular were a major source of concern, but the use of this and, later, the related herbicide 2,4-D provided a means of all but eradicating the most troublesome weeds such as the corn poppy (*Papaver rhoeas*), fat hen (*Chenopodium album*) and charlock (*Sinapsis arvensis*). Since then, the use of MCPA has declined, but other weeds, including chickweed (*Stellaria media*), black bindweed (*Polygonum convolvulus*), knotgrass (*P. aviculare*), field bindweed (*Convolvulus arvensis*) and cleavers (*Galium aparine*) have become increasingly serious problems. To some extent, these weeds can be controlled by the newer breed of broad-spectrum herbicides including 2,4-DP (dichlorprop), dicamba-MCPA mixtures and benazolin-MCPA mixtures. Today, the main weed pests are wild oats (*Avena strigosa, A. fatua* and *A. ludoviciana*), mayweeds (*Anthemis cotula*), black-grass (*Alopecurus myosuroides*), couch-grass (*Agropyron repens*) and black bent (*Agrostis gigantea*). Various herbicides may be used to treat these weeds, but to date control is far from wholly effective and, perhaps because of lack of other competition and because of reluctance by farmers to use costly, specific herbicides, infestations are growing in many areas (see, for example, Elliott 1972).

A number of animal pests attack cereal crops. In the case of birds, the effects are not all detrimental; most birds also eat insect pests and weed seeds. Pigeons (*Columba* spp.) and rooks (*Corvus frugilegus*), however, are known to eat large quantities of cereal seed, while rooks may beat down the standing crop in order to reach insects in the soil. Rabbits (*Oryctolagus cuniculus*) and rats (*Rattus* spp.) have become less of a problem as a result of habitat removal (e.g. of hedgerows and scrubland), though they still represent a threat to young crops and to stored grain.

Less obvious, but often equally important, are insect pests. The wheat bulb fly (*Leptohylemyia coarctata*) lays its eggs on the bare soil in autumn; the larvae attack the crops in the late spring, causing the central shoots to yellow and die. The leatherjacket, the larva of the cranefly (*Tipula* spp.), eats the roots and below ground stem of barley and wheat, causing patchy failure of the crop. The wireworm, the larva of the chickbeetle (*Agriotes* spp.), moves down the row eating the plants just beneath the surface.

Many of these pests can be controlled by the use of insecticides, such as HCH, aldrin and primicarb, but some, including the eelworms, are best suppressed by crop rotations or by growing resistant cereal varieties. Studies by Sykes (1979) and Evans (1979) have illustrated the effects of the nematode *Longidorus leptocephalus*. Sykes found a close correlation between nematode numbers and yields of potatoes, yields declining by 0.55 t ha^{-1} for each increase in nematode numbers of 200,000 m^{-3}. Studying an arable rotation system, Evans (1979) found a negative correlation between nematode numbers under potatoes and the yield of the subsequent wheat crop; each increase in the nematode population of 100,000 m^{-3} gave a reduction in yield of 0.53 t ha^{-1}. Infestations were controlled by soil fumigation and resulted in marked increases in the potato yields, even after two years of barley, one of the ideal host crops for *L. leptocephalus*. Similar results have been reported from the Woburn rotation-fumigation experiment (Williams *et al.* 1979). Fumigation resulted in good control of several species of nematodes and led to significant increases in the yields of potatoes. Barley and sugar beet showed smaller responses, and the effect in general varied according to the weather.

8.2.5 Harvesting and residue management

The removal of the crop during harvesting has a major impact on the soil. During its growth, the crop assimilates carbon, oxygen and other nutrient elements from the soil and atmosphere. It also gives protection to the soil and considerably modifies the microclimate of the near-ground atmosphere. Harvesting takes away the nutrients stored in the crops, leaves the soil relatively unprotected from the influence of rainfall and wind, and leads to the development of a new, often more extreme surface microclimate.

The severity of these effects depends to a great extent upon the character of the crop and the management of the crop residues. In the case of root crops, most or all of the crop may be removed with the result that the loss from the system of the nutrients taken up by the plants is almost complete. In the case of cereal crops, the losses are potentially less drastic. Although a high proportion of nutrients is concentrated in the grain of wheat and barley, considerable quantities of nutrients may be held in the straw and the stubble; where these residues are left on the soil, nutrient losses are minimised. The conservation of straw and stubble poses serious problems, however. The residues provide habitats for pests during the winter and may interfere with subsequent cultivation. For these reasons, the residues are often removed, either as straw for use as bedding material or industrial purposes, or by burning. In both cases there is a significant loss of nutrients from the system. Where straw is harvested, up to 90 per cent of the total nutrient uptake of the crop may be removed. Where it is burned, losses are less, for a large proportion of the potassium and calcium are returned to the soil in the ash. As was noted in Chapter 7, however, significant quantities of nitrogen and phosphorus may be removed in smoke, while a high percentage of the ashed nutrients may be lost by leaching.

Relatively few studies have been carried out to quantify losses in straw and stubble burning, but Biederbeck *et al*. (1980) showed that individual burns resulted in marked changes in nutrient contents of a loam soil, with a significant, though temporary, increase in exchangeable ammonium-nitrogen, nitrate-nitrogen and bicarbonate-extractable phosphorus. In the long term, however, repeated burning led to reductions in both phosphorus and potentially mineralisable nitrogen. In a long-term study in the Pacific north-west of the USA, the effects of repeated straw burning have been monitored since 1931 (Rasmussen *et al*. 1980). As a result of burning, organic matter and nitrogen contents have declined significantly over the last fifty years (Fig. 8.8). Autumn burning, following harvesting of the wheat crop, increased the loss of nitrogen but not carbon relative to losses associated with spring burning. In addition, burned plots seemed to show changes in soil colour, aggregate stability and infiltration capacity.

Experiments in Britain have shown similar, though often smaller, changes in soil nutrient contents. Hood and Procter (1961) reported a saving of 21 kg ha^{-1} y^{-1} of nitrogen by ploughing in as opposed to burning

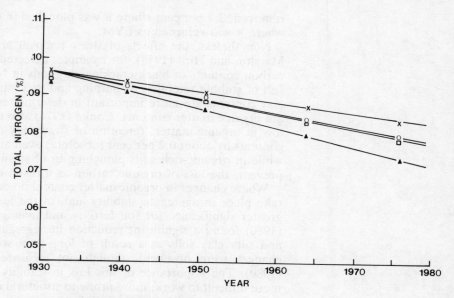

Fig. 8.8 Changes in the nitrogen content of soils under different straw disposal
treatments in a silt-loam soil, Oregon (after Rasmussen *et al.* 1980).

or removing straw. Patterson (1960), in studies at Rothamsted, found that
ploughing in of straw every second year increased nitrogen contents of
the soil by 0.005 per cent (absolute) after 18 years – an increase of about
one-seventh in the nitrogen contents. Reviewing these experiments,
Cooke (1974) comments that in many cases ploughing in of straw may
lead to reduced nitrogen availability due to immobilisation of nitrogen
during decomposition.

Changes in nutrient content due to straw and stubble management are
closely associated with changes in soil organic matter contents. Straw removal and, to a lesser extent, burning might be expected to reduce organic matter contents in the long term, and several studies have indicated
the magnitude of these losses. Bullen (1974) compared the effects of
ploughing in, burning and removing straw at four different experimental
stations for periods up to seventeen years. Organic matter contents of the
soil fell under all treatments but ploughing in of straw reduced organic
matter losses to a small degree. The changes in organic matter contents
were possibly masked, however, because Bullen sampled soils to a depth
of 16 inches (40 cm). More marked changes are likely to occur in the
upper few centimetres of the soil, and Rasmussen *et al.* (1980) and Biederbeck *et al.* (1980) reported significant declines in organic matter contents as a result of long-term straw burning. Similarly, studies at High
Mowthorpe Experimental Husbandry Farm, on chalk soils in Yorkshire,
showed a greater decline in organic matter contents when the straw was
burned or removed than when it was returned as farmyard manure or
ploughed in (MAFF 1964b). After eight years, the organic matter content
had declined from 3.96 to 3.6 per cent where the straw was burned or

removed, 3.7 per cent where it was ploughed in directly, and 3.8 per cent where it was returned as FYM.

Nonetheless, the effects of straw removal are not always significant. Marston and Hird (1978), for example, detected no difference in organic carbon contents of black cracking clay soils in New South Wales as a result of stubble retention or burning and they suggest that, in heavy soils, clay contents are more important in determining aggregate stability than are organic matter contents. Cooke (1974) also notes that in soils initially low in organic matter, retention of straw may only raise organic carbon contents by about 0.2 per cent (absolute) even after 20 years of treatment, while in organic-rich soils ploughing in of straw diminished, but did not prevent, the loss of organic carbon as a result of arable cropping.

Where changes in organic matter content do occur, related changes may take place in aggregate stability and, in the long term, these may have greater significance for soil fertility and management. Biederbeck *et al.* (1980) found a significant reduction in aggregate stability on both loam and silty clay soils as a result of long-term straw burning, and similar changes were observed, though not measured, by Rasmussen *et al.* (1980). The importance of this loss in stability is that it makes the soil more difficult to work, more prone to structural damage during cultivation or harvesting and more susceptible to erosion.

The effects of crop removal and residue management on soil erosion also operate more directly. Loss of the protective cover provided by the crop and its residues frequently results in greatly increased rates of soil erosion. As the studies of rainfall erosion by Wischmeier and others (Smith and Wischmeier 1962; Wischmeier and Smith 1965; Wischmeier and Mannering 1969; Wischmeier *et al.* 1971), and of wind erosion by Chepil and his colleagues (Chepil 1955; Chepil and Woodruff 1963; Woodruff and Siddoway 1965) have shown, crop cover plays a vital role in controlling erosion.

The protection afforded by the crop operates in a wide variety of ways. The plants intercept rainfall and store part of the moisture until it is lost by evaporation. Thus, they reduce the amount of water reaching the soil, they delay its arrival at the soil, and they thereby diminish the quantity of water available for erosion. In addition, interception results in a reduction in the velocity and, in some cases, droplet size of the rainfall reaching the soil. As a consequence, the kinetic energy of the rainfall ($0.5 M \times V^2$) is greatly reduced and the rainfall is therefore considerably less erosive (Morgan 1977). Moreover, because of the diminished kinetic energy, less compaction is caused by the raindrops, so higher soil infiltration capacities are maintained and a greater proportion of the incoming rainfall is absorbed by the soil.

In the same way, the crop provides protection from wind erosion. Wind velocity within the stem and canopy zone of the crop is reduced as a result of the greater frictional resistance and wind erosivity is reduced. Sediment which is entrained is also trapped by the plants, so that re-deposition of the eroded material is encouraged.

Given the fact that partial removal of the crop is unavoidable, it is clear that the potential for soil erosion is inevitably increased following harvesting. The management of the crop residues consequently assumes considerable importance. Retention of the straw and stubble helps maintain the protection originally afforded by the growing crop and, as many studies have shown, can lead to marked reductions in erosion. Laflen *et al.* (1978), for example, comparing six tillage treatments for continuous maize production on three Iowan soils, found that variations in the amount of residue cover accounted for between 78 and 89 per cent of the variance in rates of rainfall erosion. Similarly, Ketcheson and Webber (1978) showed that retention of the corn stover in continuous maize cropping systems reduced rainfall erosion from 0.125 cm y^{-1} to less than 0.01 cm y^{-1}. The effect of crop residues is particularly significant where the soil is cultivated into furrows (e.g. potatoes, sugar beet), for the furrows provide ready made channels for surface flow of the water. As Aarstad and Miller (1978) have indicated, even small quantities of residue placed in the furrows may reduce erosion losses considerably; 2.2 t ha^{-1} of mulch placed in the furrows (equivalent to about 0.4 t ha^{-1} overall) effectively eliminated erosion. Residues similarly inhibit wind erosion. Chepil and Woodruff (1963) presented data comparing the effectiveness of wheat and sorghum residues in reducing erosion losses. In both cases, standing stubble gave a better control than did cut residues, and wheat consistently gave bigger reductions than did sorghum (Table 8.3). More recently, Fornstrom and Boennke (1976) analysed the effects of growing barley as a mulch between sugar beet rows; losses were reduced from 110 t ha^{-1} to 37 t ha^{-1}.

Table 8.3 The effects of stubble retention on rates on soil loss by wind erosion

| Residue quantity (kg ha^{-1}) | Soil erosion (t ha^{-1}) | | | |
| | Wheat residue | | Sorghum residue | |
	Standing	Flat	Standing	Flat
0	40.2	40.2	40.2	40.2
560	7.1	21.3	32.6	36.4
1120	0.3	6.3	20.3	26.1
2240	T	0.3	9.8	13.3
3360	T	T	3.5	5.5
6726	T	T	T	0.5

Source: After Chepil and Woodruff 1963

Note: T = trace

Increases in wind erosion following crop removal reflect major changes in the surface microclimate, with increased wind shear and consequent entrainment of particles. More subtle but no less significant changes in microclimate also occur due to reduced interception of both incoming and outgoing radiation. In general, plants create an insulating layer which reduces the amplitude of temperature variations near the ground and may help to maintain higher levels of humidity. Removal of the crop destroys this insulating canopy and results in markedly greater variations in surface and soil temperatures, both on a daily and a seasonal basis. This increased variability is particularly apparent where the soil is left clear of crop residues; surface temperatures during the day may then rise several degrees above those of the air, and at night they may fall to several degrees lower than air temperature. This effect extends some way into the soil. Both the diurnal and the seasonal temperature amplitude beneath a non-vegetated surface tends to be greater than beneath a crop to a depth of at least 20 cm. Moreover, by reducing air turbulence close to the soil surface, the crop creates a zone of relatively still air with a low thermal conductivity. The effect of this is to reduce the rate at which heat is transmitted to the soil surface from the atmosphere, leading in turn to relatively slow warming of the soil during the morning or during the spring. Removal of the crop allows the soil to respond much more speedily to external temperature variations.

Loss of this insulating vegetation cover during harvesting, and the resulting change in soil temperatures, has considerable significance for subsequent plant growth. In areas of high day-time temperatures, the bare soil may reach excessive temperatures which can damage seedlings or soil organisms. Moreover, in bare soils in winter, the incidence and severity of freezing are likely to be increased. As a consequence, seeds or roots may be damaged and soil organisms killed. These effects are not wholly detrimental, however, for both weeds and insect pests may be controlled by winter freezing, while frequent freeze-thaw cycles help in the breakdown of soil aggregates and the formation of a fine, stable surface tilth. In addition, the relatively rapid warming of the bare soil in spring may be an advantage by encouraging germination and seedling growth and extending the active growing season.

Where crop residues are left on the soil, they act in much the same way as a standing crop: they intercept and dissipate incoming radiation, thereby reducing energy inputs to the soil, but the residues also intercept outgoing radiation, and thus conserve the energy which is received. As a result the amplitude of the surface and soil temperatures is reduced, evaporation is diminished and air humidity at the surface and soil moisture content increased. Fawcett (1978), for example, analysed the influence of residue treatment on soil moisture contents in cracking clay and red-brown earth soils in northern New South Wales and southern Queensland, Australia. He showed that the retention of stubble increased soil water contents for most soils, though the specific effects varied from site to site.

The combined influence of crop residue management on soil and surface microclimate, soil erosion, nutrient and moisture supply, soil structure and soil organic activity clearly has implications for crop growth and productivity. In some instances, it has been shown that residue retention increases yields significantly, though the effect often varies according to the nature of the residue management. Harvey (1959), reporting on a straw disposal experiment in which wheat, sugar beet, barley and rye were grown in rotation, found that ploughing in the straw did not affect sugar beet yields, but increased the subsequent barley yield by about 20 per cent. Similarly, Wickens (1963) demonstrated small increases in the yields of potatoes and cereals as a result of returning straw either by ploughing in or as farmyard manure, though the effects tended to disappear with increased applications of nitrogen fertiliser. Only rarely have marked increases in yield been attributable directly to the retention of crop residues, therefore, and many studies have shown no apparent effects of different methods of straw disposal on yield. Experiments at Jealotts Hill Experimental Station, for example, detected no differences in yields as a result of different stubble treatments (Hood and Procter 1961), while neither Bullen (1974) nor Biederbeck et al. (1980) found any effect upon yield of stubble burning. Russell (1977) presents data showing no effect of ploughing in straw on barley or sugar beet yields, but a small increase in potato yields. In some cases, however, retention of crop residues has been shown to reduce crop yields, either because of competition for nitrogen (e.g. Hood and Procter 1961), or because of reduced soil temperature and delayed germination (e.g. Chopra and Chaudhary 1980). In general, it seems that the main importance of stubble and straw management is in terms of the long-term influence on soil fertility: for example, by rendering the soil less susceptible to erosion, by conserving moisture (especially in areas with high soil moisture deficits), and by maintaining soil organic matter contents and thereby soil structure. It is also apparent, however, that straw and stubble burning as a management practice may have significant eccological and aesthetic implications. Indeed, in recent years a vociferous opposition to straw burning has built up among the public, arguing, among other things, that the practice destroys or damages habitats (especially hedgerows) and causes serious pollution.

8.2.6 Crop rotation and cropping sequence

Although monoculture of wheat, barley or maize is practised in some areas, the majority of arable farming systems today involves some method of crop rotation. This does not always include ley crops (see Chapter 7), but more commonly there is a break crop such as potatoes, sugar beet, sorghum, beans or oil-seed rape. Typically, these crops are grown for one or two years between longer sequences of cereals; for example, a widely practised rotation in England is to alternate winter wheat and spring barley with early potatoes and sugar beet or peas (Table 8.4). In the USA, sorghum and soybean are often rotated with maize or wheat.

Table 8.4   Typical arable rotations in Britain

Year	Crop rotations			
	A	B	C	D
1	Early potatoes/kale	Potatoes	Wheat	Sugar beet
2	Wheat	Wheat	Barley	Barley
3	Sugar beet	Wheat	Barley	Barley
4	Barley	Peas	Barley	Vegetables
5	Barley	Barley	Peas	Wheat
6		Barley	Seeds	

Source: After Lockhart and Wiseman 1978

The use of crop rotations in arable farming has a number of functions. These include the broadening of the economic base of the enterprise, the spreading of demands upon labour more evenly through the year and the production of particularly high value (but often high cost) crops as a means of increasing income. Potatoes and sugar beet are particularly significant in this respect because both have a high market value and are harvested during relatively slack periods in the farming year. In addition, however, the use of break crops may benefit soil fertility and result in higher yields of the subsequent cereal crops.

Some of the potential effects of arable crop rotations upon soil conditions are summarised by Bullen (1967) and Page (1972). They emphasise the role of break crops in pest and disease control, especially in relation to diseases such as take-all and eyespot which tend to build up during continuous cereal cultivation due to over-wintering of the organisms in the soil (Fig. 8.9). Break crops provide time for the host material in the soil to decompose before the next cereal crop is planted, so that

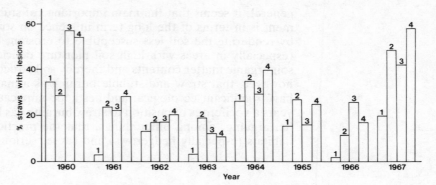

Fig. 8.9 Eyespot infection (percentage of straw with lesions) at harvest on
plots 1, 2, 3 and 4 years after fallow on Broadbalk Field,
Rothamsted, 1960–67 (data from Glynne 1969).

the cycle of the disease is interrupted. In most cases, a one-year break
is sufficient to provide control, but during particularly dry periods decomposition may be incomplete so that the cereal residues do not break down
fully and the disease may return in the first year of the cereal crop. In
addition, some pests – especially eelworms – may be encouraged by crop
rotations. Nonetheless, on the whole, break crops provide an effective
control of pests and diseases, the effectiveness increasing with the length
of breaks or their frequency. Schonrok-Fischer and Sachse (1980), for
example, investigating the effects of the cereal cyst nematode (*Heterodera
avenae*), found that increasing proportions of cereals in the crop rotation
resulted in increased infestations. They also showed that, partly as a result
of these infestations, cereal yields declined; under continuous cereal cultivation, yields of wheat were 0.9 t ha^{-1} less than when cereals accounted
for only 40–60 per cent of the rotation. These losses in yield seemed to
be a function mainly of reduced plant densities, which increased by 25 per
cent with the lower frequency of cereal cropping. The number of grains
per ear, on the other hand, was not affected, reflecting the fact that the
nematode inhibits maturation of the plants and causes die-back before ear
development. Similar results have been noted by Williams *et al.* (1979),
who found that a three-year crop rotation of barley, sugar beet and potatoes satisfactorily controlled *H. avenae*, while rotational cropping together with the use of fumigants minimised attacks by the nematode
Globodera rostochiensis.

Break crops may also have significant effects upon the structural and
chemical fertility of the soil. Relatively few studies have examined the
effects of arable rotations on organic matter contents and soil structural
stability, but it seems clear that the results of rotations are not always
beneficial. Dormaar and Pittman (1980) compared organic carbon and
polysaccharide contents of soils under various wheat-fallow rotations in
Canada and found that contents of both were reduced by the inclusion
of fallows in the rotation. Certainly both potatoes and sugar beet are often
associated with severe structural damage during harvesting, while it is
likely that the lack of residues they provide may reduce organic matter
contents. Break crops may also encourage nutrient loss, for almost the
complete crop is harvested so that returns in residues are limited. Furthermore, as Bijay Singh and Sekhon (1977) have demonstrated, shallow
root crops may result in greater leaching losses than do cereals. Compared

to the beneficial effects of ley-rotations, therefore, it seems that arable break crops contribute little to long-term soil fertility and may even be detrimental.

8.3 GROWTH OF CEREAL CROPS

The range of crops grown in arable farming systems, and the immense number of varieties of individual species – there are, for example, over 4,000 varieties of barley grown in the world (Martin and Leonard 1967) – mean that generalisations about patterns of crop growth are of limited relevance. Moreover, growth patterns depend not only on the physiological characteristics of the crop, but also on the management regime (e.g. the time and depth of sowing, fertiliser practice and harvesting) and on the soil and climatic conditions during the growing season. Many cereal crops, however, show a broad similarity in development and we can take as a simple model of crop growth the example of barley. Two main species of barley are grown in intensive, temperate arable systems: the common two-row form (*Hordeum distichum*) and the six-row form (*H. vulgare*). The main growth stages for cereals are shown in Fig. 8.10.

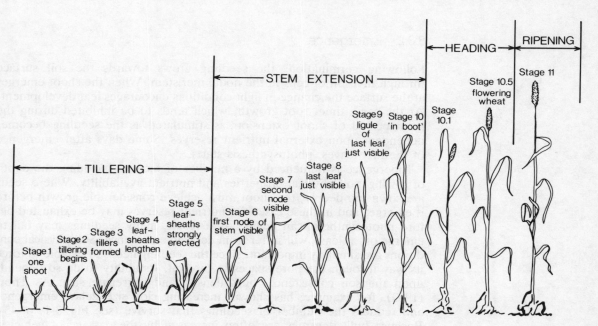

Fig. 8.10   Cereal growth stages (from Eddowes 1976).

8.3.1   Germination

Cereal crops are annual plants, growing from seed to reach maturity within a period of five to nine months. Growth commences with germination, during which first the radicle (the first seminal root) and then the plumule (the shoot) emerge from the seed. In time, often within a matter of days, additional seminal roots develop and, finally, the crown roots grow out from the elongating shoot. By this stage, the seed, which had provided food for germination, is almost exhausted and the seedling is dependent upon water and nutrients obtained from the soil.

To a great extent, the efficiency of germination is governed by the quality of the seed, and diseased or damaged seeds often show a high mortality and low germination rate. In addition, numerous edaphic factors influence germination, among them moisture status, oxygen availability and temperature. Germination only commences when the seed becomes moist, normally through uptake of water from the soil. The threshold moisture content varies from about 25 to 75 per cent for different crops, but is in the order of 40–50 per cent of the dry seed weight for many cereals. As the seed is wetted it tends to become permeable to gases and adequate oxygen supply at this stage is critical. In waterlogged conditions, oxygen may be lacking with the result that germination is inhibited. In addition, soil temperature exerts a major control; wheat, barley, oats and rye germinate only when the soil temperature exceeds freezing point, while maize requires temperatures above 9 °C. Above these threshold temperatures, however, germination improves (i.e. more seeds germinate) with increasing temperature to an optimum of about 15 °C for wheat and barley, and about 45 °C for maize (Wellington 1965).

8.3.2 Emergence

Following germination, the seedling grows towards the soil surface through division of cells in the nodal meristem. When the shoot emerges at the surface the change in light conditions encourages leaf development. At the same time, root growth, which tends to be inhibited during the main period of shoot extension, is stimulated as the seedling becomes dependent upon external nutrient reserves. Some days after emergence of the first leaves, photosynthesis starts.

Emergence is influenced by a number of factors, including the depth of sowing, soil physical properties and nutrient availability. Where seeds are sown too deeply, the shoot must achieve considerable growth before it emerges and in these cases the nutrient reserves may be exhausted before photosynthesis can begin. As a result, many seedlings may fail to emerge or, at least, will suffer from delayed emergence. Soil physical conditions are similarly important, since the seedling needs to be able to force its way through the pore spaces in the soil. Its ability to do so depends upon the soil consistence, bulk density and aggregate size. As Thow (1963), for example, has shown, increasing clod size delays emergence and reduces the number of seedlings that survive (see Fig. 3.1, p. 45). Because bulk densities are often increased by the passage of vehicles, emergence is often inhibited in wheelings. Lack of nutrients, particularly nitrogen, may also be important during seedling emergence, for it is apparent that as the internal reserves of the seed are exhausted and the plant becomes dependent upon external sources, nutrient deficiencies may severely reduce shoot growth. This has particular implications for fertiliser practice, for it is clear that during emergence the roots of the developing seedling are short and sparse, so they are able to draw on nutrients from only a small and immediately adjacent volume of soil. The need at this stage is for an abundant supply of readily available nutrients within this narrow rooting zone.

8.3.3 Tillering

After the seedling has emerged and photosynthesis has started, the growing shoot develops new leaves. The pattern of leaf development varies with the species, but as the studies by Williams (1960) have demonstrated, growth rates of the leaves depend among other things on nitrogen supply. As the total leaf area increases, the rate of photosynthesis also rises, so that growth is initially rapid. Most of the growth, however, occurs not by stem extension but by the process of tillering (Fig. 8.10).

It has been increasingly recognised in recent years that the process of tillering exerts a fundamental control on the ultimate yield of cereal crops, and many studies have been carried out examining the processes involved and the factors influencing them (e.g. Watson *et al.* 1958; Aspinall 1961; Kirby 1967, 1969, 1973; Kirby and Faris 1972; Kirby and Jones 1977; Cannell 1969a, b; Jones and Kirby 1977). Tillering commences with the expansion of vegetative buds adjacent to the leaf primordia, and over a period of weeks several tillers may develop. Not all these tillers survive to produce ears, however, and in the case of barley, tiller numbers reach a peak in the spring then fall to a minimum before ear emergence (Watson *et al.* 1958, Cannell 1969a). Several studies have shown that the earlier tillers are most likely to survive and tend to contribute most to the ultimate yield of the plant (Wellington 1965, Cannell 1969a), and it is also apparent that tillers compete for nutrients during their early growth, thereby inhibiting development of the main shoot (Kirby and Jones 1977). Thus, removal of tillers may increase yield by reducing competition for nutrients (Jones and Kirby 1977).

As this indicates, nutrient availability is one of the main controls on tillering, and thereby on crop yield. Aspinall (1961) and Cannell (1969a) have shown that the survival of tillers is improved by application of nutrients, particularly nitrogen, while nutrient deficiencies may arrest tiller development. In addition, light plays an important part, and because increased tillering causes shading of the lower parts of the plant, lack of light may inhibit the development of later tillers in closely-spaced crops. Kirby and Faris (1972), for example, found that the development of tiller buds was restricted at high planting densities, and that subsequent growth of the tillers was limited by availability of light in dense plantings. Similarly, tillering is affected by moisture availability. Tiller development is reduced by moisture stress, while death of tillers is increased (Day *et al.* 1978; Legg *et al.* 1979).

8.3.4 Reproductive growth

If allowed to develop, the tillers and the main stem of the plant enter a phase of reproductive growth soon after tillering. At this stage the energy and nutrient reserves of the plant appear to be diverted towards the apex of the growing shoots, and away from the leaves and lateral buds (Bunting and Drennan 1965). As a result, tillering ceases temporarily, leaf growth is inhibited and the main process is one of ear formation and stem extension. Flowers develop at the end of the active stems and these set to produce grain. Not all the florets within the flowers set grains and it is clear that both stem extension and ear development depend upon environmental conditions as well as physiological characteristics of the plant. Stem extension, for example, is severely affected by nitrogen and water supply, and stunted growth occurs when either are deficient. Ear formation, filling and grain growth are controlled in part by temperature, light, moisture and nutrient supply. Increasing temperature speeds up the rate of grain growth but may result in lower grain weights. The number, size and weight of ears are also affected by similar factors. Day *et al.* (1978) found that the number of grains per ear was reduced by drought during the tillering and ear formation stages, while the numbers of ears per unit area were influenced mainly during the period of stem extension; the mean weight (i.e. the size) of the grains was most affected by drought during the grain-filling stage.

As we noted earlier (section 8.2.2) the ultimate yield of the crop is greatly affected by the plant density, and this operates especially during the period of reproductive growth. Density particularly constrains the survival of tillers to produce ears, in part it seems due to competition for light. Additionally, increased tiller numbers during the ear-formation stage results in smaller ears due to both reduced numbers of grains on each ear and also lower grain size. Indeed, it is apparent that all the components of yield – the number of plants per unit area, tillers per plant, ears per plant, grains per ear and weight per grain – are closely interdependent. Complex feedback mechanisms operate throughout growth which compensate for changes in the plant population. We have seen one implication of this already: over a wide range of sowing rates, crops show limited variations in total yield. Only when the sowing density is very high is the loss in ear size and weight sufficient to compensate the higher numbers of ears; only when the sowing rate is very low is the reduction in the number of plants sufficient to offset the increased grain weight per plant.

8.3.5 Root growth

Because the cereal plants are dependent for their growth upon the nutrients and water supplied by the soil, the rate and efficiency of root growth exert a major control upon crop yield. The distribution of cereal roots in the soil depends, however, on the rate of photosynthesis and therefore internal feedback links operate between the growth of the above- and below-ground parts of the plants. In addition, soil and climatic conditions greatly influence root development.

Where environmental conditions are optimal, root growth tends to be rapid during the early stages of growth, and some 60 per cent of root development often occurs within the first 4 weeks of emergence. Most of this growth is in the upper few centimetres of the soil, and the distributions of roots at various stages in growth are illustrated in Fig. 8.11. Numerous factors influence the detailed pattern of root growth, however, including light intensity, nutrient supply, water availability and soil structural conditions.

The importance of light intensity has been demonstrated in a series of

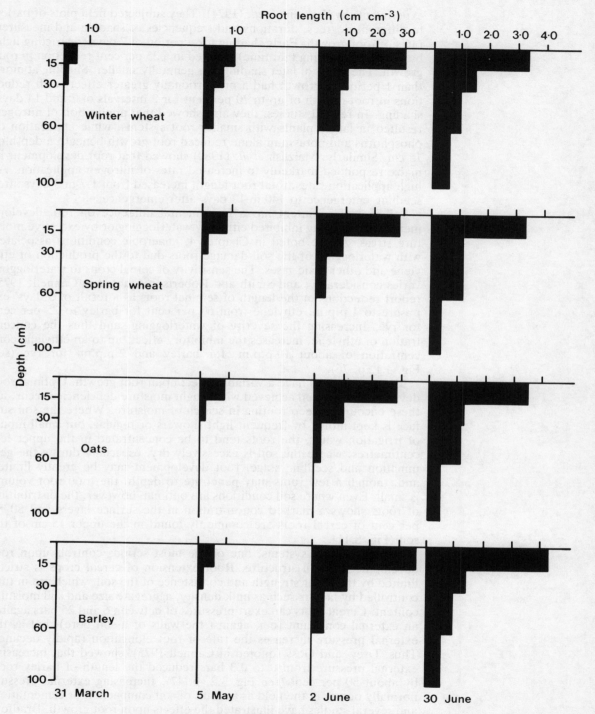

Fig. 8.11 The distribution with depth of roots of cereal crops at different stages (from Welbank 1975).

experiments by Welbank *et al*. (1974). They subjected field plots of barley to different degrees, durations and frequencies of shading and measured rates of root growth. Early shading for a period of 7 days (restricting light by 50 per cent during that time) resulted in a 25 per cent reduction in root growth. The effect of later shading was generally smaller. Shading at more than 1 period of growth had a proportionally greater effect, with reductions in root growth of up to 70 per cent for 2 intervals of 7 and 14 days' shading. In related studies, they also showed that application of nitrogen resulted in larger plants with smaller root systems, while application of phosphorus and potassium alone reduced root growth beneath a depth of 15 cm. Similarly, Maizlish *et al*. (1980) showed that root development in maize responded markedly to increased rates of nitrogen application. At high application rates, total root length increased from 1.7 m 3 days after seedling emergence to 140 m 17 days after emergence.

Soil moisture content has a fundamental influence on root development, rooting being inhibited either by waterlogging or by excessive moisture stress. As we noted in Chapter 4, anaerobic conditions associated with waterlogging of the soil damage roots due to the production of ethylene and other toxic gases. The sensitivity of cereal crops to waterlogging varies considerably, and Smith and Robertson (quoted in Cannell, 1975) report reductions in the length of seminal roots as a result of 3 days' exposure to 1 p.p.m. ethylene from 60 per cent for barley to 25 per cent for rye. Increasing the severity of waterlogging, and thus the concentration of ethylene, increases the inhibitory effect, up to an ethylene concentration of about 10 p.p.m. for barley and 2 p.p.m. for rye (see Fig. 4.1, p. 76).

Moisture deficits have a variable effect upon root growth. Optimal root development is often achieved when slight moisture deficiencies occur, for these encourage deep rooting in search of moisture. Where the soil surface is kept moist, by frequent light showers or regular, but small inputs of irrigation water, the roots tend to be concentrated in the upper few centimetres; where the soil is excessively dry, especially during the germination and seedling stage, root development may be greatly limited and, though a few roots may penetrate to depth, the total root volume is small. Even where soil conditions are optimal, however, the distribution of roots shows a marked concentration in the surface layers and 80–85 per cent of cereal roots are commonly found in the upper 15 cm of the soil (Fig. 8.11).

In many arable systems, one of the most serious controls upon root development is soil structure. Root extension of cereal crops is strictly limited by the shear strength and consistence of the soil, which are in turn controlled by factors such as bulk density, aggregate size and soil moisture content. Cereal roots can exert pressures of between 6 and 25 bars against an external constraint (e.g. against the walls of a soil pore), but as the external pressure increases the rate of root elongation rapidly declines. Thus, Drew and Goss (quoted in Cannell 1975), showed that increasing external pressure from 0 to 0.3 bars reduced the length of barley roots by about 50 per cent (see Fig. 3.2, p. 47). Increasing external pressure normally occurs in the field as a result of soil compaction or cementation, and several studies have illustrated the effects upon root growth. Bradford (1980), for example, found a correlation between root length and the penetrometer resistance of structured soils in silt-loam soils in Wisconsin, while Ehlers *et al*. (1980–81) reported that a plough-pan at 20–30 cm depth increased the root concentration in the upper 20 cm of the soil but inhibited rooting at depth. Similarly, Raghavan *et al.* (1979) compared rooting volumes in compacted and uncompacted arable soils. Compaction of the soil by 15 passes with a tractor exerting a contact pressure of 0.63 kg cm^{-2} reduced the root weight of maize plants from 5.7 mg g^{-1} to less than 2 mg g^{-1}. As Hewitt and Dexter (1979) have argued, however, root impedance is not solely a function of the physical restrictions caused by compaction; nutrient availability also plays an important role. As aggregate size increases, for example, the availability of nutrients per unit length of root is reduced due to the smaller surface area in contact with the root. This reduces the ability of the roots to obtain nutrients and thus inhibits their growth.

8.3.6 The composition of the growing crop

Throughout the life of the crop, major changes in nutrient composition occur as the plants assimilate and cycle nutrients obtained from the soil and atmosphere. The majority of nutrients are taken up during the early stages of growth, and by the tillering stage most cereals have reached 80–90 per cent of their maximum nutrient content. Indeed, during later stages of growth, nutrient contents may fall as tillers die and as roots slough off (see Fig. 5.5, p. 113). In addition, some nutrients are lost by leaching from the plant, especially during the ripening stage when a transfer takes place from the stem and leaves to the ears. Potassium, in particular, seems to be lost during the ripening stage of wheat, while magnesium may be readily leached from some crops (Russell 1973). As we noted earlier, this process of loss back to the soil means that nutrient removal during harvesting is somewhat less than it would otherwise be. It also means that nutrient analyses of the harvested crop do not provide an exact indication of nutrient requirements during growth.

8.4 SOIL CONDITIONS

The annual cycle of seedbed preparation, sowing, crop growth and harvesting which characterises arable farming systems has a major influence on soil conditions. As we have seen, organic, chemical and physical properties of the soil are all affected by these processes. In the short term, the changes which occur in these properties are generally small and reversible, but in the longer term progressive or cumulative changes may take place which greatly modify soil fertility.

8.4.1 Organic matter content and organic activity

Relatively few studies have been carried out of the seasonal variations in soil conditions under arable farming systems. Marked changes occur in the organic composition and activity of the soil, however, in response to varying rates of input and loss. The main inputs are by the more or less continuous sloughing off of dead root material and the decay of aborted tillers and leaves, and from the seasonal input of crop residues following harvest. The latter often provides the major source of organic matter and, as we noted in section 8.2.5, management of the debris controls to a significant extent the rate of organic matter replenishment in the soil. Decomposition of cereal residues tends to be slow and is carried out largely by general-purpose organisms such as actinomycetes and fungi, capable of attacking the cellulose and lignin compounds. The process tends to follow a clearly defined sequence, with the less resistant hemicellulose materials being broken down in the first days by fungi, then the pure cellulose being attacked in the following days. After a week or so, however, the rate of decomposition declines as the more resistant lignins and the lignin-protected materials are attacked. Relatively few microorganisms are able to attack these lignin compounds and much of the decomposition probably involves chemical reactions outside the living cells of the micro-organisms. In addition, the residual compounds produced during initial organic decomposition of the plant material break down only slowly, so that after the first three to four weeks the rate of organic matter loss is reduced.

Numerous factors influence the rate at which decomposition processes operate. Material left on the surface tends to break down relatively slowly, since it is not readily accessible to many of the micro-organisms which take part in decomposition within the soil. In this circumstance, breakdown depends to a great extent upon frost action; freezing distends the plant cells and causes rupture. Burial of the debris by tillage commonly accelerates decomposition, for though it protects the material from frost it does bring it into contact with the abundant mycorrhizal populations in the root zone. On the other hand, decomposition is inhibited by anaerobic conditions, for fewer soil organisms are available to attack the debris and a greater proportion of the compounds produced during decomposition are high-energy materials which are resistant to further breakdown. Thus, deep burial of the residues, or soil compaction during tillage, may reduce the rates of decomposition and lead to the survival of pockets of raw organic matter in the soil. In addition, soil temperature affects the rate of breakdown, and as temperatures fall decomposition is reduced due to the declining activity of the soil organisms.

The end product of these essentially organic processes of decomposition is the comparatively resistant compounds which constitute the soil humus; thus the processes are known as humification. The humus which is left at the end of these processes decomposes only slowly, largely through chemical oxidation but partly through the action of specific organisms which attack the polysaccharides and humic and fulvic acids that make up the large proportion of the humus. The rate at which these mineralisation processes operate is governed in general by the quantity of humus available in the soil; decomposition tends to follow the law of first order kinetics such that the rate of mineralisation is a linear function of the humus content (Russell 1973; Sinha *et al.* 1977). Variations around this general rate of decomposition occur under field conditions, however, due to changes in temperature and moisture content. Wetting and drying of the soil increases the rate of humus oxidation, as does an increase in soil temperature. Improved aeration, caused for example by loosening the soil during tillage, also encourages oxidation, and as a result rates of humus decomposition show marked variations during the course of the year. Unfortunately, few data are available on which to base detailed summaries of the seasonal patterns of mineralisation, but from theoretical considerations it is clear that maximum rates should be achieved in the autumn when the soil is being rewetted following a period of summer dryness, and when aeration is being improved by tillage. Mineralisation is probably at a minimum in the winter when the soil is wet, temperatures are low and oxygen is deficient.

In the longer term, soil organic matter contents may undergo more fundamental changes. These changes are dependent upon the ratio of organic matter inputs from crop residues to losses by decomposition, and under constant management and environmental conditions may be described by the following equation:

$$\frac{dC}{dt} = a - bC \qquad [8.1]$$

where C is the organic carbon content of the soil,
   t is time
   a and b are constants.

Integrating this equation gives

$$bC = a - (a - bC_o)e^{-bt} \qquad [8.2]$$

where C_o is the original organic carbon content of the soil. This equation states that over time the organic carbon content will approach the value of a/b asymptotically, and that in $0.693/b$ years the carbon content will have been halved. This thus describes the half-life of the soil humus. Estimates of the half-life of humus from unmanured arable fields indicate a value of about 25 years (e.g. Jenkinson 1965; Monteith *et al.* 1964), giving a value of b of about 0.28. The half-life varies considerably, however, according to management practices, and is much less when farmyard manure is applied or grass crops are sown as leys. Jenkinson (1965) found that the half-life of organic matter derived from ryegrass in the Broadbalk plot at Rothamsted was only four years. Moreover, the rate of organic matter decomposition varies for different components of the humus. Much of the fresh organic matter is stored in readily accessible locations within the soil, as coatings to aggregates or in macropores where oxidation processes are rapid. Significant quantities, however, accumulate in the smaller pores (less than about 200 μm diameter), where bacterial activity is limited and where the rates of oxidation are diminished due to the presence of relatively anaerobic conditions. Additionally, as has been noted, the final products of decomposition are highly stable humus compounds and these may take many hundreds or even thousands of years to break down. Studying the Broadbalk plots, Jenkinson (1968) found that in the unmanured arable soil the organic matter in the upper 22 cm had a half-life of about 1400 years, while that on the old, permanent grassland plots was approximately 600 years. Thus, the different components of the organic matter content of the soil – the readily decomposable plant material, the resistant plant material and the stable humus decomposition products – show markedly different half-lives. Modelling of rates of organic matter breakdown in arable soils is consequently complex, but attempts have been made by Jenkinson and Rayner (1977) and B. L. Smith (1979a, b). Jenkinson and Rayner developed a model which predicted changes in organic carbon contents on the basis of five different organic matter components. Readily decomposable plant material was assumed to have a half-life of about 0.165 years, resistant plant material 2.31 years, the

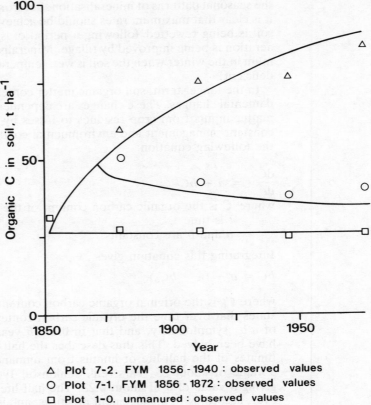

Fig. 8.12   Predicted and observed soil organic matter contents under continuous cultivation at Broadbalk, Rothamsted (after Jenkinson and Rayner 1977).

soil biomass 1.69 years, physically stabilised organic matter 49.5 years and the chemically stabilised organic matter 1,980 years. Changes in organic carbon contents of soils at Rothamsted were then simulated using this model and compared with field data from long-term field experiments. The results showed a close fit between predicted and measured organic carbon contents (Fig. 8.12).

As the analyses of Jenkinson and Rayner (1977) show, changes in the organic carbon contents of arable soils depend upon the interaction of numerous factors. Under conditions of uniform management (and thus uniform inputs and organic matter losses), however, soil organic carbon contents ultimately reach a steady-state level. Where old grassland plots have been ploughed up for arable cropping, this steady-state level is commonly less than the original organic carbon content, and may take many years to achieve. At Rothamsted, for example, the continuous barley experiment has shown that, except with high annual rates of farmyard manure application, organic carbon contents have fallen markedly. On plots receiving farmyard manure from the start of the experiment in 1852 until 1871 but none since, organic matter contents initially rose and then declined. Steady state was achieved after about 40 years, although at a higher organic carbon content than in the original, old arable soil. In plots receiving no manure from the start of the experiment, steady state seems to have been reached within 30 years. Similarly, after a period of 21 years of continuous arable cropping on silt-loam and loam soils of the Batcombe Series at Highfield, Rothamsted, organic carbon contents are only just stabilising. Over this period they have fallen from 2.75 to 1.96 per cent (Jenkinson and Johnston 1977). In a comparable study at Woburn, on sandy loam soils of the Cottenham Series, plots under a 5-year rotation of cereals and root crops showed a reduction in organic carbon contents from 1.02 to 0.88 per cent after 28 years on plots receiving no farmyard manure, and even at this stage a small annual reduction in organic carbon contents was still detectable, indicating that equilibrium had not yet been reached (Johnston 1973).

While the detailed effects of arable cropping upon soil organic matter contents depend upon a wide range of factors, including cropping sequence, fertiliser and manuring practice, soil type and climate, therefore, it is clear that in many cases the long-term effect is to reduce organic carbon contents. Significantly, though these changes undoubtedly have indirect effects upon soil fertility through their effect on nutrient supply (see below), there is no firm evidence to show that they directly impair fertility or cause reductions in yield (see, for example, Russell 1977).

8.4.2 Soil nutrients

Changes in soil organic matter contents under arable cropping have fundamental effects upon nutrient supply. Plant residues provide a major input of nutrients to the soil, and processes of organic matter decomposition exert a major control on the availability of nutrients to plants. In many cases, release of the nutrients during organic matter breakdown involves two conflicting processes: immobilisation as the nutrients are assimilated by soil organisms and mineralisation as the organisms themselves die and decay. Variations in the relative rates of these two processes lead to short-term fluctuations in nutrient availability. Using ^{15}N isotopes as a tracer, Broadbent (1978), for example, found that following the addition of fertiliser nitrogen and barley straw to the soil, net immobilisation occurred for two weeks. Gradually, however, the rate of mineralisation increased until, by the end of the second week, it exceeded immobilisation. Thus, a phase of diminishing nitrogen availability was succeeded by a period of increased availability. This pattern reflects the process noted in Chapter 5; namely that nitrogen availability often decreases for a short period after manuring of the soil due to increased nitrogen uptake by the organisms involved in decomposition. It is clear that as a result of this process nitrogen supplies in the soil vary markedly through the year in response to inputs of plant residues and farmyard manure.

Superimposed upon the effects of organic matter inputs and decomposition are several other short-term processes and responses, which together result in complex seasonal patterns of nutrient contents. Rates of nitrogen mineralisation, for example, are closely related to soil moisture contents, and Stanford and Epstein (1974), among others, have shown that between field capacity and wilting point the rate of N mineralisation is linearly correlated with soil moisture content. The effect appears in part to be a function of the increase in microbial activity as the soil is wetted from wilting point, though as it approaches saturation the activity of fungi and actinomycetes, in particular, declines (Power 1981). There is also evidence to suggest that alternate wetting and drying of the soil stimulates nitrogen mineralisation, the magnitude of the effect varying considerably, however, according to organic matter content, soil temperature and moisture content (e.g. Campbell *et al.* 1975; Campbell 1978). Similarly, mineralisation rates are closely temperature-controlled. Within the range 5 to 35 °C, Stanford *et al.* (1973) showed that increasing the soil temperature by 10 °C resulted in a 2–3 fold increase in mineralisation rate. Additionally, availability of nutrients is clearly dependent upon the input and mobilisation of fertilisers. As we saw in Chapter 5, the fate of fertiliser nutrients in the soil depends to a great extent upon the nutrient under consideration. In the case of phosphates, a large proportion of the fertiliser input is immobilised, and the release of plant-available phosphate is slow. Nitrogen fertilisers, on the other hand, tend to be much more soluble, so that a greater proportion of the applied nitrogen is taken up immediately by the plant, but more is also leached through the soil by percolating water. In this context, a critical factor is rainfall, and marked seasonal variations in leaching losses and redistribution of nitrogen occur as a result of the seasonal pattern of rainfall. Williams (1970, 1976), for example, showed major losses of nitrate in drainage waters from arable land in response to individual rainfall events.

Together, the effects of variations in weather, crop management, cultivation and biological activity cause considerable short-term variations in nutrient contents and availability. On the whole, these variations are likely to be more marked in arable than grassland systems because of the more fundamental changes in external conditions (due, for example, to tillage, harvesting and residue removal, and to the fact that, for long periods of time the soil lacks a crop cover to protect it against changes in the weather). Nonetheless, few studies have considered the nature of these short-term effects and it is, in any case, doubtful that general patterns can be identified because of the importance of specific, local conditions and management procedures.

In the longer term, changes in soil nutrient content under continuous arable cultivation show more regular patterns. Commonly, where high rates of fertiliser application have not been sustained, soil nutrient contents have fallen. In the continuous barley experiment on Hoosfield at Rothamsted, for example, nitrogen contents in plots receiving no fertiliser or farmyard manure fell from 0.34 per cent in 1852, at the start of the experiment, to 0.31 per cent in 1882, since when they have been relatively stable (Jenkinson and Johnston 1977). Similarly, in the so-called exhaustion experiments at Rothamsted, unmanured and unfertilised plots showed a reduction in nitrogen contents from 0.132 per cent in 1856 to about 0.100 per cent in 1903, since when nitrogen levels have not changed significantly (Johnston and Poulton 1977).

These changes in nitrogen contents broadly reflect the patterns of organic carbon decline in these soils, and equilibrium conditions are reached after similar periods (Fig. 8.13). This is not unexpected since cycling of carbon and nitrogen in arable systems is closely related. Other nutrients show comparable adjustments to steady-state conditions. Phosphorus contents of unmanured plots on the Rothamsted exhaustion land have changed little since 1856, but those receiving FYM between 1856 and 1872 (but none since) only reached equilibrium in 1940. Potassium contents on unmanured plots took over 100 years to stabilise. In many cases, in fact,

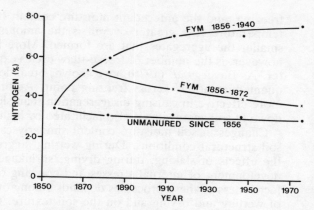

Fig. 8.13 Changes in the nitrogen content of soils under continuous arable cropping on the exhaustion land at Rothamsted Experimental Station (after Johnston and Poulton 1977).

the time taken to reach equilibrium of these nutrients seems to be higher than that for nitrogen. The slower response of these nutrients appears to be linked to their more effective retention in the soil by adsorption, chemical precipitation and fixation in stable organomineral compounds. Moreover, in the case of potassium, a greater proportion of the nutrients taken up by the crop are retained in the straw and may thus be returned to the soil in the residues. Losses of both potassium and phosphorus by leaching are also relatively low, with the result that both nutrients have long turnover times and adjust only slowly to changes in rates of input.

8.4.3 Soil structure

Cultivation, crop growth and harvesting lead to marked changes in the structure of arable soils. As we noted in Chapter 3 ploughing tends to open up pore spaces and increase the volume of macropores in the soil, particularly if tillage is carried out when the soil is in a friable state. On the other hand, ploughing of wet soils, together with the effects of compaction and smearing by vehicle wheels and implement tines, may lead to compression of aggregates and loss of porosity. Localised compaction, which is often hidden from view by loose material scraped over the surface during cultivation, results in anaerobic decomposition of organic matter and apparently leads to bacterial degradation of the polysaccharide compounds which bind the soil particles together. As a result, the stability of these aggregates may be reduced. Seasonal patterns of cultivation therefore cause related, short-term variations in many of the structural properties of the soil.

Seasonal variations in structural conditions also occur in response to climatic effects. Frost action has a major influence both in the creation and destruction of aggregates: ice-needle growth in soil pore spaces forces together the particles and encourages cohesion within the aggregates, while at the same time weakening the links between the aggregate faces. The overall effect is generally to produce smaller, more compact peds. The effects of frost action upon soil structure depend upon a number of factors, including the intensity of freezing, the frequency and rate of freezing, and the antecedent moisture content of the soil. The lower the temperature, the greater overall is the amount of breakdown and the smaller the aggregates that are formed. More important in most cases, however, is the number of freeze-thaw cycles experienced during the winter. As Baver *et al.* (1976) have shown, breakdown is encouraged by frequent fluctuations across freezing point. Additionally, rapid freezing is more effective in causing disaggregation than slow freezing, while breakdown of soil aggregates is also facilitated by higher soil moisture contents.

Changes in soil moisture content similarly cause marked variations in soil structural conditions. During wetting, aggregates break down under the effects of slaking; during drying, shrinkage of the soil leads to the development of internal stresses and cracking of the aggregates. The effectiveness of these processes depends among other things upon the rates of wetting and drying and on the soil texture. Generally, disaggregation is favoured by rapid wetting and drying while the processes tend to be most active in loamy or clayey soils, especially those containing significant quantities of swelling clays such as montmorillonite.

During the course of the year the processes of tillage, compaction by machinery, freeze-thaw, slaking and shrinkage-cracking interact to produce constant adjustments in soil structural conditions. Few studies have examined the consequences of these effects for seasonal variations in soil structure, though Wilson *et al.* (1947) assessed aggregate sizes of silt-loam soils in Iowa under continuous and rotational corn systems. They showed that aggregates under continuous corn increased in diameter between May and August, but then decreased in size during the autumn. Aggregate stability showed a similar pattern, reaching a peak in August and declining in November. The reasons for these changes were not discussed, though they presumably reflect the counteracting effects of cultivation, crop growth and atmospheric processes throughout the year.

Arable cultivation involves a number of conflicting changes in soil conditions. Tillage loosens the soil, reduces aggregate size and increases macroporosity (Djeniyi and Dexter 1979). Smearing by vehicle wheels and implement tines, and compaction by machinery increases bulk density, increases aggregate size and reduces macroporosity (Greenland 1977, 1981). The loss of organic matter, due to crop removal and tillage effects, results in lower structural stability and a higher soil bulk density (Hamblin and Davies 1977). The long-term changes in soil structure under arable cropping thus depend upon the relative rates of these conflicting processes. Many authors have argued that arable production leads to structural damage (e.g. Low 1955; Martel and MacKenzie 1980; Grieve 1980; Saini and Grant 1980), and Russell (1973) has noted the loss of 'mellowness' seen in old arable fields. The most serious long-term effects, however, appear to be the loss in structural stability which takes place during continued arable cultivation. Low (1972), for example, monitored structural stabilities in soils of the Hanslope series under old grassland, recently ploughed out grassland and old arable land for a period of five years. In the old grassland the structural stability remained at 73–78 per cent. In the old arable plot it was only 12–17 per cent. During the first year after ploughing, on the ploughed out grassland, the structural stability fell from 76 per cent to about 35 per cent, after which it declined more slowly to 25 per cent. In the same plots, at the end of the 5-year period, the air-filled porosity (i.e. the volume of soil filled with air at field capacity) was 20.2 per cent in the old grassland, 7.8 per cent in the ploughed out grassland, and only 5.0 per cent in the old arable soil.

As these results indicate, the decline in structural stability and macro-porosity of the soil is relatively rapid under arable cropping, and presumably relates to the loss of readily decomposable organic compounds. Conversely, the rates of recovery may be very slow, especially when damage occurs at depth (Low 1955; Greenland 1977). The effects vary, however, according to soil conditions, tillage and cropping practice and stubble management. As mentioned earlier, changes in structural stability in clay soils tend to be less than those in lighter soils, possibly because organic matter is less important in controlling stability (Marston and Hird 1978). Vorob'ev and Saforov (1977) found that the monoculture of potatoes reduced the proportion of agronomically significant water-stable aggregates by 50 per cent compared with wheat or barley. Hewitt and Dexter (1980) noted that while tillage increased macroporosity by 30 per cent, stubble burning resulted in a loss of voids and a reduction in aggregate size compared with stubble retention; they concluded that differences due to stubble management were at least as important as those due to tillage. Zero tillage also affects structural changes under arable systems, and commonly results in higher bulk densities in the surface layers and significantly greater structural stabilities than conventional tillage (e.g. Soane 1975; Pidgeon and Soane 1978).

As these examples indicate, structural effects of arable farming vary considerably, but it is nevertheless apparent that in the long term, on many soils and under a wide variety of arable farming systems, significant soil structural changes occur. In some cases these lead to detectable reductions in yield (Fig. 8.14; see also Table 3.1, p. 44). More often their impact is indirect, in that they make the soil more prone to compaction, more difficult to cultivate to a fine tilth, more susceptible to erosion, and thus more difficult to manage. It is for these reasons that the impact of arable farming on soil structure is a major concern in many areas, especially on soils which are inherently sensitive to damage (e.g. MAFF 1970a). Similarly, it is partly for these reasons that increasing attention

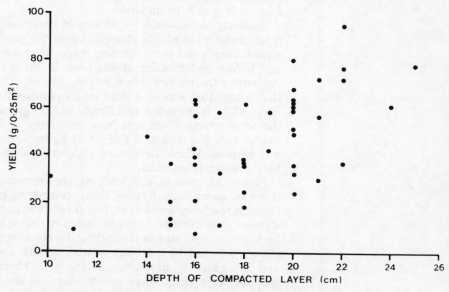

Fig. 8.14 The effect of soil compaction by machinery on the yield of spring barley in a single field in Devon (from Briggs 1977).

is being devoted to cropping systems – such as rotations – and tillage practices – such as zero tillage – which help to conserve soil structure.

8.5 NUTRIENT CYCLING IN ARABLE SYSTEMS

Compared to grassland systems, the nutrient cycles of intensive arable systems are characterised by high throughputs of nutrients but relatively simple and minor internal circulations. In general, the cycles are dominated by the annual removal of nutrients in the harvested crop, and their replenishment in fertilisers. Other losses include leaching, erosion and denitrification, while small inputs of nutrients occur via atmospheric deposition and rainwash, weathering and – to a very minor extent – nitrogen fixation.

The importance of harvesting within arable systems is illustrated by data from farms in western Europe and USA (Frissel 1978) which indicate that between 65 and 85 per cent of nitrogen and phosphorus taken up by the crop is lost in harvests, and this accounts for 60–100 per cent of the total loss of these nutrients. The majority of the nutrients are removed in the grain (in cereal crops), or tubers (in root crops), but significant quantities are also lost in the crop residues. Consequently, harvesting regimes and residue management methods have a major influence on the nutrient cycle and on the condition of nutrient reserves under steady state. Possibly the most important factor is the policy of residue management. As we have already noted, up to 30 per cent of nitrogen and phosphorus may be stored in the straw of cereal crops, while the tops and leaves of sugar beet may contain up to two-thirds of the total nitrogen and half of the total phosphorus in the plant. The removal or burning of crop residues markedly influences nutrient losses as the examples in Figs. 8.15 and 8.16 illustrate.

Similarly important is the timing of harvesting. As explained in section 8.3.6, nutrient contents change during the growth of the crop, reaching a peak during the early ripening stage, after which leaching from the ears and leaves and transfer of nutrients to the stem and roots reduces the content of the harvestable material. Russell (1973), for example, showed that wheat lost almost a third of its potassium between 18 June and 16 July, while Widdowson and Penny (1965) found that almost a quarter of the total nitrogen was lost between earing and harvesting in wheat fertilised with nitrogen at a rate of 91 kg ha^{-1}, but none when no nitrogen was applied. Where harvesting is delayed, therefore, nutrient losses may be significantly reduced.

Losses by erosion and leaching are commonly considerably less than those by harvesting. Brady (1974), comparing quantities of nutrients lost through leaching, erosion and crop removal in humid regions in the USA showed that average nitrogen losses by harvesting were six times those by leaching, and almost twice those by erosion. Similarly, almost no phosphorus was lost by leaching, compared with about 20 kg ha^{-1} by erosion and 25 kg ha^{-1} by harvesting (Table 8.5). These data refer to continuous cropping of maize; where arable rotations are practised, or where other cereal crops are grown, losses by erosion are likely to be significantly lower. Nonetheless, arable farming undoubtedly increases the potential

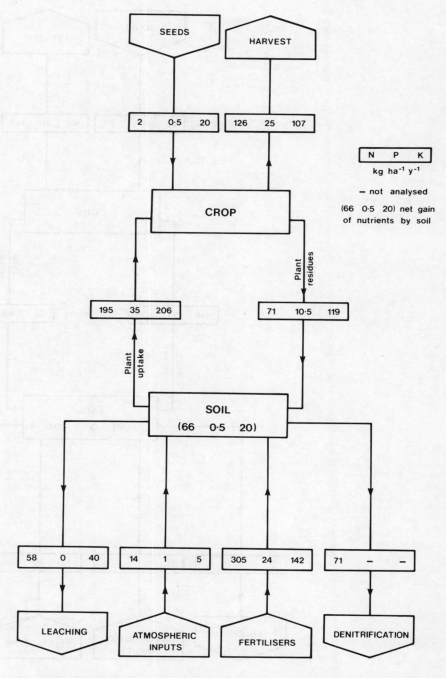

Fig. 8.15 The nutrient cycle of an intensive arable farm in which crop residues are retained on the land: potatoes – wheat – sugar beet rotation on clay soil in the Netherlands (data from Henkens, in Frissel 1978).

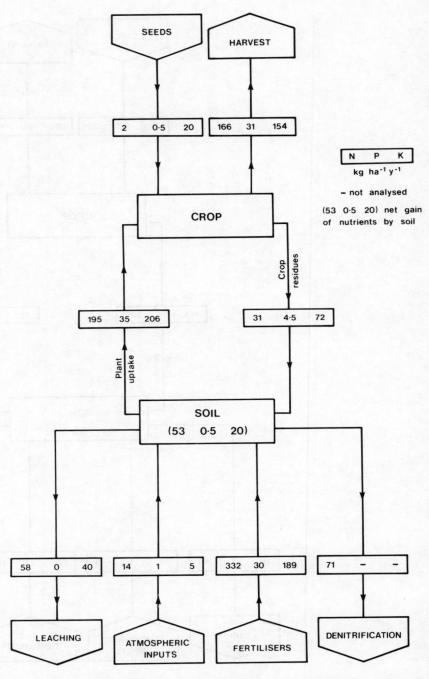

Fig. 8.16   The nutrient cycle of an intensive arable farm in which crop residues are removed from the land: potatoes – wheat – sugar beet rotation on clay soil in the Netherlands (data from Henkens, in Frissel 1978).

Table 8.5   Nutrient losses from intensive arable systems in the USA

	Nutrient losses (kg ha^{-1})			
	N	P	K	Ca
Removed in harvest	132	25	112	44
Leached	22	T	28	112
Eroded (continuous maize)	73	20	671	242
Eroded (rotation)	29	9	235	94

Source: Data from Brady 1974

for nutrient losses by leaching and erosion compared to such losses in grassland or forest systems. The lack of plant cover, relative reduction of rainfall interception and reduction of transpiration increases the soil moisture content, diminishes the rate at which nutrients are drawn up from the subsoil against the direction of leaching, and increases the amount of water available for drainage. As a consequence, leaching rates tend to be higher than under grass; Cooke (1977), for example, presents nitrogen economies for the UK, showing that 120 kt N was lost by leaching from arable crops, compared with 10 kt from temporary grasslands and 25 kt from permanent grassland (Table 8.6). Allowing for the relative areas of each crop type, this indicates leaching rates on arable land some 3–3.5 times greater than the rates on grassland. Similarly, lack of plant cover, loosening of the soil surface by tillage and related subsurface compaction increase the potential for erosion. As Massey and Jackson (1952) and Lal (1976) have shown, the eroded material tends to consist mainly of fine-grained particles, and especially organic matter. In addition to the nitrogen lost in organic matter, the removal of these material results in the preferential loss of adsorbed potassium and of phosphorus bound in precipitates on the soil particles and held in the colloidal structures. As a consequence, nutrient loss by erosion is generally greater under arable farming than under grasslands.

The major input of nutrients to arable systems occurs in the form of fertilisers; often 90 per cent or more of the total inputs are derived from fertilisers (Table 8.6). Because of this, and because of the high rates of nutrient loss, fertiliser regimes exert a fundamental control on nutrient cycling in arable systems. Increasing fertiliser applications increases also

Table 8.6   Nitrogen budgets for farming systems in England and Wales

	Arable	Temporary grass	Permanent grass
Area (million ha)*	4.95	1.69	4.64
Inputs (kt y^{-1})			
Fertiliser	370	220	200
Farmyard manure	30	—	500
Natural sources	370	450	540
Total	770	670	1240
Outputs (kt y^{-1})			
Harvesting	460	440	650
Leaching	120	10	25
Total	580	450	675

* From Lazenby and Down (1982)

Source: Data from Cooke 1977

the throughput of nutrients, for crop growth and nutrient removal in harvested material are encouraged, while leaching losses are also greater. Thus, as comparisons of the data in Figs. 8.16 and 8.17 show, losses increase more or less proportionately with the rate of fertiliser application. These data also indicate that as fertiliser inputs are increased, the efficiency of nutrient usage (i.e. the amount taken up by the crop as a percentage of the total input) tends to decline. Cooke (1977) suggests that in the case of nitrogen, efficiency is no better than 50 per cent, implying a considerable wastage of fertiliser inputs. Power (1981), reviewing nitrogen use by all cultivated crop-land in the USA, quotes an efficiency of only 36 per cent and argues that this shows the potential for improved efficiency in the use of nitrogen from all sources. Data presented by Frissel (1978) also suggest that in many arable systems, nitrogen surpluses occur, the inputs to the systems exceeding the outputs (Table 8.7). Although these data cannot be treated uncritically, since some components of the nutrient budget (e.g. inputs from weathering and losses by denitrification) have not always been measured, they do again indicate that the nutrient cycles in arable systems are not in a state of equilibrium. Given the possible environmental effects of accumulating nitrogen and phosphorus levels in the soil, this is clearly a matter for some concern.

Table 8.7   Nitrogen budgets for arable farming systems

Location	System	Fertiliser	Other inputs	Primary production	Other outputs	Balance
UK	Wheat	98	22	77	28	+15
Netherlands	Roots/wheat	305	16	126	129	+66
Netherlands	Roots/wheat	332	16	166	129	+53
USA	Maize	112	10	85	46	− 9
USA	Wheat	34	6	36	14	−10
USA	Soybean	0	133	90	41	+ 2
USA	Potatoes	168	6	80	94	0

Source: Data from Frissel 1978

8.6   PRODUCTIVITY IN INTENSIVE ARABLE SYSTEMS

The actual productivity of arable farming systems depends to a great extent upon management factors: in particular, the choice of crop and cropping sequence, fertiliser practice, irrigation policy, sowing dates, tillage systems and methods of crop protection. Crop management, however, operates within the framework of a relatively fixed environment, and it is the nature of this environmental framework which constrains – though it may not actively control – agricultural productivity. It does so not only by exerting direct limitations on crop growth – for example, through the control of moisture availability or radiation inputs – but also by influencing the day-to-day and longer-term decisions and responses of the farmer.

At a relatively simple level, it is therefore possible to detect relationships between environmental conditions and productivity of arable systems: notably, between crop yield and climate or soil conditions. Briggs (1981), for example, analysed yield data for spring barley in England and Wales and showed that yields were generally greater on brown earths,

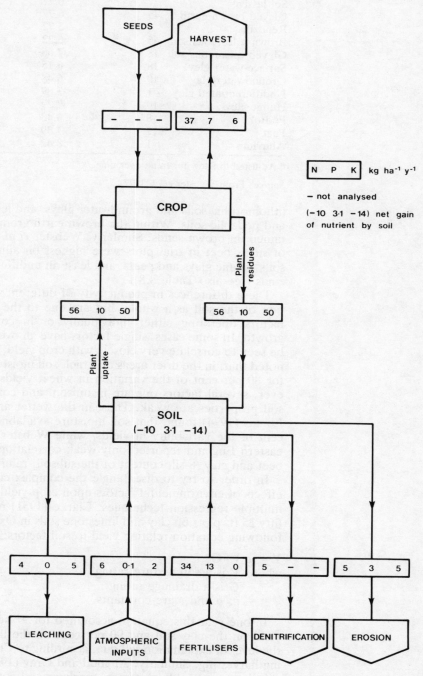

Fig. 8.17   The nutrient cycle in a wheat farm in central Kansas, USA (data from Thomas and Gilliam, in Frissel 1978).

Table 8.8   Yields of sugar beet (t ha^{-1}) on major soil groups in England

Major soil group	Number of plots	Mean yield*	Standard error of mean
Brown earth	91	6.41	0.14
Sol lessivé	27	6.77	0.28
Gleyed brown earth	45	6.77	0.21
Rendzina	8	5.72	0.53
Brown calcareous soil	65	6.22	0.17
Gleyed calcareous soil	18	7.66	0.34
Surface-water gley	18	6.15	0.35
Groundwater gley	19	6.48	0.33
Undifferentiated gley	1	3.89	1.50
Humic gley	19	7.29	0.35
Podsol	8	6.43	0.52
Peat	17	7.39	0.37
Alluvium	1	8.42	1.52

* Adjusted to allow for trend over time.

Source: From Webster *et al.* 1977

lithomorphic soils and groundwater gleys, and less on surface-water gleys and podzolic soils. Within the brown earth group, lowest yields were obtained on brown sands. Similarly, Webster *et al.* (1977) found that yields of sugar beet in trial plots were highest on alluvium, gleyed calcareous soils, humic gleys and peats, and least on undifferentiated gleys and rendzinas (see also Table 8.8).

These differences in productivity of different soil types are not a function of the soil as a whole, but are due to the influence of specific soil factors operating either individually or in consortium to affect crop growth. In some cases, single factors have an overriding influence and can be seen to correlate very closely with crop yield; Lewin and Lomas (1974) noted that, in the drier areas of Israel, soil moisture availability accounted for 80 per cent of the variation in wheat yields. More commonly, however, several factors operate in unison, and correlations with individual soil properties are weak. Thus, in the wetter areas of Israel, Lewin and Lomas (1974) found that soil moisture availability explained only 25 per cent of the variability in yields, while Webster *et al.* (1977) working in eastern England reported only weak correlations between yield of sugar beet and clay + silt content of the subsoil, rainfall and altitude.

In order to try to disentangle the complex cumulative and interactive effects of environmental factors upon crop yield, many studies have used multiple regression techniques. Clarke (1951) measured barley yields of fifty 25 ft^2 plots on clay and limestone soils in Oxfordshire and devised the following equation relating yield to soil factors:

$$Y = c \, V^a \, G^b \qquad\qquad [8.3]$$

where V is a soil texture rating
 G is a drainage rating
 a, b and c are constants.

In one year this equation accounted for 97 per cent of the variation in yield on the clay soil and 90 per cent on the limestone, but in the next the level of explanation fell due, according to Clarke, to bad weather during harvesting. Similarly, Allgood and Gray (1978) analysed the relationship between yields of wheat, sorghum and cotton lint and selected soil properties. In the case of wheat, the following relationship accounted for 83 per cent of the variation in yields on 35 plots:

$$Y = -31.38 - 7.80G + 5.25C - 0.33S - 0.09C^2 + 0.01S^2 + 0.21Ca + 1.55A - 0.26T + 7.23H \qquad [8.4]$$

where Y is yield (cwt/acre)
  G is gradient (%)
  C is clay content (%)
  S is sand content (%)
  Ca is calcium carbonate content (%)
  A is topsoil thickness (inches)
  T is solum thickness (inches)
  H is hydrogen ion concentration (pH).

Other studies have included both soil and climatic factors in similar analyses. Briggs (1981), for example, analysing data from 100 sites in England and Wales, related the following:

Y: yield of barley (kg ha^{-1})
S: subsoil sand content (%)
R: available rooting depth (cm)
M: potential summer soil moisture deficit (mm)
W: available water capacity to 60 cm (cm)
T: accumulated temperature during the growing season (°C)

thus:

$$Y = 8.790S - 0.012S^3 + 8.627R - 0.001R^3 - 20.530M + 613.870 \log W + 6.400T - 1579.711 \qquad [8.5]$$

The relationship, however, accounted for only 25 per cent of the variation in crop yield. In Canada, Williams *et al.* (1975) found stronger relationships between yields of wheat, barley and oats and environmental factors including rainfall, potential evapotranspiration, soil texture and topography and trend over time. Heapy *et al.* (1976a, b), using pooled data for 17 plot years of experiments, explained up to 57 per cent of the variation in barley yields in central Alberta on the basis of fertiliser input, nitrogen and phosphorus availability and soil moisture stress. Similarly, Haun (1975) analysed yield-environment relationships for potatoes.

The results of these analyses illustrate a number of interesting features. One of the main points to emerge is that in many cases very similar environmental factors are found to correlate with yield: in particular, soil texture, available water capacity, soil moisture stress and soil drainage status. All of these appear to be surrogates for the effect of moisture availability upon crop yield and illustrate the contention by Austin (1978) and G. W. Cooke (1979) that yields of cereals in Britain are constrained mainly by droughtiness and inadequate moisture availability. They also reflect the findings of the National Wheat Survey in 1956 (quoted in Mackney 1969) that yields of wheat were greatest on moderately to imperfectly drained soils (where water supply was adequate), and least on poorly drained profiles (where water supply was excessive) or on freely drained soils where droughtiness limited productivity. Briggs (1981) reported similar relationships for spring barley.

Nevertheless, the value of these equations is limited for a number of reasons. They are essentially empirical formulae, and are not based on any explicit, preconceived theory of environment-crop relationships. They cannot, therefore, be extended unmodified to new areas, crops or farming systems. They are also based in most cases on the assumption of an additive relationship between the dependent variable (yield) and the independent variables (the environmental factors). As we have noted, this is rarely valid and in most instances the various environmental factors interact in a complex manner.

Because of the inherent limitations of multiple regression approaches, it appears that a more rigorous understanding of productivity in arable farming systems depends upon the development of more deterministic models of the relationships between crop growth and environmental factors. A common starting point for such analyses is the relationship between crop growth and the supply of energy and water. In broad terms, it can be argued that the supply of solar radiation to the crop provides an upper limit to the amount of biomass that can be produced by photosynthesis. The crop's ability to make use of this energy and to build plant tissue, however, is constrained by the supply of water and its associated nutrients from the soil. Where moisture is lacking, actual biomass production tends to fall below the potential rate. One way of modelling this interaction between energy supply and water availability is through the assessment of actual evapotranspiration, which is itself largely a function of energy and moisture supply and which has been shown to correlate closely with biomass production.

One of the first to recognise this relationship was Thornthwaite (1948), and since then various attempts have been made to quantify the relationship for a wide range of natural and agricultural ecosystems. Leith and Box (1972), for example, defined the following equation:

$$P = 3000 \ (1 - e^{-0.0009695 \ (E\text{-}20)}) \qquad\qquad [8.6]$$

where P is net primary productivity (g m^{-2} y^{-1})

$\quad E$ is total evapotranspiration (mm)

$\quad$e is the base of the natural logarithm.

This relationship has been applied by, among others, Moss and Strath Davis (1982) in a study of agricultural productivity in southern Ontario, Canada. Using empirically determined coefficients to express soil suitability for each crop, they adjusted values of P to provide a measure of potential net primary productivity for each county. Estimated values of potential net primary productivity were correlated with measures of actual productivity (Fig. 8.18), though in absolute terms actual productivity was seen to exceed greatly the estimated potential productivity.

The discrepancy between actual and potential productivity found in this study demonstrates that environmental factors alone do not determine crop yield; management practices such as fertiliser inputs and irrigation clearly allow yields to be increased considerably. Consequently, general relationships between environmental conditions and crop yield are likely to provide only a partial picture of productivity in arable systems. As we saw in section 8.3, processes of crop growth are influenced by a range of different environmental factors at different growth stages, and also by characteristics of the crop itself (Fig. 8.19). The major determinants of yield are thus the crop characteristics, such as rate of seedling emergence, tillering and ear formation per unit area. As mentioned earlier, these characteristics are partly physiologically determined, but partly controlled also by management factors such as sowing density. In addition, it is likely that those soil factors which are correlated with individual yield components do not necessarily influence crop growth directly but operate indirectly through their effect upon the efficiency of management practices such as seedbed preparation and sowing.

If productivity in arable systems is to be predicted, therefore, much

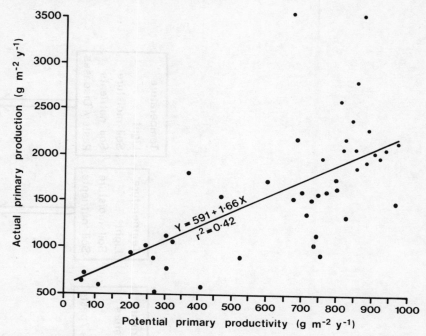

Fig. 8.18 Relationship between potential and actual primary productivity for southern Ontario agro-ecosystems (from Moss and Strath Davis 1982).

more complex, dynamic models are required. The development of such models is clearly a daunting task, and relatively few attempts have been successful. Possibly the most effective has been that devised by de Wit and his colleagues at Wageningen in the Netherlands (e.g. de Wit and Goudriaan 1974; Centre for World Food Studies 1980). This relates the yield of the crop to the available radiant energy, soil moisture and nutrient availability. From data on the geographical position of the site and the position of the sun, estimates are made of total irradiance for monthly periods. Taking account of the rate at which the crop assimilates CO_2 (a function of crop type, canopy cover and leaf characteristics), these irradiation values are then used to assess the gross CO_2 assimilation for each month. These assessments are then further modified to allow for water availability, which is itself predicted on the basis of a moisture budget equation:

$$\triangle S = P + I + VF - D - LF - ET \qquad [8.7]$$

where P is the effective precipitation (after allowing for interception)
   I is the effective irrigation
   VF is the input from groundwaters
   D is drainage
   LF is the lateral flow from the soil
   ET is the actual evapotranspiration.

The effect of nutrient availability is incorporated through the evaluation of four factors: the slope of the yield-nutrient uptake curve, the maximum potential yield of the crop (both these are crop-specific factors), the nutrient availability at zero fertiliser application rate (a soil factor) and the slope of the nutrient uptake-fertiliser application curve. On the basis of

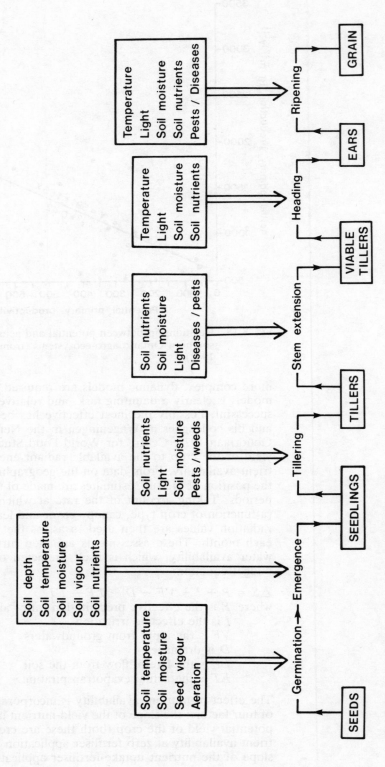

Fig. 8.19   Factors affecting the growth of cereal crops.

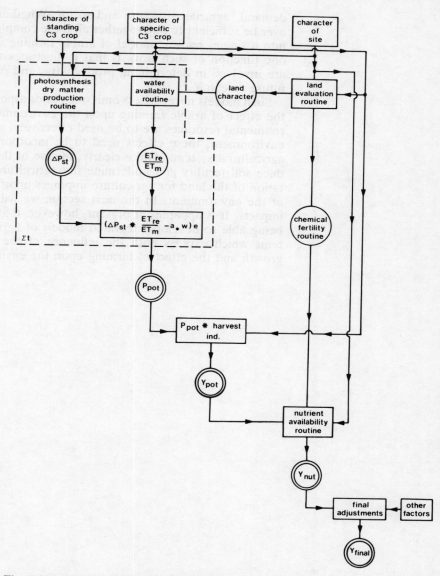

Fig. 8.20 The 'Wageningen model' of crop production (after Centre for World Food Studies 1980).

all these factors, the potential yield of the site is determined (Fig. 8.20).

It is apparent that this approach to modelling productivity of arable systems is highly complex. Even so, it represents a gross simplification of the real system and can provide only a general picture of the relationship between the environment and crop growth. Research is still proceeding, therefore, to develop the model further and to take account of other factors such as weed and disease effects, harvest and storage losses and crop protection. In addition, the model is designed to be linked to related socio-economic models which take account of government policies, public demand, agrarian structure and so on. Whether our understanding will ever be sufficient to pull together all these complex and conflicting factors into a single, realistic model of arable farming systems is debatable, but one function of such work is that it illustrates the range of factors that are involved in determining productivity, and provides a framework for future research.

Such models nevertheless omit one major aspect of the system: namely, the effect of arable farming upon the environment. In time, if the environmental resources are to be used effectively, yet without damaging the environment, these effects need to be incorporated into our models of agricultural systems. Quite clearly, misuse of the land can ultimately reduce soil fertility and undermine the agricultural systems. Also, exploitation of the land for agriculture impinges upon other uses and functions of the environment. In the next section we will examine some of these impacts. It is apparent at present, however, that we are a long way from being able to develop integrated models of agriculture-environment systems which express both the influence of the environment upon crop growth and the effect of farming upon the environment.

PART 4

ENVIRONMENTAL IMPACT OF MODERN AGRICULTURE

9 HYDROLOGICAL IMPACTS OF AGRICULTURE

9.1 INTRODUCTION

The high yields characteristic of modern, intensive farming systems in temperate areas reflect man's ability to modify the agro-ecosystem in such a way as to remove or diminish natural limitations upon productivity, and to provide a more favourable environment for crop growth. As we have seen in earlier chapters, these modifications include the introduction of new species or varieties of crop plants and livestock, the elimination of competing plants and other organisms by the use of pesticides and the removal of habitats, the use of high levels of fertilisers, manipulation of soil structural conditions by tillage, and control of soil moisture by irrigation and drainage. Inevitably, these practices have a fundamental influence on the wider environment. We have already discussed the ways in which they affect the soil, and have noted that in many cases continuous intensive farming leads to long-term, accumulative changes in factors such as soil structure, organic matter content, soil nutrient contents and organic activity. In later chapters we will consider the ecological implications of such farming practices. It is also clear, however, that many of the procedures involved in modern farming have major implications for hydrological processes. In this chapter we will examine some of these hydrological impacts and assess the extent to which modern farming systems pose a threat to water resources.

It is common to represent the hydrological cycle as a closed system comprising major storage components such as the atmosphere, oceans, ice caps, soil, groundwaters and streams. In the context of agricultural impacts, however, we can focus attention upon a limited part of this cycle, which we may refer to rather loosely as the terrestrial hydrological cycle. This consists of four main components – the soil, groundwaters, streams and seas – linked by various transfers, such as drainage, throughflow, runoff and seepage. The main inputs to this system are by precipitation, though small quantities of water may also be derived from deep (fossil) aquifers. The main losses occur by evapotranspiration and, to a much lesser extent, by deep percolation (Fig. 9.1).

This part of the hydrological cycle concerns us because it is almost all vulnerable to the effects of agriculture, in particular through the impact of farming practices upon the transfer mechanisms. These impacts influence not only the rate of water flow through the system, but also its rout-

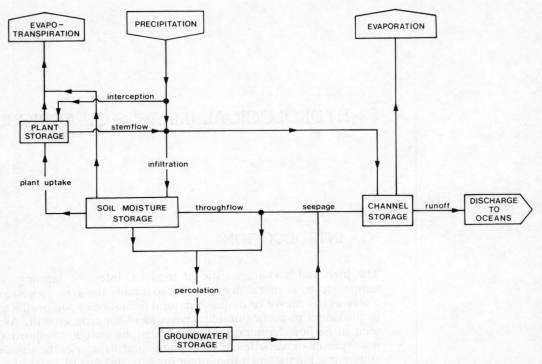

Fig. 9.1   The terrestrial hydrological cycle.

ing and its quality. Thus, to a considerable extent, agriculture may control the quantity of water intercepted by the vegetation and the amount reaching the soil surface. It also affects the routing of this water at the surface, by influencing the infiltration capacity of the soil and hence the volume of surface runoff. Within the soil, the rates of downward percolation and lateral flow are affected to a large degree by drainage and tillage practices, whilst the quality of the water percolating into the bedrocks or escaping by seepage back to the surface is influenced by factors such as fertiliser regimes, pesticide usage and livestock management. Through these various effects, agriculture also influences both stream discharge and stream water quality. Additionally, the character of the crops and the methods of soil tillage control in part the rate of evapotranspiration. Ultimately, many of these effects are transferred to coastal and marine waters, for changes in the hydrological processes operating on the land result in changes in erosion rates and sediment transport, while agricultural pollutants are frequently carried downstream into the sea. It is apparent, therefore, that agriculture may have far-reaching impacts upon hydrological processes and conditions (Fig. 9.2), and, because water represents a vital resource for man and an important habitat for wildlife, concern has been increasing in recent years about the potential effects of modern farming systems (e.g. Gasser 1980; IAHS 1980). To assess these effects it is useful to consider separately the impacts upon the main route-ways for water flow within the system, and the quality of the water as it passes through the system.

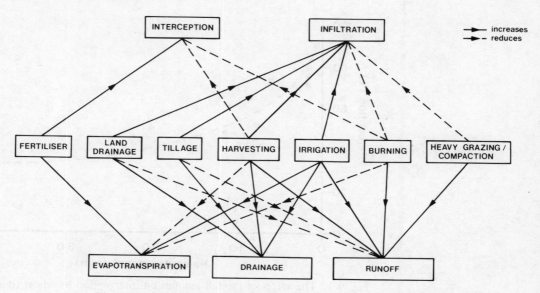

Fig. 9.2 The effects of farming practices on the components of the hydrological cycle.

9.2 INTERCEPTION

Under natural conditions, the vegetation cover intercepts a considerable proportion of precipitation during any individual storm. The intercepted rainfall is held by surface tension on the leaves and stems or trunks, collects in crevices or depressions on the plant surface, and may to a very small degree be absorbed through the stomata into the plant itself. Much of the moisture held on the surface of the plants is subsequently lost by evaporation, partly during the storm but mainly after rainfall has ceased. A proportion, however, ultimately reaches the ground, either by leaf-drip or by flow down the stem. This rerouting results in delayed inputs of rainfall to the soil, and also redistributes the precipitation at the surface to produce areas of concentration, particularly around the base of the stem.

The amount of precipitation intercepted by the vegetation depends upon the characteristics of the storm and of the crop, as well as factors such as antecedent moisture conditions, wind speed and post-rainfall temperature and humidity. In general, the gross amount of rainfall intercepted by the vegetation increases with rainfall amount, though the proportion falls (Fig. 9.3). Many attempts to generalise these relationships have been made, and equations often take the form:

$$I = a + bP^n \qquad [9.1]$$

where I is gross rainfall interception (mm)

P is rainfall amount (mm)

a, b and n are constants which depend upon crop characteristics.

Interception also varies with precipitation intensity. For storms of equal rainfall amount but different duration (and thus different intensity) both

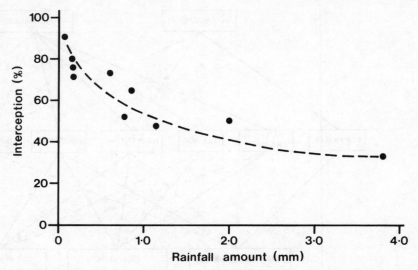

Fig. 9.3   The effect of rainfall amount on interception by wheat (data from Gray 1973).

the amount of rainfall intercepted and the proportion this represents of the total storm precipitation, decrease with increasing intensity. This is due in part to the increased kinetic energy of the raindrops which are able to splash off the leaves and also shake loose moisture already held on the surface. It is also due to the rapid wetting of the vegetation, producing water films thick enough to overcome the surface tension effects and allowing stem-flow and leaf-drip to occur before evaporation can take place.

A third characteristic of the storm controlling interception is the type of precipitation. Interception from snowfall is almost invariably greater than that from rainfall of the same water equivalent. Snow adheres more effectively to the plant surfaces and also is more cohesive than water. As a result thick layers of snow can accumulate, even on vertical surfaces. Some of this snow may be blown or shaken off the leaves, or ultimately melts and reaches the soil as leaf-drip or stem-flow, albeit after a considerable delay, but much is lost by sublimation (direct evaporation of the snow). The relative rates of melting and sublimation depend amongst other things on air temperature and humidity after snowfall; sublimation, for example, is generally at a maximum in dry, relatively cool air, while melting is encouraged by warm, moist conditions.

The main vegetation factors affecting interception are canopy density, leaf shape and orientation, and vegetation height. Increasing the canopy density increases the surface area of the plants available to intercept rainfall and to store the moisture retained. As canopy density increases, therefore, both the gross amount and the percentage of rainfall which is intercepted increase. This effect continues well beyond an actual ground cover of 100 per cent, for retention of rainfall involves secondary interception of leaf-drip from higher parts of the canopy by lower strata. In this context the concept of leaf area index (LAI) is useful. This is the ratio of the surface area of the vegetation to that of the ground beneath. Relatively few studies of interception by agricultural crops have been carried out, but by analogy with the numerous studies on forests (summarised, for example, by Courtney 1981), it seems that interception increases up to a leaf area index of at least three, before stabilising.

At constant LAI, interception by vegetation varies markedly with leaf shape and orientation. Plants with wide, flat leaves tend to retain more moisture than those with acicular leaves, simply because they provide a greater area of horizontal surfaces and a greater volume of depressions in which to trap water. Similarly, interception is increased in vegetation with leaves aligned horizontally, and is diminished where leaves are oriented more vertically. Vegetation height is also important, in part because tall vegetation tends to be multi-layered but, in addition, because it affects the rate of loss from the vegetation surface by evaporation. All else being equal, evaporation is more effective from tall vegetation because of the tendency for higher wind speeds and lower relative humidity at height.

As has been noted, few studies of interception by agricultural crops have been published, not least because of the practical difficulties in measuring interception by low-standing plants. Nevertheless, it is apparent that agriculture has a profound effect upon interception rates: the choice of crops, sowing rates, weeding, harvesting and residue management all determine fundamentally the character of the vegetation cover and thus its capacity to intercept rainfall. These effects are likely to be most significant in arable systems where, for long periods of the year, the soil may be bare, with the result that at that time interception is negligible. Moreover, these periods often coincide with the winter rainfall season, with the result that the effects on interception are accentuated. Some indication of the magnitude of differences in interception rates under different management systems is given in Fig. 9.4, which compares effective rainfall (rainfall − interception) under maize, soybean and oats. Similar results were found in the USA by Lull (1964) and in Russia by Kontorshchikov and Eremina (1963); as the data in Table 9.1 illustrate, however, interception

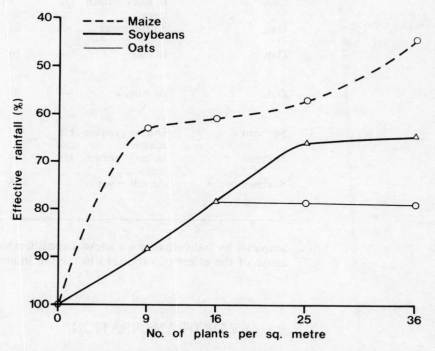

Fig. 9.4 Effective rainfall (rainfall − interception) for three agricultural crops (after Ward 1975).

Table 9.1 Rainfall interception by agricultural crops

Crop	Description	Rainfall (mm)	Interception (%)	Source
Alfalfa	During growing season	275	36	Lull (1964)
Alfalfa	In early growth stage	275	22	Lull (1964)
Blue-grass	In Missouri	—	17	Musgrave (1938)
Heather	In Aberdeenshire, Scotland	—	35–66	Aranda and Coutts (1963)
Mixed grasses	*Avena/Stipa/Bromus*	80	26	Kittredge (1973)
Bluestem	1-hour storm	25	47	Clark (1940)
Western wheatgrass	30-minute storm	6	50	Clark (1940)
Prairie cordgrass	30-minute storm	12	55	Clark (1940)
Prairie cordgrass	30-minute storm	3	72	Clark (1940)
Maize	During growth season	181	16	Lull (1964)
Maize	In early growth stage	181	3	Lull (1964)
Maize	At full cover	—	40–50	Wollny (in Ward 1975)
Oats	During growing season	171	7	Lull (1964)
Oats	In early growth stage	171	3	Lull (1964)
Oats	At full cover	—	23	Wollny (in Ward 1975)
Oats	In July	—	16	Kontorshchikov and Eremina (1963)
Oats	In August	—	23	Kontorshchikov and Eremina (1963)
Soybean	During growing season	158	15	Lull (1964)
Soybean	In early growth stage	158	9	Lull (1964)
Soybean	At full cover	—	35	Kontorshchikov and Eremina (1963)

amounts by individual crops show a considerable range, presumably because of the effect of variations in rainfall intensity and crop cover.

9.3   EVAPOTRANSPIRATION

Evapotranspiration represents one of the main processes of loss from the hydrological system. As we saw in Chapter 4, the potential for evapotranspiration depends primarily upon environmental conditions, including air temperature, atmospheric pressure, air humidity, wind speed and turbulence. Potential evapotranspiration (PT) thus varies over both time and space in response to variations in meteorological conditions.

Under conditions of constant water supply, the actual rate of evapotranspiration from field soils closely approximates to potential evapotranspiration (as measured, for example, by evaporation from an open water body). Plants act for the most part as passive conducting bodies for water moving from the soil to the atmosphere in response to the atmospheric demand for water. Where moisture supply from the soil is limited, however, actual evapotranspiration (ET) falls below the potential rate due to the inability of the plant roots to extract sufficient water from the soil, and the consequent closure of the stomata in the leaves (see Ch. 4).

Various attempts have been made to define general relationships between actual and potential evapotranspiration, on the basis for example of soil moisture content and available water capacity (Fig. 9.5). It is apparent, however, that the ratio of actual to potential evapotranspiration (ET/PT) depends in detail upon crop conditions, such as the stomatal characteristics of the plants, leaf area and rooting depth and volume, as well as upon soil conditions such as texture and structure (see Barry

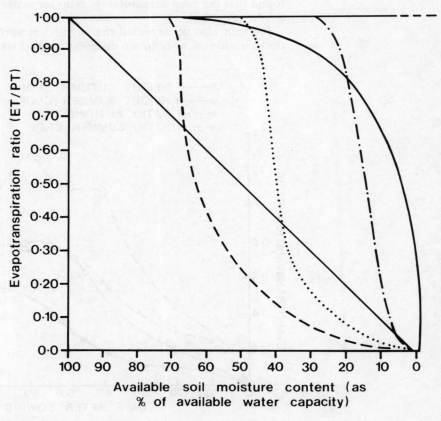

Fig. 9.5 General relationships between soil moisture content and evapotranspiration rates (after Baier and Robertson 1967).

1969). As a result, variations in crop cover and soil management may significantly influence actual rates of evapotranspiration.

Marked differences in evapotranspiration rates occur throughout the growing season due to changes in crop height, leaf area and root distribution during growth. These differences are, of course, superimposed upon the seasonal and shorter-term variations in evapotranspiration resulting from changes in meteorological conditions and soil moisture availability, and are thus not always apparent. By expressing actual evapotranspiration as a proportion of potential evapotranspiration, however, these external factors can be allowed for. Denmead and Shaw (1959), for example, argued that the ratio of actual to potential evapotranspiration for maize increased sigmoidally, from 0.36 at planting to 0.81 at silking (i.e. the development of the tassle); thereafter, the ratio declined. More recently, data provided by Lomas *et al*. (1974) have confirmed this general relationship, though they also showed that the ratio varied significantly between autumn- and spring-planted crops. The average ET/PT value during the growing season for spring-planted maize was 0.62; that for autumn-planted maize during the same growing period was 0.85. This difference presumably reflects the greater height and leaf area of the autumn-sown crop early in the subsequent spring. Similarly, it has been shown by Stanhill (1965) that the amount of water transpired by alfalfa increases with increasing crop height, whilst Lomas *et al*. (1974) found that the evapotranspiration ratio for maize was correlated with crop height (Fig. 9.6).

It might also be expected that significant variations in rates of evapotranspiration occur between different crops. Plants with a higher leaf area

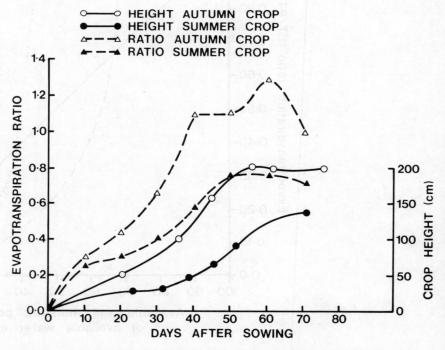

Fig. 9.6   Changes in the evapotranspiration ratio and crop height of autumn- and summer-sown maize in Israel (from Lomas *et al*. 1974).

or a longer growing season may be expected to use more water than plants with a small surface area of leaves or with a very restricted growth period. Penman (1963), reviewing a large number of early studies, nevertheless detected few differences between different crops, though he noted that comparisons were made difficult by the inherent variations in experimental design both within and between individual studies. Blad and Rosenberg (1974), however, have shown that broad-leaved plants tend to transpire more water than do grasses, and Rosenberg (1974) argues that taller vegetation will have higher rates of evapotranspiration than short crops because of the more efficient energy exchange between taller plants and the ambient air. Under field conditions such variations tend to be masked, it appears, by the host of other factors which affect evapotranspiration – for example, planting density, soil treatment and rooting characteristics.

The effects of plant density may be marked. The effects operate in a variety of ways. Increasing plant density increases the total leaf area index and thus provides a greater surface area from which transpiration may occur. Brun et al. (1972) have shown that, in soybean and sorghum fields, the ratio of actual to potential evapotranspiration is correlated with leaf area index, rising from about 0.50 at a leaf area index of 2, to 0.95 at a leaf area index of 4. Thus, it might be anticipated that higher planting densities will result in proportionally higher rates of evapotranspiration. Shading also increases with increasing plant densities, however, so that energy receipts by individual plants are reduced. As a consequence, the rate of increase of ET (or ET/PT) with increasing plant density levels off beyond a LAI of about four to five, and the evapotranspiration rate of individual plants falls as plant density increases. These effects are clearly illustrated by Espinoza (1979). He compared evapotranspiration from maize grown at densities of 20,000, 40,000, 60,000 and 80,000 plants per hectare in the Cerrado district of Brazil. Average rates of daily evapotranspiration were similar at all four densities during the first 70 days after planting, but by the end of the growing season (140 days after sowing) showed marked differences between the four treatments. Total water losses after 140 days were 369.3 mm, 339.9 mm, 443.8 mm and 531.1 mm for the four plots respectively. This represented rates of water use of 180, 104, 77 and 64 litres per plant respectively.

Similar effects are produced by variations in the spacing and orientation of the crop. Because shading influences energy exchanges between the plant and the atmosphere, and because changes in canopy roughness affect turbulence and wind speed, differences in planting geometry result in marked variations in evapotranspiration. The effects are complicated, however, by interactions with atmospheric conditions, such as wind direction and solar angle and azimuth (compass direction). McCauley et al. (1978) illustrated the complexity of these interactions in a study of evapotranspiration by Spanish peanuts (Arachis hypogaea) and grain sorghum (Sorghum bicolor). They compared crops planted in north–south and east–west rows, spaced at narrow (30 cm) and wide (90 cm) intervals. No significant differences in evapotranspiration were detected when potential evapotranspiration was low (i.e. under cool or cloudy conditions). During periods of high potential evapotranspiration, however, marked differences were seen. With a south wind, radiation was less in the narrowly spaced N–S rows due, apparently, to the higher aerodynamic roughness of this plot, causing greater energy losses by advection and lower net radiation inputs.

As this example indicates, management practices which modify the morphology of the crop canopy may cause differences in evapotranspiration rates. Similar differences may also be caused by variations in soil management (*e.g.* Ekern *et al.*, 1967). Increased applications of nitrate fertilisers, for example, may result in higher growth rates and higher rates of evapotranspiration (e.g. Singh 1978), though the effect probably varies according to the placement of the fertiliser and the effect on root distribution. Irrigation also increases evapotranspiration, both by improving growth and by maintaining soil moisture contents closer to field capacity. Data from North Dakota quoted by Penman (1963), for example, show increases of 30–90 per cent in water use by irrigated maize compared with unirrigated plots. In addition, tillage practices appear to affect evapotranspiration through their influence on root growth and moisture retention. In field trials at Gottingen in Germany, Ehlers (1976) found no significant differences between tilled and untilled plots for either winter wheat or sugar beet. Weatherly and Dane (1979), however, showed that moisture uptake by maize was greater under conventional tillage systems than under direct drilling either with, or without, subsoiling. Similarly, Massee and Cary (1978) demonstrated that ET may be reduced by maintaining mulches on a fallow soil or by chemical treatments to reduce the capillary rise of water to the surface.

Variations in the input of water to, and the loss of water from, the ground surface, clearly have major significance for all parts of the terrestrial hydrological cycle. It is apparent from what has been said that these inputs and outputs are subject to influence by agricultural practices. A general indication of the nature of these influences is given in Fig. 9.2 (p. 251). As this shows, many of the effects of agriculture are potentially counteractive, in that individual practices may decrease both inputs and outputs. On the other hand, processes such as vegetation removal, by reducing interception and evapotranspiration, lead to marked increases in the amount of water within the system, with consequent effects on processes such as drainage, runoff and leaching. In the next sections, we will consider the movement of water through the terrestrial hydrological cycle and examine the ways in which agricultural activity impinges upon the processes involved.

9.4 INFILTRATION

9.4.1 Infiltration processes

The precipitation which reaches the soil surface is directed along one of two routes: it either drains into the soil and enters the soil moisture store, or it remains on the surface and contributes to surface storage and flow. The relative proportions of precipitation following each of these routes depend to a large extent upon the infiltration capacity of the soil. In general, only that portion of the rainfall not able to enter the soil through infiltration remains at the surface, though some of the water held in surface depressions may infiltrate later. As a consequence of control on surface routing of water, however, infiltration exerts a fundamental influence on hydrological processes, and agricultural effects on infiltration assume considerable importance.

Infiltration involves the movement of water across the air-soil interface into the pore spaces of the soil. The ability of the soil to absorb water in this way is known as its infiltration capacity. This depends in part upon relatively permanent properties of the soil, such as the texture and structure which affect the pore volume of the soil, and in general it is seen that light textured (i.e. sandy) or well structured soils have higher infiltration capacities than heavy or poorly structured soils. During an intense storm, however, the infiltration capacity tends to decline as the surface layers of the soil become saturated; water can then infiltrate only as the surface pores empty by drainage to depth. The rate at which drainage occurs is largely a function of pore diameter, connectivity and tortuosity, and these factors determine the longer-term, saturated infiltration capacity of the soil. In addition, however, infiltration capacity may decline during rainfall due both to expansion of clay particles as they absorb water and also to clogging of soil pores by slaking of particles from aggregates. Thus the infiltration capacity is not constant for any single soil type and may vary according to storm characteristics such as rainfall intensity and duration.

For the same reason, marked spatial variations are seen in infiltration rates during a storm. On footslopes, for example, infiltration rates are often limited by the saturation of the soil due to downslope movement of water in the soil. This results in the preferential development of surface storage and runoff in these areas, while upslope the soil is still able to accept rainfall.

9.4.2 Agricultural influences on infiltration

Agriculture affects infiltration processes in a wide variety of ways. Through its control on vegetation cover, it influences interception and thus the amount of water available for infiltration. Tillage generally loosens the soil surface and increases the volume and connectivity of pores. Compaction by machinery or animals leads to closure of pores and reduced infiltration capacities. Subsurface compaction (e.g. the development of a plough-pan) encourages water retention in the topsoil, leading to reduced rates of saturated infiltration, while subsoiling, moling or pipe drainage speed up the removal of water and increase the saturated infiltration capacity. In the longer term the effects of cultivation, crop removal and manurial treatments on soil organic matter contents and aggregate stability influence the structural characteristics of the soil and consequently affect infiltration capacity.

Numerous studies have illustrated these diverse effects. Chow (1967), for example, presented representative infiltration curves for various cropping systems showing that, in general, infiltration capacities declined in the order: old permanent pasture – moderately grazed pasture – tilled arable land – heavily grazed pasture – ploughed arable land – bare arable land (Fig. 9.7). The effects of tillage have also been demonstrated by many experiments on direct drilling and ploughing. As we noted in Chapter 3, direct drilling characteristically leads to higher bulk densities in the surface layers of the soil, but lower bulk densities at depth (e.g. Selim and Voth 1980), but in the longer term the effects are often reversed. Improved pore connectivity due to deeper and more extensive rooting, a more extensive network of earthworm channels, and improved structural stability, together lead to a progressive increase in infiltration capacity on directly drilled soils (e.g. Goss et al. 1978).

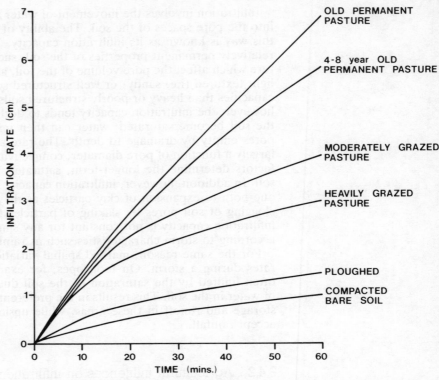

Fig. 9.7   Infiltration rates under different cropping systems (after Holtan and
          Kirkpatrick 1950).

Similar effects are displayed by other management techniques which
likewise improve soil structural conditions. The application of soil stabil-
isers commonly increases infiltration capacity and reduces surface runoff;
Hartmann *et al.* (1981), for example, found a 94 per cent increase in infiltration capacity as a result of treating arable soils with polyurea soil
conditioners. Application of FYM or retention of crop residues also increases infiltration capacity (e.g. Aarstad and Miller 1978). On the other
hand, straw burning may reduce infiltration capacity, not only because it
leads to a loss of organic matter but also because ash is washed into the
surface pore spaces (Biederbeck *et al.* 1980). For similar reasons the
choice of cropping system significantly influences infiltration capacity.
Crops which leave the soil bare for long periods encourage surface slaking
and pore sealing; those which add large amounts of organic matter to the
soil, produce extensive root systems, and give a good surface protection
against rain-splash all increase infiltration capacity. Zabek (1979), for example, found that infiltration capacities were least on bare fallow soil and
greatest under lucerne or an annual crop rotation. The effect varied with
soil type, however, and infiltration capacities were consistently two to
three times higher on sandy soils than on loams.

As the data in Fig. 9.7 show, infiltration rates on grassland are generally higher than those on arable land. Nevertheless, marked variations
occur in grassland soils due, in particular, to differences in sward age,
composition and grazing intensity. In general, infiltration capacity increases as the pasture gets older due to the accumulation of an organic
mat at the surface and the development of an extensive root system and

stable soil structure. Different types of grass also have different effects, partly because of the way in which they affect soil structure, but also because of their varied resistance to animal trampling (Gifford and Hawkins 1978).

The most important factor controlling infiltration capacity in grassland is grazing intensity. Reviewing the hydrological effects of grazing, Gifford and Hawkins (1978) concluded that light-moderate grazing may reduce infiltration capacities by about 25 per cent compared with ungrazed pasture, while under heavy grazing infiltration capacities fall to about 50 per cent (Fig. 9.8). Within a field, these effects show marked spatial varia-

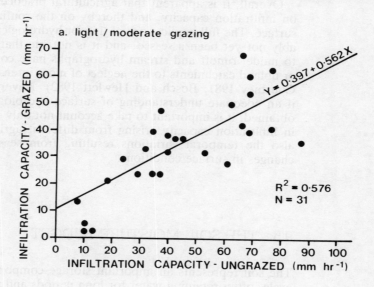

a. light / moderate grazing

$Y = 0.397 + 0.562 X$

$R^2 = 0.576$
$N = 31$

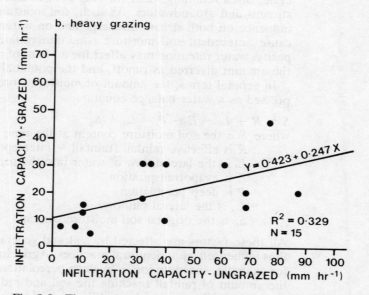

b. heavy grazing

$Y = 0.423 + 0.247 X$

$R^2 = 0.329$
$N = 15$

Fig. 9.8   The relationship between infiltration capacities on ungrazed and grazed pastures under light/moderate and heavy grazing (after Gifford and Hawkins 1978).

bility as a result of the behavioural patterns of grazing animals. Typically, infiltration capacities fall to their lowest levels in gateways, around feeding or water troughs, and in the trackways created by the animals as they move around the field. Briggs (1978), for example, comparing infiltration capacities in different areas of a single field on clay soils, showed that in the most heavily trampled areas infiltration rates were zero, whereas in the least trampled, grazing areas the infiltration capacity was 7.6 cm h^{-1} Similarly, Selby (1972) noted that in New Zealand, grazing may cause severe compaction which reduces infiltration capacity, promotes surface runoff and encourages soil erosion.

Overall, it is apparent that agricultural practices have a major impact on infiltration capacity, and thereby on the routing of water at the soil surface. The full implications of this for hydrological processes have probably not yet been assessed, and it is notable that many of the attempts to model runoff and stream hydrographs have concentrated on forested or upland catchments to the neglect of more intensively farmed land (e.g. Courtney 1981; Bosch and Hewlett 1982). Nevertheless, it is clear that if an adequate understanding of surface hydrological processes is to be obtained, it is important to take account not only of the spatial variations in infiltration capacity arising from different agricultural land uses, but also the temporal variations resulting from seasonal and shorter-term changes in surface conditions.

9.5   THE SOIL MOISTURE BUDGET

The soil represents an important storage component in the hydrological cycle, often retaining water for long periods and releasing it gradually to streams and groundwaters. As such, soil moisture storage has a marked influence on both stream discharge and on water quality. Moreover, because antecedent soil moisture conditions influence the infiltration capacity, water retention may affect the amount of rainfall entering the soil, the amount diverted as runoff, and the potential for soil erosion.

In general terms, the amount of moisture stored in the soil can be expressed as a water balance equation:

$$S = R + L_i - E - P - L_o + S_o \qquad\qquad [9.2]$$

where S is the soil moisture content at the time under consideration

R is effective rainfall (rainfall − interception)

L_i is the lateral flow of water into the soil

E is evapotranspiration

P is deep percolation

L_o is the lateral outflow

S_o is the original soil moisture content.

All these factors are affected by soil and vegetation conditions, and are thus influenced by agricultural practices. Agricultural drainage and tillage, for example, speed up losses by deep percolation; crop removal increases the amount of rainfall reaching the soil and reduces evapotranspiration; changes in soil structure due to tillage, residue management, crop rotation or use of FYM affect rates of percolation, evapotranspiration and lateral flow. Several studies have illustrated these effects, although it is apparent that, because of the temporal and spatial variability of all the factors involved, it is not easy to quantify the overall impact of agriculture on the soil moisture budget.

One of the main controls is exerted by crop cover, for this influences both inputs to and losses from the soil moisture store. Russell (1973) quotes early work by R. K. Schofield showing that after a prolonged drought fallow soils retain more water at all depths than does land carrying barley – a reflection of the effect of crop transpiration. Similarly, French (1978), reported that soil moisture storage at sowing was 28 mm greater in soils subjected to a preceding 9–10 month fallow period than in soils receiving preparatory cultivations for the preceding 2 months. The effect, however, was seen to vary in relation to soil type: differences were minimal in coarse-textured soils, but considerable in fine-textured soils.

Comparable effects occur as a result of weed removal. Weeding reduces water losses by evapotranspiration and increases soil moisture contents. Lavake and Wiese (1979), for example, found that, on a clay-loam soil in the southern Great Plains of the USA, delaying weeding by 17–24 days after emergence significantly reduced the average soil moisture content in the top 120 cm of the soil. Retaining crop residues also reduces evapotranspiration and helps conserve soil moisture. Unger (1978) found that soil moisture contents at planting time increased from 12.3 to 21.4 cm as a result of retaining 12 t ha^{-1} wheat-straw stubble. As a consequence, in the subsequent growing period, evapotranspiration also increased and water-use efficiency (the amount of moisture removed from the soil per unit volume) rose from 55.6 to 115 kg ha.cm^{-1}

As this last example indicates, agricultural effects on water storage are complex, and gains at one stage in the year may be offset by equal or larger losses at other periods. Thus, farming practices do not necessarily affect the total quantity of water held in the soil in the long term (the average soil moisture content), but they do affect the pattern of retention throughout the year. In addition, numerous interactive effects complicate the picture. Techniques which conserve moisture may encourage greater crop growth and therefore increased uptake of soil water by the plants. Practices which increase evapotranspiration may, as a result, reduce losses by lateral outflow or deep percolation. Thus Keuren et al. (1979) found that summer-grazed pastures had higher rates of evapotranspiration than winter-grazed plots (possibly because of improved sward recovery and growth) but less surface runoff and subsurface outflow.

The overall effects of cropping practice are thus best illustrated by continuous monitoring of plots to show either net changes in soil moisture content over the year, or variations in the inputs and outputs. Such an approach was adopted by McGowan and Williams (1980a, b) in a study of the water balance of an agricultural catchment at Kingston Brook, Rempstone in Nottinghamshire. Potential evapotranspiration was calculated from standard daily weather records using the Penman formula (MAFF 1967); soil moisture contents were measured using a neutron probe; runoff was monitored at the outflow from the catchment; and rainfall was measured at initially eight (and subsequently nine) sites. The catchment comprised a range of soil types, including clays (Ragdale, Hanslope and Worcester series) and sands (Newport and Arrow series). Sites in woodland, grassland and spring barley were analysed.

Monitoring of soil water contents at different depths throughout the growing season showed the way in which root extension into the subsoil caused progressively deeper drying of the soil. The period during which

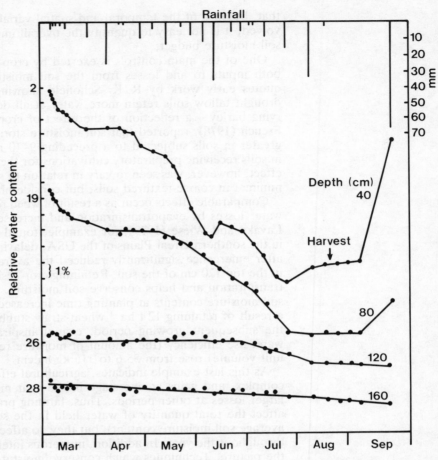

Fig. 9.9   Soil moisture contents at selected depths in a sand soil (Wick series)
under barley at Rempstone, Nottinghamshire, 1972 (from McGowan
and Williams 1980a).

roots reached any single layer in the soil was marked by a relatively sudden reduction in soil water content, as losses by evapotranspiration began
to have an effect. Superimposed upon this seasonal pattern of evapotranspiration at each depth were shorter-term changes caused by rainfall and,
in the case of barley, by harvesting (Fig. 9.9). McGowan and Williams
(1980b) also plotted the maximum depth at which drying could be detected throughout the growing season and compared the patterns in different soils and under different crops. Discrepancies between the depths
of drying in the sand and clay soils were small during relatively wet years,
but during 1971 and 1972 which were both affected by prolonged drought,
drying was found to proceed more deeply in the sand soil (Fig. 9.10).

Water deficits under the two crops varied more or less in sympathy
throughout the year, though moisture contents were generally lower (i.e.
deficits greater) under the grass crop than under barley (Fig. 9.11). The
differences were small, however, for whilst evapotranspiration from
spring barley tended to lag behind that from grass in the early part of the
growing season (when leaf cover in the cereal crop was incomplete), later
in the summer the greater height and rougher canopy of the barley crop

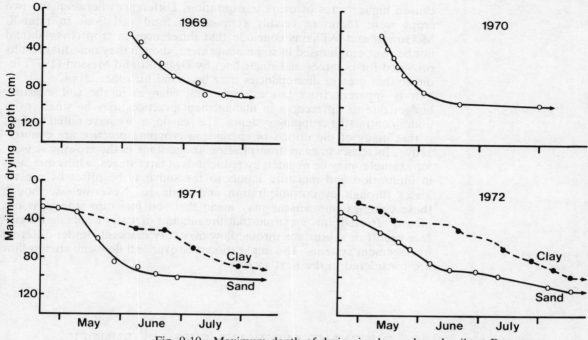

Fig. 9.10   Maximum depth of drying in clay and sand soils at Rempstone, Nottinghamshire (from McGowan and Williams 1980b).

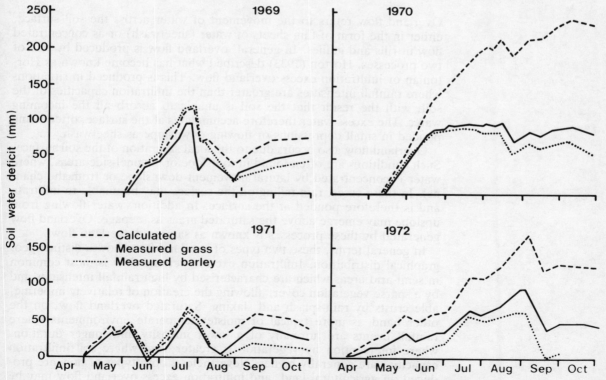

Fig. 9.11   Soil moisture deficits under pasture and barley at Rempstone, Nottinghamshire (from McGowan and Williams 1980b).

caused higher rates of evapotranspiration. Differences between the two crops were therefore readily eclipsed by local variations in rainfall. McGowan and Williams conclude that differences in crop cover should not be over-emphasised in soil management, though they note that results provided for potatoes and sugar beet by Draycott and Messon (1977) indicate that greater discrepancies may be found for other crops.

It is apparent from this example that changes in the soil moisture budget due to differences in management practices may be small, even under contrasting cropping systems. The reason, as we have noted earlier, is that many of the effects of changes in cropping practice are counter-active. Increases in evapotranspiration at one stage in the growing season, for example, may be negated by reductions at later stages, whilst increases in infiltration and moisture inputs to the soil may be offset by greater losses through evapotranspiration and drainage. Nevertheless, though these complex interactions may mean that soil moisture storage is not greatly affected, it is apparent that the amount of water available for surface runoff or subsurface throughflow may vary markedly under different management systems. The implications for overland flow and stream flow are considered in the next section.

9.6   OVERLAND FLOW AND SURFACE RUNOFF

9.6.1   The generation and character of overland flow

Overland flow refers to the movement of water across the soil surface, either in the form of thin sheets of water (sheetwash) or as concentrated flow in rills and gullies. In general, overland flow is produced by one of two processes. Horton (1933) described what has become known as Hortonian or infiltration excess overland flow. This is produced in situations where rainfall intensities are greater than the infiltration capacities of the soil, with the result that the soil is unable to absorb all the incoming water. The excess water therefore accumulates at the surface, often being ponded in small depressions or flowing downslope as sheetwash.

Overland flow also occurs due to the local saturation of the soil surface. Such conditions are often found in footslope or channel-side areas, where water is concentrated by lateral movement downslope or from the channel. In these cases, rain falling on the surface may be unable to infiltrate and is therefore ponded at the surface. In addition, water flowing from upslope may emerge above the saturated areas as seepage. Overland flow generated by these processes is known as saturated overland flow.

In general terms, these two types of overland flow show a distinct geographical distribution. Infiltration excess overland flow is most common in semi-arid areas which are characterised by high rainfall intensities and by a sparse vegetation cover, allowing the creation of relatively impermeable crusts by rain-splash and slaking. Saturated overland flow, on the other hand, is more typical of moister temperate environments where rainfall events are normally of a lower intensity and longer duration, where interception by vegetation is greater, and where soil infiltration capacities are generally higher. Nevertheless, both types of flow are produced on agricultural land, and infiltration excess overland flow may be common on exposed or compacted arable soils with low infiltration capacities or on soils which are inherently susceptible to crusting (e.g. de Ploey 1981). As this indicates, agricultural practices may exert a major control on the generation of overland flow. Such flow has considerable significance, both agriculturally and more widely. Because it may be highly erosive and may remove not only soil but also seeds, fertilisers and pesticides, overland flow may represent a major risk on farmland. Moreover, because it provides a major input of water, sediment and solutes to streams it plays a vital role in controlling stream discharge and stream water quality.

The generation of overland flow, in fact, does not always occur spontaneously. In many cases, flow is preceded by a period of surface storage, when the ponded water accumulates in depressions on the soil surface. With further rainfall, these depressions fill until the water spills out of them into neighbouring depressions, and in time they become linked and a more or less continuous flow downslope is initiated. The time-lag between the appearance of surface ponding and the generation of overland flow depends upon the relative rates of rainfall and infiltration, but also upon the depression storage capacity of the soil surface. Storage capacity is itself related to surface roughness, and on arable land, for example, is greatest in newly ploughed soils, then declines as the surface is smoothed by the effects of frost action, rain-splash and settling of the soil (Fig. 9.12). Thus, marked seasonal variations in surface storage capacity are seen at a single site.

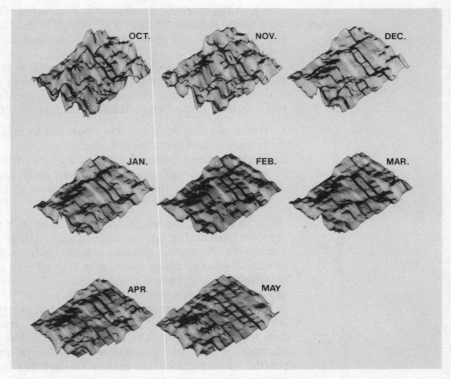

Fig. 9.12   Changes in surface morphology of a clay soil five years after ploughing up from old pasture (from Reid 1979).

Surface roughness, together with slope angle and water depth, also controls the velocity of overland flow. Velocity profiles in runoff waters typically show maximum velocities at or near the water surface and a decline towards the base. This profile results from the drag exerted by roughnesses on the bed which produce a zone of eddying. The larger these protrusions – that is, the rougher the soil surface – the greater is the drag, the thicker the layer of eddying and the lower the mean flow velocity. Average velocities of overland flow thus tend to decline with surface roughness.

Overland flow occurs in two main forms. On smooth surfaces, sheetwash tends to develop, flow occurring as a thin veneer of water a few millimetres in depth. Velocities in sheetwash may be relatively high (often in the order of 0.1 m s^{-1}) because of the lack of frictional drag, but the flow has limited erosive potential because of the lack of turbulence and the high cohesive strength of the soil. Moreover, overland flow often ceases before it reaches the stream channel, as the water infiltrates the soil, whilst much of the sediment is deposited as the water reaches more gentle footslopes. Consequently, sheetwash rarely transports significant amounts of sediment into streams. Sheetwash is rarely stable for long periods or over great distances, however, for splash from raindrops falling into the flow, together with local irregularities in the soil surface, causes turbulence and encourages the initiation of sediment transport. When this happens, localised scour occurs leading to the development of rills. These concentrate flow and increase the erosive potential of the water. The rills thus expand and ultimately may form gullies (e.g. Cooke and Reeves 1976). Although gully formation is associated with an increase in surface roughness, the concentration of flow into relatively deep channels leads to high depth/roughness ratios and consequently to higher mean flow velocities.

Flow velocity is an important parameter in relation to runoff because it affects the time taken for water to enter the permanent stream network, and thus the response time (flashiness) of the stream system. In general, rapid stream responses to rainfall are favoured by:

1. Rapid generation of overland flow during rainfall – i.e. low infiltration capacities and high rainfall intensity;
2. High overland flow velocities – e.g. due to steep slopes and the development of well-defined and deep gully systems;
3. Short travel distances for overland flow to the stream channel – i.e. high stream densities.

These conditions are commonly associated with upland areas, where valley sideslopes are steep, rainfall is relatively high, and soils are thin or impermeable. As a consequence, streams in these areas are often flashy with stream discharge rising quickly in response to rainfall, and then falling relatively quickly when rainfall ceases. This leads to characteristically peaked storm hydrographs (Fig. 9.14a).

Where soils are more stable, where vegetation cover is greater, or where rainfall intensities are generally low, the generation of overland flow is greatly limited. The main process by which water moves to stream channels is then by throughflow – the lateral movement within the soil profile. Movement occurs mainly through the larger structural voids, including pipes created by roots and burrowing animals (e.g. Jones 1981). Even so, the average rate of throughflow is characteristically much less than rates of overland flow, often no more than 0.5 m h^{-1}. As a result it takes many hours for water from more distant parts of the stream catchment to reach the channel. Storm hydrographs in these circumstances are typically less peaked than in streams fed by overland flow, and the lag between rainfall and stream response is greater.

9.6.2 Effects of agriculture on overland flow

It is clear that agriculture exerts a major control on processes of overland flow, both through its effect on crop cover and rainfall interception and its effect on infiltration capacity, surface roughness and surface moisture detention. In addition, agricultural drainage practices may modify the rate at which water moves through the soil to the stream channel.

In general, practices which increase rainfall interception and the infiltration capacity of the soil may be expected to reduce the incidence and magnitude of overland flow. Similarly, practices which encourage water depletion by evapotranspiration or drainage diminish losses by overland flow. Consequently, overland flow tends to decline with increasing crop cover or with greater retention of plant residues (Hauser and Hiler 1975; Adams *et al.* 1978).

Early research on the effects of cropping practices on overland flow in the USA has been summarised by Glymph and Holtan (1969). Among other studies, they quoted experiments at Hastings, Nebraska, in which peak rates of overland flow were measured from plots under small grain crops (e.g. wheat), maize and a rotation of maize, oats and wheat. Under both the rotation treatment and the maize crop, peak runoff rates were greatest with straight-row tillage, intermediate with subsurface tillage and least with contour ploughing. In the small grain plots, no difference was detected between the latter two treatments, though both produced less runoff than straight-row tillage. The effects of different cropping systems were also demonstrated by studies at three different experimental stations, in Wisconsin, Oklahoma and Iowa. These showed that, expressed as a percentage of rainfall, runoff was least from continuous grass, intermediate from rotation crops, and greatest from continuous maize (Fig. 9.13). The results thus indicate that runoff is inversely related to vegetation cover and directly related to the frequency of cultivation.

Similar effects have been shown by more recent research. Verma *et al.* (1979), for example, found that runoff from cropped land was greater than that from uncropped land, due to the greater infiltration capacity and evapotranspiration under a crop. They also demonstrated that both deep tillage and mulching reduced overland flow. Freebairn and Boughton (1981) compared different cropping systems in the eastern Darling Downs in Australia and showed that double cropping reduced overland flow relative to either winter cropping or summer cropping alone. In Italy, Chisci and Zanchi (1981) recorded the effects of different cropping, cultivation and drainage practices on runoff and soil loss from silty clay soils in the Vicarello area, near Pisa. Overland flow was greatest from continuous wheat plots subjected to minimal tillage, intermediate from conventionally tilled wheat plots and least from grass. Drained plots had consistently lower rates of overland flow than undrained plots (Table 9.2), but, significantly, soil erosion was greatest under conventional tillage, apparently due to the reduced cover of crop residues and the less stable soil structure on these plots.

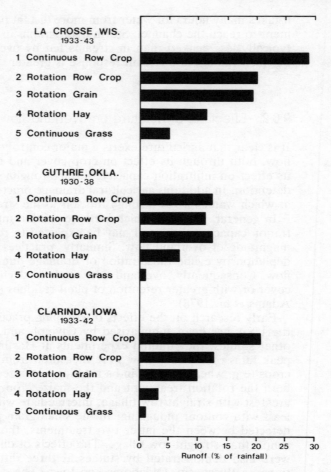

Fig. 9.13 Runoff from 0.01 acre (0.004 ha) plots under 5 cropping systems in
Wisconsin, Oklahoma and Iowa (after Glymph and Holtan 1969).

Table 9.2 Overland flow and soil loss from silty clay soils in the Vicarello area
of Italy under different cropping systems

| | Undrained | | | Drained | | |
| | Runoff | | Soil loss | Runoff | | Soil loss |
	(mm)	(%)	(t ha⁻¹)	(mm)	(%)	(t ha⁻¹)
Conventional tillage	25.8	3.64	4.05	17.9	2.53	3.72
Minimal cultivation	41.1	5.81	1.61	26.9	3.81	1.52
Grassland	21.1	2.98	0.18	15.9	2.24	0.15

Source: From Chisci and Zanchi 1981

As several of these studies have indicated, overland flow from grass
crops is generally less than that from arable land. A grass sward provides
a more or less continuous vegetation cover which intercepts rainfall and
impedes any overland flow which does occur. The improved rooting and
organic matter accumulation in grassland soils also means that infiltration
capacities tend to be higher than in arable soils (see Fig. 9.7, p. 260).

Nevertheless, soil structural damage caused by trampling or vegetation removal due to over-grazing may allow overland flow to take place, and in some cases serious erosion may be initiated. Costin (1979) illustrates these effects in a comparison of plots under a range of grazing regimes, from moderate to heavy stocking, in the southern Tablelands of New South Wales, Australia. Higher grazing intensities resulted in lower vegetation cover and higher rates of overland flow and soil loss. Similarly, use of herbicides to control rangeland weeds has been shown by Richardson and Bovey (1979) to reduce evapotranspiration by 80 mm y^{-1} and increase runoff by 10 per cent.

9.6.3 Effects of agriculture on streamflow

Although the effects of agriculture on overland flow are well defined, the implications for streamflow are less clear. In part this results from the greater complexity of the system under consideration: streamflow represents a response not only to surface controls of infiltration and overland flow, but also to the effects of soil moisture storage and throughflow. In addition, analysis of relationships between agricultural practices and streamflow are considerably more difficult due to the greater scale of the system. Nevertheless, several studies have shown that both peak stream discharges and the flashiness of streams increase as the proportion of cultivated land in the catchment increases (e.g. Ursic and Dendy 1965). McGowan *et al.* (1980) have also suggested that a change in land use in the Kingston Brook catchment in Nottinghamshire, England from 56 per cent pasture and 30 per cent cereals to 86 per cent cereals would increase streamflow by 10 per cent. Similarly, more detailed effects of agricultural practices have been detected. Yates (1971), for example, found that oversowing of pasture grass or top-dressing grass with fertiliser diminished overland flow, increased surface detention, reduced the number of days on which overland flow occurred, and led to marked changes in the character of streamflow. Although no difference in the hydrograph rise time was noted, the lag between rainfall and stream response was increased, less flow occurred before the peak discharge, both the peak and the total stream discharge were reduced, and the falling limb of the hydrograph was increased. These effects were greater under a light grazing regime than under heavy grazing. Nonetheless, such effects are not always noted. Lvovitch (1958), amongst others, has shown in Russia that streamflow and flood discharges may actually be reduced by cultivation. In the Transvolga area of European Russia he estimated that cultivation of former forest lands had halved the average streamflow of the rivers, whilst the peak discharges of the Don, 70 per cent of the catchment of which is under cultivation, had decreased by 25–30 per cent as a result of agricultural activity.

The discrepancies between results from the different studies of agricultural impacts on streamflow reflect the wide range of factors which control stream discharge. Variations in catchment morphology (e.g. slope angle and distribution), in catchment size and shape, in soil conditions and in average climatic conditions all affect streamflow production and make comparisons of different catchments difficult. Moreover, year-to-year variations in rainfall and evapotranspiration rates may mask the effects of short-term changes in farming practices. Ideally, long-term monitoring of agricultural catchments, with suitable control sites, is required to assess reliably the effects of factors such as tillage, cropping or fertiliser practices upon streamflow, but relatively few studies of this nature have been carried out. As a consequence, some doubt remains about the relationship between agricultural land use and streamflow processes.

This uncertainty is seen most acutely in the debate about the effects of agricultural drainage on stream discharge. It has often been argued that improvements in drainage lead to increased flashiness of streamflow and higher peak discharges. Howe *et al.* (1966), for example, analysing trends in stream discharges in the Severn and Wye catchments in Wales between 1911 and 1964, claimed that field drainage had increased the risk of flooding. Similarly, McDonald (1973) concluded that agricultural drainage in areas of upland peat resulted in higher peak streamflows and a more flashy flow regime. On the other hand, several studies have found no effect of land drainage on streamflow, and Trafford (1978) suggests that in most cases the impact of drainage is negligible in the long term. Yet other workers have found that improving agricultural drainage reduces peak discharges (e.g. Green 1979).

The conflict between these various interpretations can be traced back to the wide range of factors affecting streamflow, and the inherent experimental differences between different studies. Logically, it might be anticipated that the effects of land drainage would vary according to local soil and topographic conditions, according to the rainfall regime, and according to other management practices. In catchments comprising largely waterlogged soils which do not drain effectively between storms, streamflow is likely to be supplied largely by surface runoff. In these circumstances, streamflow responses are likely to be rapid and peak discharges high. Installation of land drainage will tend to reduce discharge and increase the lag time to rainfall, by helping to lower the water table in the catchment and provide higher soil moisture storage capacities at the start of the storm. Conversely, in catchments in which streamflow is provided mainly by throughflow, drainage improvements may accelerate the movement of water to the stream channel by providing a more effective network of conducting pipes within the soil. The consequence will be higher peak discharges and a reduced lag time between rainfall and stream response.

Evidence is now accumulating to suggest that the effects of drainage can be rationalised in this way. Ryecroft and Massey (1975), in a study of mole drainage of clay soils in the Shenley Brook End catchment in Buckinghamshire, found that drainage either reduced peak streamflow or caused no change, depending on specific rainfall and soil conditions. They also re-analysed data from Southern Ireland and showed that mole drainage reduced peak flows and increased the lag time of stream responses for high intensity storms. In conclusion, they state that:

1. There is no evidence that land drainage increases flooding.
2. Mole drainage reduces peak flows from catchments for heavy storms which would otherwise give rise to flooding.
3. Mole drainage lowers the soil water table and thereby increases the strorage capacity at the start of the subsequent storm.
4. Undrained clay catchments have limited storage which results in runoff during storms.
5. An undrained, waterlogged clay catchment remains wet for a long time after a storm, thus providing no buffer against further rainfall.

Similar conclusions are drawn by Beven (1980) in a study in the River Ray catchment in Oxfordshire. This catchment comprises soils derived mainly from Jurassic clays and is mainly under permanent pasture (70 per cent), arable land and mixed deciduous woodland. Beven compared outflows from a drained and an undrained plot for a period of two years, and showed that hydrograph response from the drained plot was slower than that from the undrained site, but that the timing and magnitude of the peak flows were similar for both plots. On the other hand, the recession limb from the drained plot was higher than that from the undrained site (Fig. 9.14). He concluded that:

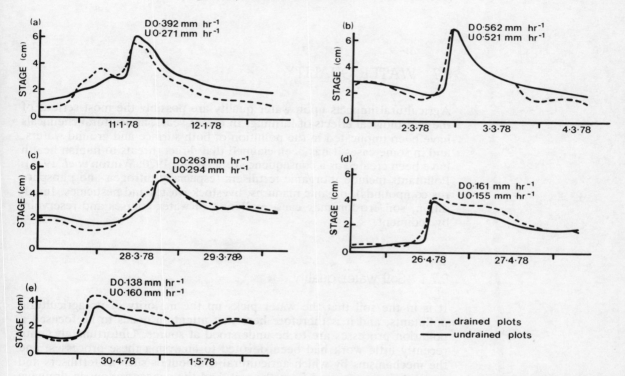

Fig. 9.14   Outflow hydrographs for drained and undrained plots for selected storm events at Grendon Underwood (from Beven 1980).

1. There was no evidence that improvements in field drainage since 1964 had affected the flow of the River Ray.
2. The effects of drainage on peak discharge varied according to antecedent soil moisture conditions.
3. Recession discharges were always higher from the drained than the undrained plot and led to a higher overall water yield.
4. Drainage eliminated the development of surface ponding in winter; thus, although the hydrographs from the drained and undrained plots were similar, the processes of response were probably different.
5. Movement of the water to mole drains takes place along macropores in the soil; drainage may help to maintain these macropores by draining them between storms and reducing the degree of clay expansion around them.

Overall, it is apparent that agricultural drainage may have significant effects on streamflow, but the effects depend upon catchment characteristics such as soil type, relief and climate, as well as the nature of the drainage system itself. Before these effects can be predicted with any certainty, more long-term, controlled experiments will be needed under a wider range of environmental and management conditions. Certainly the topic needs further study, for agricultural impacts on streamflow have significance not only for flow regimes but also for erosion, reservoir and bridge design, channel management and ecological processes within the streams (see Ch. 13).

9.7   WATER QUALITY

Agricultural impacts upon water quality are possibly the most serious of the hydrological effects of farming. In recent years, agricultural chemicals have been implicated in the pollution of both surface and ground waters, and in some cases it has been claimed that direct threats to human health have been created as a consequence (e.g. WHO 1970; Winton *et al.* 1971). Pollutants include inorganic fertilisers (especially nitrogen and phosphorus compounds), organic manures, livestock wastes and pesticides. In addition, soil erosion may cause pollution of water courses and reservoirs by sediment.

9.7.1   Soil water quality

It is in the soil that the water picks up the majority of its agricultural pollutants, and it is therefore here that attention needs to be focused if pollution processes are to be understood at source. Unfortunately, until recently little work had been devoted to analysing these processes, and the mechanisms by which agricultural compounds such as fertilisers and pesticides are removed from the soil, and the concentrations in drainage waters, remain uncertain.

A major factor controlling solute losses is the solubility of the compound. As we noted in Chapter 5, nitrates tend to be more soluble than phosphates or potassium compounds and they are thus lost more readily from the soil. Consequently, nitrates often pose the main threat to water quality. Similarly, pesticides which are neither strongly adsorbed nor rapidly lost by volatilisation are likely to be available for leaching and to contribute to pollution more readily than highly volatile pesticides (e.g. parathion, carbaryl) or pesticides which are actively held by clay or organic colloids, such as the triazine and phenylurea herbicides (e.g. simazine, atrazine, propazine). On the other hand, adsorbed pesticides may be carried on eroded sediment, and Walling (1980) notes that many of the more persistent pesticides, including DDT, dieldrin and diquat, may enter streams attached to sediments.

Related to this factor is the sorptive capacity of the soil. In the case of cationic compounds, retention is encouraged by a high cation exchange capacity, which is in turn a function of the presence of clay and organic compounds. Sorption of phosphates, on the other hand, is dependent mainly on iron, aluminium and, possibly to a lesser extent, calcium compounds; ferric hydroxide, in particular, seems capable of forming stable films of phosphate. Nitrates are held mainly by organic processes. Overall, therefore, the sorptive capacity of the soil is related to its texture, organic matter content, pH and mineralogy, and in general retention of fertilisers and other compounds is seen to be most effective in clay or peat soils of moderate to high pH. Soil moisture content also plays an important role, however, for this affects both the sorptive capacity of the soil and the solubility of the compounds. Phosphates and nitrates are readily reduced in anaerobic conditions to more soluble forms which are then available for leaching, and Sawhney (1979) noted that, while agriculturally applied phosphorus accumulated in the upper 30 cm of the soil well beyond the level explicable by its sorptive capacity, losses did occur when the water-level was high and reduction of the phosphorus compounds was possible. Similarly, several studies have shown that pesticides applied to wet soils, or subjected to heavy rainfall soon after application, are readily mobilised by reduction and may be leached from the soil (e.g. Lichtenstein 1958; Helling et al. 1971).

Because of these effects, it is apparent that rainfall conditions greatly influence the potential for leaching losses and thus for pollution. It is for these reasons, indeed, that autumn or winter applications of fertilisers are deterred, for otherwise considerable leaching may occur before the crops can make use of the fertiliser (see Ch. 5). Even so, the correlation between rainfall and solute concentrations in drainage waters is not strong for two main reasons. Firstly, as rainfall increases, drainage amounts also increase and solute loads are diluted; thus, although the total load may be greater, the concentration is less. Secondly, a critical factor in determining the quantity of solutes washed from the soil is the residence time of the soil water. Simply, water which flows rapidly through the soil has little opportunity to pick up solutes, while what which is stored for longer is able to come into equilibrium with the soil matrix and thus to accumulate high concentrations of solutes. On the whole, rapid flow is confined to the larger soil pores associated with fissures between aggregates and with earthworm or root channels (i.e. structural and bio-pores). Conversely, water flowing through the finer pores within the aggregates (the textural pores) moves only slowly. The relative proportions of soil water following these two routes, and thus the solute concentrations of the drainage waters, depend upon rainfall intensity and antecedent soil moisture conditions as well as the pore size distribution of the soil. During an individual storm, however, there is a tendency for the initial drainage to be of relatively solute-free waters moving rapidly though the larger pores. This is followed by the appearance of the more solute-rich water which has been displaced from the textural pores. Finally, as these pores drain to equilibrium, solute levels again fall. It is thus apparent that many factors influence solute losses and the quality of soil drainage waters. Agricultural practices affect may of these factors, however, and thereby exert a major control on water quality. One of the most important effects is the application of fertilizers. Theoretically, losses from fertilisers should be small, for if they are applied at the appropriate time and at recommended rates there should be almost no residues in the soil available for leaching. As Cooke (1977) notes, such losses represent a wastage of agricultural inputs as well as a potential environmental hazard. In the UK, however, it has been shown that potatoes often leave considerable residues of nitrogen in the soil after harvest, and these may be leached by winter rain-fall. On this basis, Gasser (1980) estimates that 200,000 ha of land in the UK are susceptible to intense N losses each year (some 3.5 per cent of the total arable land area). In the Netherlands, fertiliser application rates are considerably higher, but even so Kolenbrander (1972) calculated that fertilisers account for only 5 per cent of the NO_3-N load of drainage waters.

A more important source of solutes (especially N and P) in many cases is livestock residues. Faeces are returned to the soil throughout the grazing season in grassland systems, and these may still be decomposing during the winter when soil moisture contents increase, nutrient uptake by plants declines, and rainfall is able to flush the solutes from the soil. Under highly intensive livestock systems, the problems are exacerbated. Where cattle are housed for long periods, for example, slurry must be disposed on the land; the available pasture is often limited, excessive dressings of slurry are applied, and solute losses in drainage water may be considerable. Seepage from slurry silos is also a problem while intensive feedlot systems result in the concentration of cattle for long periods in small feeding areas, producing excessive quantities of wastes in the soil.

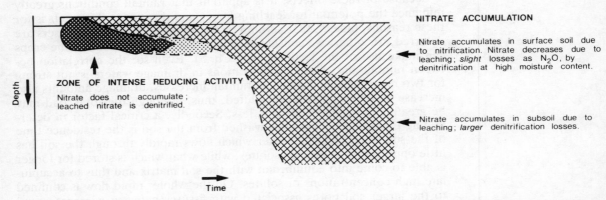

NITRATE ACCUMULATION

Nitrate accumulates in surface soil due to nitrification. Nitrate decreases due to leaching; *slight* losses as N_2O, by denitrification at high moisture content.

ZONE OF INTENSE REDUCING ACTIVITY
Nitrate does not accumulate; leached nitrate is denitrified.

Nitrate accumulates in subsoil due to leaching; *larger* denitrification losses.

Fig. 9.15   Model of nitrate redistribution from surface applied livestock wastes (from Burford *et al*. 1976).

Terry *et al*. (1981) found no effect of different rates of feedlot waste application on the nitrate concentrations in a shallow, underlying aquifer, but several studies have identified runoff from feedlots as a major source of pollution (e.g. Keller and Smith 1967; Stewart *et al*. 1968; Koelliker *et al*. 1971; Riskin 1973). A general model of the effects of animal wastes on soil nitrate levels has been developed by Burford *et al*. (1976) (Fig. 9.15). Haghiri *et al*. (1978) found a significant effect of soil type on leaching losses from surface applied manures. They compared losses over four years from plots on three soils previously treated with four different rates of manure. The results showed that losses of nitrate, potassium, calcium and magnesium and sodium all increased with increasing rate of manure application. Over time, however, losses were also greater from the well-drained Wooster silt-loam than from the moderately drained Celina silt-loam, and least from the poorly drained Hoytville silty clay (Fig. 9.16).

Several other practices influence the rate of solute losses from agricultural land. Ploughing up grassland characteristically produces a flush of

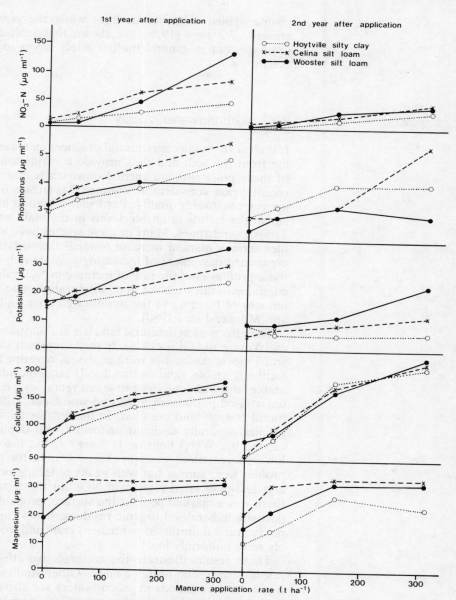

Fig. 9.16 Losses of nitrate-nitrogen, phosphorus, potassium, calcium and magnesium from surface applied manures on three soils. (Data from Haghiri *et al.* 1978).

NO_3-N in drainage waters as nitrate is released from the rotting vegetation. Losses from leguminous crops also tend to be greater than those from grass or arable crops, a difference which is accentuated when the legumes are ploughed in. Low and Armitage (1970), for example, found that total N losses in drainage waters from lysimeters carrying clover were 30 kg ha^{-1}, while those for grass (meadow fescue) were only 2.5 kg ha^{-1}. Agricultural drainage also affects leaching losses, though the effect cannot easily be predicted. Improved aeration in the subsoil reduces the rate of denitrification, but increases the amount of water moving through the

profile. Thus, solubility declines while the potential for transport increases. Williams (1976) has shown that leaching losses increase after
drainage, but in general the net effect seems to depend on specific soil
conditions.

9.7.2 Groundwater quality

It is clear that many agricultural practices increase the rate of solute leaching from the soil, and thus provide a major source of pollutants. Many
of these pollutants are carried downwards into the groundwater and in
recent years considerable concern has arisen over the possible effects
upon groundwater quality. Particular attention has been devoted in Britain to the trends in nitrate levels in the main aquifers of the Chalk and
Triassic sandstones. Many of these studies have indicated that nitrate profiles show a marked increase towards the surface, reflecting the gradual
downward movement of increasingly solute-rich waters. The character of
these profiles (and the rate of increase in N_3 levels) depends upon climatic
conditions, thickness of the unsaturated zone above the aquifer, the
amount of fissuring in the unsaturated zone and also on land use (Gray
and Morgan-Jones 1980).

The effects of agricultural land use are illustrated by Young *et al.* (1976)
and Young and Gray (1978). In the latter study thirty-seven sites on Chalk
and Triassic sandstones were analysed, covering five main land use types:
fertilised arable (and horticultural) land including short leys; fertilised
arable land with long (four to seven years) leys; fertilised permanent grass;
unfertilised permanent grass and woodland. Different NO_3 profiles were
found in each land use class. In the fertilised arable land with short leys,
profiles generally declined uniformly with depth. Above 15 m they exceeded the WHO limit of 11.3 mg l^{-1}, but below that level declined to
low concentrations at 25 m (Fig. 9.17). In the arable-long ley sites, the
profiles were similar but with peaks reflecting occasional ploughing up of
leys and associated release of NO_3. In the fertilised permanent pasture
there was a marked peak at the surface, probably due to the presence of
partially mineralised organic residues; below, nitrate levels declined. Finally, in the unfertilised permanent grass and woodland profiles, NO_3 levels were uniformly low.

These results illustrate the accumulative effect of fertilisers and crop
residues on groundwater quality. Other studies have demonstrated that
nitrate concentrations in groundwaters are also markedly affected by organic manures and animal wastes. Gray and Morgan-Jones (1980), for
example, analysed three wells in the Chalk west of London. At one, a
marked peak in N levels was identified which was attributed to the use
of slurry for agriculture since 1953. Wellings and Bell (1980) compared
nitrate levels in the Upper Chalk near Winchester under slurry and fertiliser application. In the control plot, which received neither slurry nor
fertiliser, NO_3-N concentrations were consistently low. Similarly, no effect was detectable in the plot receiving 376 kg ha^{-1} N as ammonium nitrate, indicating that the nitrogen was being effectively taken up by the
ryegrass crop. In contrast, the plot receiving slurry equivalent to excreta
from 10 cows ha^{-1} y^{-1} had NO_3-N levels 10 times those in the control,
while the plot with slurry applications equivalent to 40 cows ha^{-1} y^{-1} had
N concentrations 50 times those in the control.

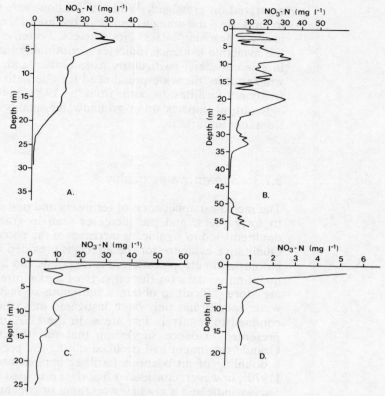

Fig. 9.17 Nitrate profiles in groundwaters under different cropping systems (a) fertilised arable land including short leys (b) fertilised arable land with long leys (four to seven years) (c) fertilised permanent pasture (d) unfertilised permanent pasture. (Note different scales.) (From Young and Gray 1978.)

Despite the occurrence of locally high nitrate concentrations such as these in groundwaters, and the accumulation in some cases of nitrate levels well above the recommended 'safe' limit of 11.3 mg l⁻¹, it is far from certain that any general threat to groundwater resources exists. Within the groundwaters, diffusion and slow mass flow act to reduce these local concentrations, and it has also been argued that denitrification occurs as the waters percolate into the saturated zone. Whitelaw and Rees (1980), for example, found substantial populations of nitrate-reducing bacteria in the vadose zone of the Chalk, and denitrification was found to be occurring within a profile beneath a permanent grassland site at Deep Dean, Sussex. Foster *et al.* (1982) dismiss the likelihood of rapid attenuation within the unsaturated zone beneath arable soils, but concede that such denitrification may be significant in aquifers beneath pasture. In this case, the 'bulge' characteristically seen in many nitrogen profiles beneath grassland soils may not be actively moving downwards, but may represent a balance between rates of input at the surface and loss by denitrification at depth.

Future impacts upon groundwater quality are likely to be dependent upon changes in fertiliser usage. Gasser (1982) predicts an increase of about 50 per cent in the use of N fertiliser in Britain by 1990, mainly concentrated on grassland. Wild (1977), however, notes that current trends in land use – including a decline in the use of short-term leys – may result in lower leaching losses. Nonetheless, Foster *et al.* (1982) conclude that the available evidence indicates a continuing increase in levels of nitrate in groundwaters, particularly from arable land, and they argue that this will result in 'the widespread need for alternative water supplies or water treatment facilities, or both, from the 1990s onwards'. Clearly, therefore, agricultural impacts upon groundwater quality remain an issue of considerable importance.

9.7.3   Stream water quality

The increased application of fertilisers and pesticides during the last thirty to forty years, and the increases seen in grazing intensities, have undoubtedly led to significant increases in the potential for stream pollution. Qualitative evidence for the effects is provided by the occurrence of algal blooms and eutrophication in rural ponds and lakes (Lee *et al.* 1978), but quantitative data on the impacts of agriculture on stream water quality are more difficult to obtain. Continuous or regular monitoring of stream water quality has only been instigated in recent years, and even now is confined to relatively few areas. In the USA, Bower and Wilcox (1969) presented evidence suggesting that nitrate levels in parts of the Rio Grande catchment had declined during the previous thirty years, despite a doubling of nitrogenous fertiliser applications. Harmeson and Larsen (1970), however, considered that data provided by the Illinois State Water Survey indicated a gradual worsening of stream water quality since 1945. More recently, Hill (1978) in a study of fourteen catchments in southern Ontario, found a close correlation between the mean spring (February to April) load of nitrate and nitrite nitrogen and the percentages of watersheds under crops.

In Britain, data for nitrate contents in seventeen rivers in the Trent River Authority area for the period since 1950 were summarised by Tomlinson (1970). These showed a progressive increase in nitrate concentrations in six rivers, a decline in one and no trend in the others. Owens (1970) compared nitrate levels in a range of rivers over a four- to ten-year interval, but detected no significant differences in solute levels, although rates of fertiliser application had risen markedly. Subsequent data for nitrate concentrations for twenty-nine rivers in Britain collated by Owens *et al.* (1972), however, show a general relationship with the rate of nitrate fertiliser application. Green (1979) also notes that nitrogen concentrations in lake waters and rates of fertiliser application have risen more or less in tandem in both Loch Leven (Scotland) and Lough Neagh (Northern Ireland). Such apparent relationships, of course, are not necessarily indications of a direct link between the two factors.

One of the most thorough long-term studies of nitrate concentrations in stream waters is provided by Edwards (1974) in a study of the River Stour in Essex. Between 1948 and 1964 there was no trend in NO_3-N concentrations, but since 1965 there has been a considerable linear increase in nitrate levels, such that by 1971 the mean annual nitrate-nitrogen concentration was over three times the average prior to 1951. This trend is paralleled by the change in rates of nitrogen fertiliser application, though Edwards emphasises that other factors, such as increased tile drainage and deep ploughing, may also have contributed to the reduction in stream water quality.

As all these results indicate, long-term trends in stream water quality are far from universally clear. In many cases this reflects the wide range of often contradictory processes affecting stream systems. These processes include changes due to climatic effects (e.g. Walling and Foster 1978), inputs of industrial effluents, and inputs of sewage. In recent decades there has been some improvement in industrial effluent treatment and in the sewerage infrastructure (e.g. Farrimond 1980), but in general the treated effluents now have higher nitrate concentrations as a result of action to reduce the biological oxygen demand of the waste water (Wild 1977). Given that Cooke (1977) estimates that 40 per cent of the nitrogen entering water courses in Britain is derived from urban sewage effluents, and that industrial wastes probably account for much of the remainder, it is apparent that the effect of changing agricultural practices may be masked by these non-agricultural trends. Thus, it remains impossible in most instances to isolate the agricultural impact upon stream water quality. This problem is to some extent common to all attempts to evaluate the effect of modern farming on the environment.

9.8 CONCLUSIONS

Modern agricultural effects upon the environment are only the most recent in a long continued sequence of environmental impacts. In the case of effects on the hydrological cycle, probably the most fundamental changes have been associated with forest clearance and the initial conversion of woodland to range or arable land. This is a process which has been going on in Britain for several thousand years, and which is still continuing in many parts of the temperate world as additional land is reclaimed for agriculture. The effects upon hydrological processes are wideranging. As the studies by Likens and others at Hubbard Brook have shown, forest clearance results in major reductions in rainfall interception and evapotranspiration, increases in soil moisture contents, leaching and runoff, and consequent increases in erosion and solute losses in stream waters (e.g. Likens 1970; Likens *et al.* 1967, 1977; Bormann and Likens 1979). Compared to these changes, the effects of developments in farming practices are often small, but with larger human populations to support and with more limited reserves of natural resources, the potential implications for man are nonetheless considerable. It is clear that much has still to be learned about the trends in hydrological conditions, and the relationships between farming systems and hydrological processes are still not well understood. This is somewhat unsatisfactory and it is not surprising that concern with these questions is increasing. A need for the future is undoubtedly to make more explicit the links between agriculture and hydrology, and to evaluate modern farming practices in the light of their impact upon water resources.

10 THE IMPACT OF AGRICULTURE ON INDIVIDUAL PLANT AND ANIMAL SPECIES

10.1 INTRODUCTION: AGRICULTURE AND NATURAL SYSTEMS

Since its inception, agriculture has involved intervention and modification of natural ecological systems. The ways in which agriculture can affect natural systems are extremely complicated and, in many respects, still not well understood. In many temperate areas, such as the British Isles, various ecosystems are in such close geographic proximity and are so intimately interdependent that modification of one system (or a part of a system) can have an immediate and significant impact on adjacent, and even on more remote, systems.

As was noted in Chapter 1, British agricultural practice is very highly developed. For a range of historical and economic reasons, it is generally more mechanised and less labour-intensive than in mainland Europe. The ecological impact of agriculture has therefore, in many ways, been relatively greater in Britain. Furthermore, in very general terms, natural ecological systems tend to be geographically larger on mainland Europe and on other continental temperate areas, than they are in Britain. Hence, again, agricultural impact on natural ecosystems has tended to be greater in Britain than in many other areas – simply because there is a great diversity of systems in a small area. Current trends towards intensification of agriculture within the European Community (often supported by regional development funds or by money from the Common Agricultural Policy) are, however, increasing the threat to natural and semi-natural ecosystems throughout Europe.

A general problem is that the effect of any 'external' factor – of which man's agricultural activity may be considered one – on an ecosystem varies greatly according to the scale of examination. To understand this problem it is worth considering some general examples. (The examples quoted will be considered in more detail in the succeeding two chapters.)

The Fenland areas in south Lincolnshire, Cambridgeshire and west Norfolk were mainly drained from swamp and fen-carr by Dutch drainage experts in the seventeenth century. After drainage they were, at first, used mostly for cattle grazing and, more recently, almost exclusively for arable crops. There would appear to be little in common, ecologically speaking, between the present open-field landscape and the swamps and marshes that existed before drainage took place. At the scale of the habitat as a whole, therefore, the changes have been such that there is no longer any realistic comparison between the present system and its original state.

In contrast there are some remnant wetland habitats, still retaining a similarity to the pre-existing Fens, where important ecological change is continuing, largely as a result of modern agricultural influences. Examples in Britain are the Somerset Levels and the Yare Marshes, Norfolk, where drainage activity has recently increased. Here there is clearly a need to understand the character of these influences and to predict their longer-term effects.

On a smaller scale, a feature characteristic of the last few years has been the increase in clearance of ditches and stream sides by mechanical excavators. This has been undertaken to try to increase water runoff from farm and other land and thus to improve land drainage and to minimise the possibility of flood damage. The vast majority of ditches, streams and river-banks in Britain are kept far freer of subaquatic, aquatic and terrestrial vegetation than they were, say twenty years ago. Whilst this is beneficial in drainage terms it involves significant ecological change at the local scale. So, even in areas such as the Fenland, although there may be little recent large-scale ecological change there is, nevertheless, still substantial continuing small-scale ecological change. In addition, the various developments in pesticides – particularly in the early 1960s – wrought a considerable change in the ecology of the plants and animals which inhabit the field areas.

From this discussion it is apparent that one useful way to evaluate the effect of agriculture on natural systems is to look at the impact at various scales. With most ecosystems, change is more significant at certain particular levels of scale and the relative significance of different scales varies from system to system. It must also be recognised that, as with all ecological studies, there are a multiplicity of linking factors and small-scale ecological changes may build together to promote more widespread ecological effects. Also, as already noted, changes in one habitat may rebound on other habitats and at different scales. Thus any attempt to simplify an ecosystem in order to understand individual links should only be a first step in beginning to grasp the complexities of the whole.

10.2 EFFECTS ON SINGLE SPECIES: SOME GENERAL CONSIDERATIONS

In recent years there has been a tendency towards the employment in ecology of the 'indicator species' concept. In simple terms this involves measurement of population density (or, at its simplest level, evaluation of presence or absence) of a single species as an indicator of more general ecological changes. One of the best known examples of this approach is the use of lichens, which have been employed in at least two ways. Firstly, Hawkesworth *et al.* (1973) and others have used them to indicate air pollution. Secondly, Rose (1974) for example used them in Savernake Forest, Wiltshire, to demonstrate the continuous presence of trees from ancient times.

Although these two examples of use of the indicator species concept are not directly concerned with agriculture, they are, nevertheless, valid and useful. There are, however, dangers in concentrating on too restricted sections of the system, the principal one being that one may fail to notice other important effects. This is discussed in relation to air pollution by Edwards *et al.* (1975) and in a more general ecological context by Helliwell (1977) who uses the analogy of the radio receiver. As he suggests, 'a high fidelity receiver with accurate fixed tuning will provide excellent reception on a given frequency but other signals will be missed'. Thus, in studying the effects of agriculture on single species, one has to be very careful not to miss other, perhaps more significant effects.

There are other serious problems when studying the effect of agriculture on individual species. The amount of research undertaken on particular species varies considerably, firstly, from species to species and, secondly, from habitat to habitat and from place to place. These difficulties are particularly acute with the smaller organisms. The first aspect is illustrated by work on Diptera (two-winged or true flies) where over the last thirty years many species have been recorded which are either new to Britain, or new to particular localities or, in some cases, altogether new species. So, for example, as recently as 1978, a new species of scuttle fly was recorded from Gloucestershire (Disney 1978). Quite obviously, if knowledge of the diversity and nature of species is incomplete it is impossible to assess accurately the effect of any changing external factor upon them.

The second aspect of the problem is illustrated by Ratcliffe (1977) who, in his assessment of the chalk grasslands of southern England comments: 'Almost no entomological work has been done in south Wiltshire, except on the Lepidoptera (butterflies) so that there is practically no information about the insects of the south-west Wiltshire Downs.' If there is no information on either the past or the current status of particular groups of organisms, then, again, it is impossible to make observations regarding changes in status or effects of external factors.

However, despite these problems, there have been several plants and animals which have been subjected to detailed long-term assessments. So far as agricultural effects are concerned there are three important groups of ecological factors:

1. The effect on an organism of the physical destruction of its natural habitat. As will be seen in later sections, some organisms are more capable of adaptation than others.
2. The effect on organisms of artificial chemicals. Broadly, this usually involves pesticide application but fertiliser may also have an effect. There are many ways in which organisms may be affected and, in many cases, they may not be the direct target of the pesticide application.
3. Secondary effects on organisms resulting from changes in their food chain. This can manifest itself in many ways, including:
   (a) the eradication of some members of the chain;
   (b) progressive uptake of pesticide concentration through the chain.

The first of these effects – habitat changes – is considered in more detail in Chapters 11 to 13 (since a change in habitat will usually affect more than one species at a time). Subsequent sections of the present chapter will be concerned with the effects of these various factors on particular species. Again it must be stressed that research has been rather patchy in terms of concentration on particular species or groups of species: this inevitably affects the general conclusions that one is able to draw.

10.3 HERBICIDES: EFFECTS ON WEEDS AND OTHER SPECIES

It has been suggested that between 150 and 200 species of plant can be described as weeds in arable land in Britain (Ratcliffe 1977). Of these, about ninety species do not frequently occur elsewhere and are mainly dependent on arable land. Because weeds are of economic significance they have been the subject of considerable research.

Many weeds are now infrequent or are decreasing in abundance and distribution. Among those that have declined considerably are darnel (*Lolium temulentum*), corncockle (*Agrostemma githago*) (see Fig. 10.1), cornflower (*Centaurea cyanus*) and corn marigold (*Chrysanthemum sagetum*), all of which Ratcliffe notes were listed among the most common weeds of the sixteenth century. Species such as the brome grass (*Bromus interruptus*) and the Composites *Arnoseris minima* and *Filego spathulata* ('spathulate cudweed'), which have been dwindling over the last fifty years have now reached a state where less than a dozen localities are known for each.

Much (but not all) of this decline is a result of the increase in the use of selective weedkillers or herbicides which commenced in the early 1930s (see Ch. 6). There is considerable pressure on farmers to reduce the numbers and varieties of weeds in crops. Hill (1977) suggests that there are nine main effects of weeds in crops:

1. Weeds may be parasitic upon crop plants.
2. Weeds may be poisonous.
3. Weeds may be unpalatable, nutritionally poor or may cause tainting of animal products, even if they are not actually poisonous.
4. The physical characteristics of some weeds may be a problem (they may tangle with animal coats).
5. They may cause damage to, or at least interference with, machinery at harvest or other times.
6. Even if the yield of a crop is not reduced, its value may be seriously affected by the presence of weeds.
7. Weeds may act as hosts for diseases and pests which affect crop plants.
8. Some weeds may be important in blocking drainage ditches and irrigation channels.
9. Weeds originating from crops may affect many man-made environments other than farmland (e.g. roads and railways).

The importance and relative significance of these factors clearly varies according to the crop, the weed, the location, the time of year and many other factors. However, the list is significant in that it clearly demonstrates the considerable and understandable pressure for the continued use of effective herbicides.

As already noted, herbicide use developed in the early 1930s, and it is since that time that substantial changes in the weed flora have taken place.

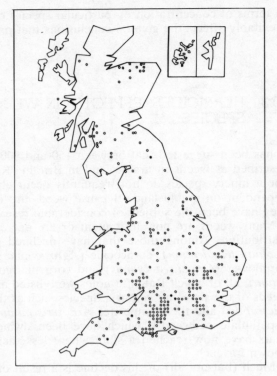

Corncockle
(*Agrostemma githago*)

o Recorded before 1960
  but no longer present.

● Recorded before 1960
  and subsequently, or
  first recorded since
  1960 and presumed
  present before that date.

Dependent on cornfields
(unmodernised)

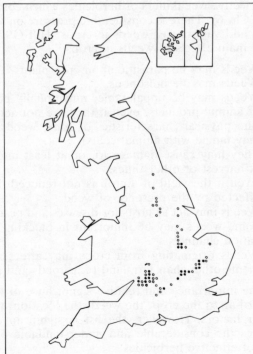

Pasque flower
(*Pulsatilla vulgaris*)

o Recorded before 1950
  but no longer present.

● Recorded before 1950
  and subsequently, or
  first recorded since
  1950 and presumed
  present before that date.

Dependent on chalk
downlands and
limestone pasture.

Fig. 10.1   Distribution of the corncockle (*Agrostemma githago*) and pasque
flower (*Pulsatilla vulgaris*) in Britain (from Nature Conservancy
Council 1977).

Among the earliest herbicides used were significant quantities of dinitro compounds related to 2,4-dinitrophenol. (A common example is DNOC.) Although these compounds were used mostly as herbicides to control weeds in cereal crops, as Duffus (1980) suggests, 'to call them herbicides gives a false impression of their specificity'. Their pseudo-specificity as killers of broad-leaved dicotyledonous weeds in cereal crops was mainly because they ran off the narrow-leaved surfaces of the crop plants, which tend to have poor wetting properties. In fact, the dinitro compounds are general biocides and human fatalities have been caused through incorrect handling in application. They act as uncouplers in respiration and so interrupt the energy cycle necessary in all aerobic life.

Fortunately DNOC is rapidly broken down and does not appear to leave toxic residues. Hence it does not accumulate in food chains. In any case, DNOC was largely replaced in the mid-1940s by the phenoxyacetic acids or 'hormone weedkillers', common examples being MCPA and 2,4-D. These latter compounds act as growth regulators and plants thus treated commonly develop stem elongation and other forms of abnormal growth. These chemicals are far less toxic than the dinitro compounds and are not considered harmful to animals (Mellanby 1981).

Very large quantities of herbicides such as MCPA are now used in agriculture in Britain. Fryer (1977) calculated that between 10.4 and 13.5 million acres (4.2 and 5.5 million ha) of agricultural land was treated annually with herbicide, comprising over 90 per cent of the arable land in the country and of which about 40 per cent was treated more than once a year. Fryer estimates that only about 10 per cent of grassland was treated.

An important question is whether modern farming techniques (including herbicide use) are causing the extinction of particular weeds or, alternatively, whether they are simply being reduced in numbers and distribution. Ratcliffe (1977) considers that populations of some species are now reaching critical levels and Moore (1977b) believes that some arable weeds are now close to extinction. Three years earlier Fryer and Chancellor (1974) thought, however, that weeds were simply reducing in number and localities of occurrence. The latter considered that the major effect of herbicides, in particular, has been to reduce rather than to eliminate weed populations. They suggested that even the most sensitive weed species such as charlock (*Sinapsis arvensis*) occur just as widely as they used to although in fewer numbers.

Some weeds have not declined in number as a result of herbicide use but for other reasons associated with modern farming developments. An example is the corncockle (*Agrostemma githago*) (Fig. 10.1), which almost disappeared before the advent of major herbicide use, because of improved methods of seed cleaning. Similarly, liming had probably been the main factor responsible for the diminishing occurrence of the corn marigold (*Chrysanthemum sagetum*) (Fryer and Chancellor 1974).

The approach to herbicide use may be at least as important as the type of herbicide used in determining the changes in weed flora (Fryer 1977). So, for example, previously important crop rotation techniques designed to reduce weeds have in some cases become unnecessary. In some of these cases break crops have been eliminated, giving rise to continuous cropping of the most profitable crop. This, in turn, has produced new weed problems and neccessitated further herbicide development. It is inevitable that at some stage financial considerations cause farmers to seriously question continuous use of particular herbicides.

A particular problem arose in the mid-1960s with some of the most successfully competing weeds of arable crops such as wild oat (*Avena fatua*) and black-grass or black-twitch (*Alopecurus myosuroides*). These are, of course, morphologically and physiologically very similar to the crop plants. Black-grass is a weed of cereals on heavy soils. It may be that it is able to compete with wheat on soils low in potassium because this particular weed has, itself, a low requirement for potassium. Species such as wild oat and black-grass can now be controlled by herbicides such as barban (a carbamate) and triallate (a thiolcarbamate), although this treatment is relatively expensive (Moore 1977b).

Improved seed-cleaning techniques have also affected the populations of some arable weed species. In the European Community, for example, the grass *Bromus grossus* is considered to be endangered. This species is endemic to Belgium, where it is today confined to a few localities at the edge of the Ardennes Plateau and in the Beauraing region. Here it occurs in association with crops of spelt wheat (*Triticum spelta*). Improvements in seed cleaning, however, seem to be causing a marked decline in its population (Smith 1973) and there have been no confirmed records since 1970. Aymonin (1974) believes that the species is now near to extinction, and although casual populations are found in France, Luxembourg, Germany and Italy these too seem to be rapidly declining in numbers (P. M. Smith 1972, Aymonin 1977). The same is true of the Casphyllacea *Silene linicola* which is found in flax fields in southern Germany, and more sporadically in northern Italy and France (Aymonin 1977). The populations appear to be declining both as a result of loss of habitats as flax production decreases, and as a result of improved control of seed quality and purity.

Substantial reductions in plant numbers were caused in the 1960s by the spraying of road-side verges by local authorities. Although farmers contended that areas acted as reservoirs for weed populations, most of the species occurring on verges are grassland species which are not capable of surviving arable conditions (Way 1970). As a result both of pressure by conservationists and also because of increased herbicide cost, in recent years there has been a considerable reduction in verge spraying (Way 1978).

Moore (1977a) points out that weeds can sometimes have beneficial effects on crops. Not only can weeds assist in the early stages of crop growth by providing shade and shelter, but weeds may harbour predators which feed on crop pests. He cites the example of growth of caterpillars of the cabbage white butterfly (*Pieris brassicae*) which can be reduced by ground-living predators such as the beetle, *Harpalus rufipes*, and the arachnid, *Phalangium opilio*. Dempster (1968) reported a study in which spraying a Brussels sprout crop with DDT not only killed the cabbage white larvae but also the predators. Thus if immigrant butterflies later lay more eggs on the sprouts then this later generation is less subject to control. The situation is exacerbated after a second year of spraying because of persistence of DDT in the soil. Moore extends the argument to suggest that if the weed level could be manipulated then an optimal level of predators and larvae could be maintained.

Most modern herbicides are not highly toxic to vertebrates and few are persistent. In the past few years there has been considerable public concern about one compound in the chlorophenoxy acid group – 2,4,5-T. This compound was used by the United States in the Vietnam war as a forest defoliant and the chemicals used in Vietnam appear to have contained appreciable quantities of dioxin, as an impurity (Westing 1977). Dioxin is the name given to a family of chemicals related to 2,4,7,8-tetra-chlorodibenzo-p-dioxin (TCDD) – which, although not phytotoxic, are strongly teratogenic and probably carcinogenic, causing the skin disease chloracne. An additional problem with dioxin is that it is environmentally very stable and may accumulate. The principal use of 2,4,5-T is to control woody growth, rather than as an agricultural herbicide, and its use as a crop-related herbicide has now been banned in the USA. The amount of dioxin impurity in 2,4,5-T available in Great Britain is very small indeed and thus fears about its use are probably overstated (Fryer 1977).

Although not toxic to vertebrates, many herbicides destroy the non-vertebrate fauna. This may, in turn, have a secondary effect on the vertebrates. In one of the few studies in this field, Potts (1971) has examined the effect of changes in the arthropod fauna (resulting from herbicide use and other factors) on young chicks of the partridge (*Perdix perdix*). The main habitat of the partridge, which has declined considerably in numbers in recent years, is arable farmland. Potts showed that at times of adverse conditions, reductions in arthropod numbers could be critical to chick survival. It is therefore suggested that increased use of herbicides is at least partly responsible for the general decline in partridge numbers.

In upland areas there has been a significant recent increase in the use of asulam (sulphonyl carbamate) and similar compounds to control bracken (*Pteridium aquilinum*). Asulam is also used to control dock (*Rumex* spp.) in pastureland and was first marketed in 1968 (Worthing 1980). The compound is sprayed on the fully grown plant and attacks the rhizome, which rots away. Long-term eradication can be almost total if the spray is correctly applied. Although of very low toxicity to animals and having a minimal effect on other plants, it clearly has a profound effect on the overall ecology of the area. This is because of the dominant role of bracken within the plant community. Also, the removal of the bracken allows more intensive grazing and the possibility of other grassland developments which can substantially further alter the ecosystem. (This aspect is considered more fully in Chapter 11.)

10.4 FUNGICIDES: EFFECTS ON 'WILD' FUNGI AND ON ANIMALS

The importance of fungicides in arable crops has been discussed in Chapter 8. However, there is very little understanding of the effects of fungicides on wild fungi. Mellanby (1981) suggests that this may be more important than is at present realised: the wild fungi, after all, act as the gene pool from which new varieties originate.

Both arable crops and fruit trees would suffer considerable damage from fungal infestations if they were not treated with fungicides. The use of organomercurial seed dressings is common and in the 1960s there began the introduction of a range of general fungicides and those with action specific to powdery mildew. During the 1970s the most important developments were the introduction of systemic fungicides and the organofin fungicides (Sly 1977).

Most modern fungicides have a relatively low toxicity, although there have been cases of mercurial poisoning resulting from human consumption of dressed seed in developing countries. Environmental contamination of wild animals by mercury has been reported from Japan, Sweden, Holland, Canada and elsewhere (Newton 1979). Industrial effluent has been a major source of mercurial pollution but increases in concentration have at least partly resulted from the use of a large range of mercural compounds as seed dressings. The main source of mercury in the terrestrial environment has been shown to be the 'alkyl-mercury' compounds, which include methyl-mercury.

These compounds are often so lethal that they kill birds outright. The effect of sub-lethal doses is less easy to assess, partly because mercurial contamination tended to coincide with contamination by organochlorine insecticides (see following section). Whatever the effect it is clear that average mercury levels increased dramatically in the late 1950s and early 1960s (Fig. 10.2). The amount of mercury in the environment has declined since the early 1960s and it also seems that birds are better able to cope with sub-lethal doses of heavy metals than they are with organochlorines: this may be because the former occur naturally in the environment, albeit at lower concentrations (Newton 1979). Similar problems have not occurred in Britain, probably because different chemicals are used (phenyl-mercury rather than methyl-mercury compounds).

It is clear that even modern target-specific fungicides *do* affect other organisms, though to a variable and usually limited extent. Thus, for example, Graham-Bryce (1977) comments that both nematodes (pests) and earthworms (not pests) may be affected by the benzimidazole group of specific fungicides.

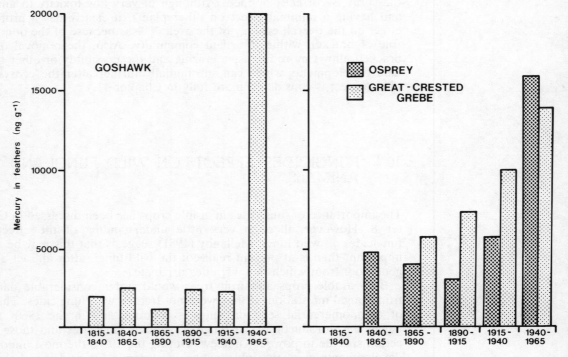

Fig. 10.2 Average mercury levels in the feathers of museum specimens of goshawks, ospreys and great crested grebes, collected at different periods in Sweden (from Newton 1979).

A further problem with fungicides is that fungi tend to become resistent, necessitating rapid research development in order to keep pace. Research efforts have therefore tended to concentrate on attempting to understand the mechanism of fungicide resistance (Graham-Bryce 1977).

10.5 INSECTICIDES: EFFECTS ON PLANTS AND ANIMALS

During the Second World War two main groups of insecticides started to come into use:

1. The organochlorines or chlorinated hydrocarbons. These include DDT and cyclodienes such as aldrin, dieldrin, endrin and heptachlor.
2. The organophosphorus group. These include examples such as 'metasystox' (demeton-S-methyl).

Newton (1979) suggests that there are three reasons why the first group – the organochlorines – are particularly damaging to wildlife:

1. They are chemically extremely stable;
2. They dissolve in the body fat of vertebrates; and
3. They can become dispersed over wide areas through the bodies of migrant animals.

Stability is demonstrated by the fact that DDT has a half-life of over 10 years whilst aldrin and dieldrin require $2\frac{1}{2}$ years in the soil for 95 per cent degradation (Duffus 1980). Beyer and Gish (1980) showed that dieldrin remained at potentially hazardous concentrations in the soil for several years through persistence in earthworms and, by implication, other soil fauna. It remained at a level of over 8 p.p.m. (critical for the woodcock, *Philohela minor*) for 3 years if originally treated with 2.2 kg ha^{-1} and for 11 years if treated with 9.0 kg ha^{-1}. Although these examples derive from more recent work, it was these kinds of problem which caused the United States administration to ban most organochlorines from use.

Serious worries about organochlorine pesticides began to develop in Great Britain in the late 1950s. As will be detailed in the following section, in the late 1950s significant numbers of dead birds were found with accumulations of organochlorines. This was traced to seed corn which had been injected with aldrin or dieldrin to protect the subsequent seedlings from attack by the larvae of the wheat bulb fly, *Leptohylemyia coarctata*. Use of these chemicals as seed dressings was reduced in the United Kingdom in 1962 and has now ceased. Under the Agricultural Chemicals Approval Scheme most chlorinated hydrocarbon insecticides are now banned. Dieldrin may still be used in dip as a protection against cabbage root fly, *Erioischiae brassicae*. Although DDT could be used until recently against some glasshouse pests and in certain other (mainly non-agricultural) circumstances (MAFF 1981); however, it has now been totally banned.

The organophosphorus compounds in use soon after the Second World War were often extremely poisonous and thousands of operatives have been killed throughout the world (Mellanby 1981). In Britain the more toxic organophosphorous compounds, such as parathion and TPP have very limited and carefully controlled use. Much less toxic organophosphorous compounds such as malathion are in common use. These have the advantage, from the point of view of wildlife, that they break down very rapidly and are not accumulated in the bodies of vertebrates.

The organophosphorous compounds are not, however, entirely without danger to wildlife. Geese, in particular, seem to be susceptible to carbophenothion seed-dressing. Bailey *et al.* (1972) report the killing of large numbers of greylag geese (*Anser anser*) in Perthshire, Scotland, and Pashby (1976) describes a similar incident with pink-footed geese (*Anser brachyrhynchus*) in Humberside, England.

Moore (1977b) and Mellanby (1981) both suggest that heavy reliance, as at present, on one group of insecticides is, itself, a dangerous thing. Pest resistance will inevitably increase and there will then be a demand for a return to more dangerous pesticides, which the organophosphorous compounds have themselves replaced. Indications that this is already happening are apparent with the introduction of some of the more toxic carbamates, such as aldicarb, oxamyl and carbofuran. Aldicarb, in particular, is increasingly used against nematodes and aphids on sugar beet. Moore (1977b) points out that the range of the rare stone curlew (*Burhinus oedicnemus*) overlaps with the sugar beet area. The total population of the stone curlew is only about 400 pairs so any potential threat is very serious.

10.6   EFFECTS OF PESTICIDES ON WILD ANIMALS: SOME CASE-STUDIES

A considerable amount of work has now been undertaken on the effects of the various organochlorines on birds and mammals. The effect on animals near the top of various food chains is particularly acute. This fact received much attention after the publication of Rachel Carson's emotive book *Silent Spring* (Carson 1962). There is no doubt that predators will absorb organochlorines from their prey but Moriarty (1977) has pointed out that it is not simply a matter of accumulation, even with chronic exposure. Prediction of ecological effects of pesticides becomes complex because significant and largely unpredictable amounts of pesticide may become 'lost' in the system.

10.6.1   Effects on predatory birds

Insecticide concentrations calculated from the results of two separate studies are given in Table 10.1. The first study (Hickey *et al.* 1966) is concerned with DDT in the lake bottom and in organisms in the food chain in herring gulls at Lake Michigan, North America. The second study (Hunt and Bischoff 1960) is from Clear Lake, California, and shows concentrations of DDD, a somewhat less toxic derivative of DDT. In this latter case DDD was sprayed over the lake to kill gnat larvae. The concentration in the grebes (less than 1600 p.p.m.) in Hunt and Bischoff's study was lethal. (Hickey (1968) has noted that only 5.3 to 9.3 p.p.m. of

Table 10.1   Concentration of pesticides in the food chains of birds

	Pesticide concentration (p.p.m.)	
	(a) DDT	(b) DDD
Birds	3177 (herring gulls)	up to 1600 (grebes)
Fish	3.0–8.0	40–2500
Small invertebrates	0.41	(not given)
General environment (lake bottom sediments)	0.0085	0.02

(a) Lake Michigan (Hickey *et al.* 1966)
(b) Clear Lake, California (Hunt and Bischoff 1960)

dieldrin and heptachlor epoxide was sufficient to cause death of Larmer and peregrine falcons.) The figures shown in Table 10.1 are best interpreted in a qualitative rather than a quantitative way. Because different pesticides, different organisms and different environments are concerned they cannot be directly comparable. What is clear is that organochlorines

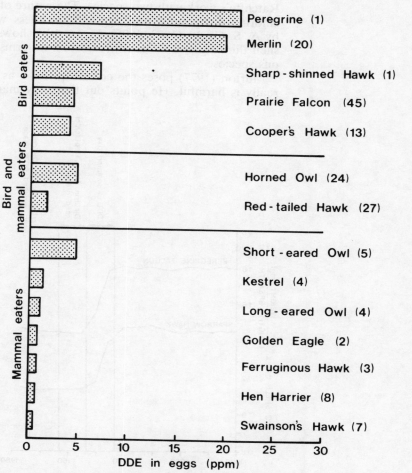

Fig. 10.3   Mean levels of DDE in the eggs of various predatory birds in the Canadian prairies during the 1960s (numbers in brackets = total sample size) (after Newton 1979).

do built up in food chains and that a certain level of dose – different for each organism and for each compound – is lethal. Figure 10.3 shows the different levels of DDE in various predatory birds in the Canadian prairies in the 1960s. It shows that the highest concentrations are suffered by those birds that are at the highest trophic levels in the food chain, namely the bird eaters.

Various alternative approaches to evaluating concentrations in the food chain are assessed by Moriarty (1977), such as laboratory models of ecosystems. At present none of the approaches is without considerable problems.

Whilst the lethal dose of pesticides may be relatively easy to establish (see section 6.4), what is by no means clear is the effect of sub-lethal doses on organisms. Ratcliffe (1967a, 1970) clearly established that there was a link between eggshell thickness of birds of prey and increased use of DDT and other organochlorine pesticides. He derived an eggshell index (basically an estimate of relative weight) and applied this to large numbers of peregrine (*Falco peregrinus*), golden eagle (*Aquila chrysaetos*) and sparrowhawk (*Accipter nisus*) eggs. Figure 10.4 shows the results of Ratcliffe's work with two species. The nature of the link between organochlorine compounds and eggshell thickness was further demonstrated by A. S. Cooke (1973, 1975). Figure 10.5 shows the relationship between shell thinning and DDE concentration in terms of fresh weight, for various species.

Murton (1977) poses the critical question as to whether shell thinning really is harmful. He points out that very many research workers have

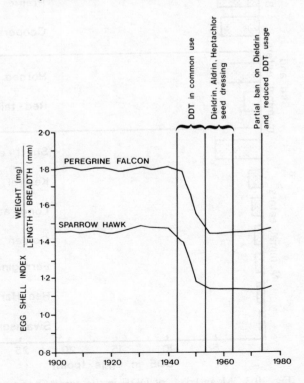

Fig. 10.4   Relationship between eggshell thinning and usage of organochlorine insecticides (from Brown 1976).

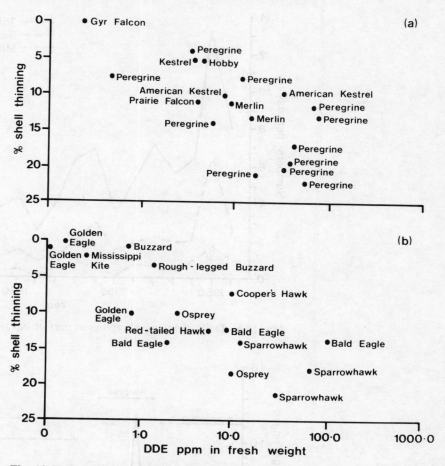

Fig. 10.5　Relationship between DDE levels in eggs and percentage eggshell thinning in different raptor populations (a) falcons (b) others (from Newton 1979).

tended to assume that if a population of birds lays eggs with thin shells and is, at the same time, declining, then the two facts must be directly linked. Nevertheless, Brown (1976), among others, argues strongly that the significant correlation must reflect a causal relationship. That there was a decline in many birds of prey at this time cannot be denied (Fig. 10.6 and 10.7). Newton (1979) cites many pieces of research which demonstrate the fairly obvious fact that weaker shells lead to more frequent breakages and thus less breeding success. He also quotes a number of laboratory studies where birds have been fed various concentrations of pesticides with similar result: the cause and effect therefore now seems clearly established.

In the United States the effect on populations has in some cases been extreme. The brown pelican (*Pelecanus occidentalis*) has virtually become extinct in its former North American range through inability to breed. Pelican eggs in California were found to contain 2600 p.p.m. of DDE (Keith *et al.* 1970). The source of DDE was thought in this case to be factory effluent. However, the fact that the peregrine falcon has now become extinct east of the Rockies, where it was formerly plentiful is more

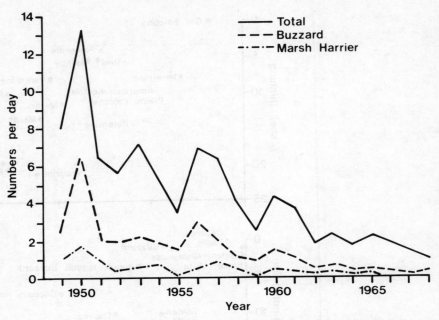

Fig. 10.6   Daily sightings of raptors in part of Bavaria (after Newton 1979).

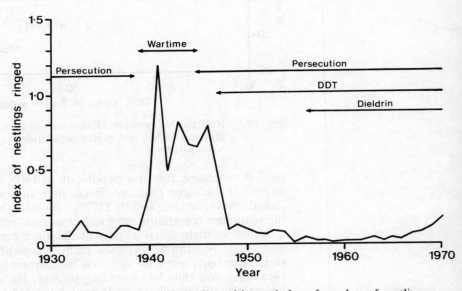

Fig. 10.7   Annual birth rate (as indicated by an index of number of nestlings ringed) of sparrowhawks in Britain, 1930–70 (from Newton 1979).

likely to be due to agricultural use of organochlorines (Newton 1979).

Other environmental factors of course interact with pesticide accumulation and in some cases it may be the other factors which precipitate critical changes in population. However, in birds of prey, in particular, it is clear that organochlorine pesticides have a significant detrimental effect.

The complex interactions involved in the response of higher organisms to pesticides is illustrated in a very detailed field study of the sparrowhawk centred on Annandale, south Scotland (Newton *et al.* 1979). All known nest sites for 5 years within a 700 km^2 area were examined. The number of nests declined year by year from 110 in 1971 to 82 in 1975. Newton *et al.* showed that most aspects of breeding were better on lower-altitude mixed farmland than on higher-altitude sheep range or plantation forest. This was indicated by the earlier laying dates on farmland, larger brood sizes, faster growth of young and higher proportion of successful nests. Similarly it has been shown elsewhere in Scotland that sparrowhawks are more successful breeders in valley forests than in hill forests (Newton 1976). An earlier paper (Newton and Bogan 1978) had demonstrated the correlations between DDE concentration and eggshell thinning in the sparrowhawk. In the Annandale study it was found that clutches which succeeded contained significantly less DDE, PCB (industrial polychlorinated biphenyl) and HEOL (from aldrin and dieldrin). Examples of the number of egg breakages in sparrowhawks are given in Table 10.2. The much higher breakage levels in the 1956–70 sample were attributed by Newton to organochlorine pesticides reducing shell strength.

Table 10.2   Sample number of egg breakages in nests of the sparrowhawk (*Accipiter nisus*)

Dates	Sample (no. of clutches)	No. of egg breakages
Prior to 1947	635	2
1956–70	160	31

Source: Data from Newton 1974

The golden eagle (*Aquila chrysaetos*) suffered in Scotland, particularly in the early 1960s as a result of the use of dieldrin in sheep-dip (Table 10.3). During the years of high use of dieldrin (1961–63: it was banned from 1966) the breeding success of eagles in Wester Ross fell from about 70 per cent to about 30 per cent (Lockie and Ratcliffe 1964). Similarly it fell from about 60 per cent to about 30 per cent in Argyll (Lockie *et al.* 1969). Brown (1976) points out that, at a 30 per cent success rate, each bird would have to breed for at least 28 years to replace itself. He suggests that it is unlikely that many birds would survive in the wild for this length of time. Fortunately breeding success seems to have recovered remarkably quickly after the discontinuance of dieldrin use (Everett 1971).

Indications of pesticides causing eggshell thinning and reducing breeding success are reported by Cramp and Simmons (1980) for various other birds of prey. These include the osprey (*Pandion haliaetus*) in Sweden and from Britain the kestrel (*Falco tinnunculus*) and the merlin (*F. columbarius*).

Table 10.3   Breeding success of golden eagle (*Aquila chrysaetos*) in Scotland

Dates	Total no. of pairs	No. of pairs raising young	% of pairs raising young
1937–60	36	26	72
1961–63	39	10	29
1964–68	48	34	71

Source: Data compiled by Cramp and Simmons (1980) from Everett (1971), Lockie *et al.* (1969) and Lockie and Ratcliffe (1964)

The apparent effect of pesticides on the peregrine falcon (*F. peregrinus*) has been particularly well studied. The decline of the peregrine population by half from an estimated total 820 occupied territories in Britain and Ireland in 1930–39 to 413 territories in 1971–72 has been carefully documented in a series of papers by Ferguson-Lees (1951), Treseaven (1961) and Ratcliffe (1962, 1963, 1965, 1967b), all of which have been collated and summarised by Brown (1976). Decline has been greatest in southern England and Wales where peregrines preyed on the birds of arable farmland. There has been some recovery since the lowest numbers were reached in Britain in 1963, but this has been in northern England and southern Scotland rather than in southern Britain. The common symptoms of adult egg-breaking and birds eating their own eggs was recorded as early as 1948–49. The eggshell thickness study of Ratcliffe (1967a), already referred to, used the peregrine as one of the case-studies (see Fig. 10.5).

There have been other factors involved in the peregrine decline, apart from pesticides, including an extermination campaign during the Second World War when it was thought that birds might interfere with carrier-pigeons bringing military messages. Also, over many years, the peregrine has suffered from persecution by, among others, gamekeepers, egg collectors, falconers and pigeon fanciers. Nevertheless, as Brown (1976) points out, peregrines had successfully withstood these various other pressures in the past without substantial declines in numbers and it would appear likely that organochlorine pesticides were the main cause of the steep decline in the early 1960s. The increase since 1963, coinciding with the reduction in use of organochlorines, again appears significant.

Even in the early 1960s, pesticides were not of course the only adverse agricultural factor affecting predators and there are detailed studies of various species where it has been shown that pesticides were a less than dominant control. Discussions of population changes in the buzzard (*Buteo buteo*) and the red kite (*Milvus milvus*), both of which have been particularly well documented, will be given in the following section (10.7).

Population decline of the magpie (*Pica pica*), which is omnivorous rather than predatory (its food consisting of insects, small mammals, small birds, eggs, molluscs, seeds, nuts, fruits and carrion), has been studied in eastern England (A. S. Cooke 1979). In some counties, such as Norfolk, there has been a gradual long-term decline, but in other areas, such as Cambridgeshire and west Suffolk, there was a sudden marked decrease in the late 1950s and early 1960s. Cooke considers various possible reasons for this, such as hedgerow removal, but again the conclusion is that it related to the use of organochlorines. Cooke notes that hedgerows have also been removed in areas where the magpie has actually increased. Perhaps more convincing is the fact that, like the birds of prey, numbers of magpies have somewhat recovered since the withdrawal of dieldrin as a seed-dressing. However, many areas that would appear to have suitable habitats for magpies are still devoid of them, which indicates that their habitat requirements are not yet fully understood.

10.6.2   Effects on mammals

The effects of pesticides on birds has been more fully documented than similar effects on mammals, but there are now a number of studies which examine this latter aspect. It has, for example, been shown that dieldrin is lethal to the fox (*Vulpes vulpes*) at very low levels although Blackmore (1963) pointed out that increased mortality of foxes due to organochlorine pesticides, which appeared to reach a peak about 1960, probably had little permanent effect on the fox population. Rothschild (1963), however, reported serious local effects.

In a more recent study of fox populations in Scotland, Kolb and Hewson (1980) considered the various measures of fox control at present in use, which include trapping and shooting by clubs, farmers, gamekeepers and Forestry Commission rangers. They concluded that these, similarly, are not limiting the fox population as a whole: it is clear that foxes are particularly resilient to adverse conditions. (The recent large-scale increase in numbers of urban foxes should be noted.)

Population fluctuations of stoats (*Mustela erminea*), weasels (*M. nivalis*) and hedgehogs (*Erinaceus europaeus*) have been studied in estates at Hexton, Hertfordshire, and Damerham, Hampshire (Jefferies and Pendlebury 1968). In the former case records of gamekeepers' 'vermin bags' were examined for a period from 1943 to 1965 and in the latter between 1947–48 and 1959–60. Stoat numbers dropped to a low post-myxomatosis level in 1953–54 and remained low thereafter, whilst the weasel population remained at about the same level that it had maintained during the 1940s and 1950s. Hedgehogs reached a peak in numbers in 1959. Jefferies and Pendlebury concluded that in none of the cases did the data they examined show any evidence of a detrimental effect of organochlorine pesticides. This was despite the fact that the three species were known to be contaminated in the areas examined. It is important to note that these animals are much lower in the food chain than birds of prey and this, in itself, will tend to reduce relative levels of concentration. On the other hand, lethal doses might also be assumed to be less in the smaller animals and in those with high metabolic rates, such as the hedgehog.

One of the main factors controlling the population of weasels has been shown to be the numbers of one particular food source, the vole (*Microtus agrestis*). Tapper (1979) studied a 25 km^2 area of farmland on the South Downs in West Sussex in which the vole numbers showed a 4-year cycle of abundance whilst weasel numbers varied annually, with peak numbers of weasels being over twice the minimum number. Relative changes in weasel numbers were shown to lag behind vole numbers. Tapper reported that female weasels usually failed to breed during the year of lowest vole numbers. During these years the main alternative food source for the weasels became birds. Tapper's study demonstrates that population changes among predators are complex phenomena. Single factors, such as toxic pesticide applications, add to the interplay of the factors, but relatively rarely seem to be wholly responsible for large long-term changes. This may be because the effects of such pesticides have usually been realised before they become critical.

Nevertheless, most animal individuals are affected by certain direct pesticide applications. Thus, for example, Mellanby (1981) notes reports of hares (*Lepus capensis*) emerging from fields sprayed with paraquat and displaying symptoms similar to myxomatosis – that is, with swollen eyes and nostrils. Some of these animals, he notes, died soon afterwards. However, there is a distinction between individual deaths and controls on the population as a whole and Mellanby suggests that we do not understand how these relate to hares, in particular. There was a marked increase in the hare population between 1953 and 1960, whilst rabbits were suffering from myxomatosis. Since 1960 hare numbers have fluctuated, although they now appear to be below the pre-1953 level (see, also, Fig. 11.8, p. 328). Clearly pesticides are only one of many factors, which include shooting, coursing (hunting with dogs) and food availability.

In most of the cases described in this section pesticides appear to have had at least a temporary effect on population size and in some cases serious long-term deleterious influences. This seems particularly the case with birds of prey, the effect of organochlorine pesticides in the early 1960s being especially acute. The following section examines various other cases of individual species where populations have fluctuated in recent years apparently mainly as a result of changes in agricultural practice – other than pesticide use.

10.7 FARMLAND MANAGEMENT AND PREDATORY BIRDS

A number of birds, mammals and invertebrates appear to flourish better on less well managed agricultural land and thus suffer as farmland is more intensively developed. The nature of this development will be more fully discussed in Chapter 11. Most of the present section refers to studies of two individual raptors that have been particularly exhaustively examined, the buzzard (*Buteo buteo*) and the red kite (*Milvus milvus*). The reasons why these two particular species have been studied in particular depth will be explained later but it is clear that studies of birds are at a much more advanced state than those of most other animals.

10.7.1 The buzzard

The buzzard has been studied in Britain and on continental Europe, whilst other *Buteo* species have been examined in North America. This bird feeds on a wide diversity of prey, which consists mainly of small mammals (especially voles and rabbits) but also includes small birds, reptiles, amphibians, large insects and earthworms (Cramp and Simmons 1980). Because it feeds mainly on small mammals rather than on birds, Brown (1976) suggests that its distribution has been much less affected by organochlorine pesticides than has that of other birds of prey. The changes of distribution of the buzzard are thus of particular interest.

It is thought that the buzzard bred throughout Britain in the beginning of the nineteenth century (Tubbs 1974). By 1900 the bird was restricted to parts of western Scotland, the Lake District, Wales and south-west England and also the New Forest, apparently through human persecution, particularly by gamekeepers. From 1914–54 the buzzard largely recovered although the onset of myxomatosis in the mid-1950s caused a subsequent decline. Moore (1957) estimated that total numbers of buzzards fell from about 20,000–30,000 before myxomatosis to about 12,000 subsequently. There has since been more recovery and the numbers now total between 15,000 and 25,000 (Brown 1976). There has also been some dispersion since 1954. Moore (1957) considers that the main control on buzzard dis-

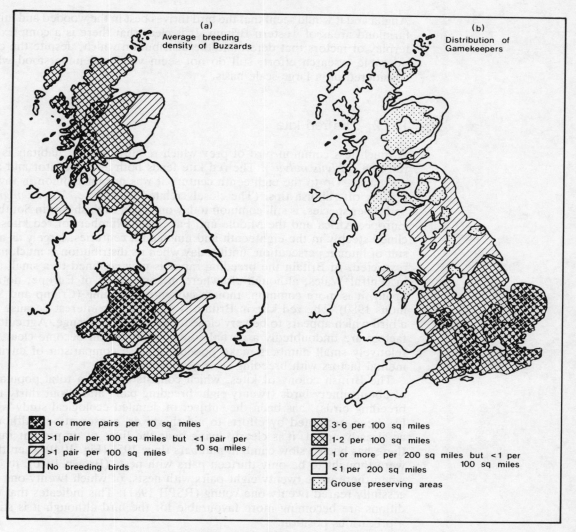

Fig. 10.8   Distribution of breeding buzzards in Britain in 1954 (a), compared
with the distribution of gamekeepers (b) (from Newton 1979).

person is keepering for game preservation. Figure 10.8 compares the
1954 distribution map for the species with the contemporary distribution
of gamekeepers. Moore's view is shared by Brown (1976) who suggests
that there is no other reason why the bird should be able to establish itself
in part of Morayshire or Hampshire but not in Norfolk. Although this
may be the case, it is difficult to explain the bird's absence from southern
Ireland (it has re-colonised Ulster) on these grounds. Elsewhere in Europe the buzzard has generally experienced a gradual decline, which has
been partially arrested in recent years except in countries such as France
where hunting pressure is intense (Cramp and Simmons 1980). In some
areas of Czechoslovakia it is suggested that the decline is closely related
to intensification in agriculture and, also, increases in afforestation. Notwithstanding Brown's contention regarding gamekeepers, it is difficult to
envisage the buzzard becoming common on the arable lands of East

Anglia and it would seem that the bird thrives best in the wooded and mixed farmland areas of western Britain. It is clear that there is a complex interplay of factors that determine the distribution which, despite the considerable research effort, still do not seem very well understood when considered on a large-scale basis.

10.7.2   The red kite

A much less common bird of prey which requires similar habitats is the red kite (*Milvus milvus*). The red kite feeds both as a predator and as a scavenger. Up to the eighteenth century it was not uncommon in towns, feeding off rubbish tips. (The closely related black kite, *Milvus migrans* and other species, is still common today on urban rubbish tips in Southern Europe, Africa and the Middle and Far East.) Numbers of red kites declined steeply in the eighteenth and nineteenth centuries, largely as a result of human persecution, until today when its distribution is much more restricted. In Britain the breeding range is now confined to a small area in central Wales, although elsewhere in some parts of Europe, notably Spain, it is more common, though currently declining (Cramp and Simmons 1980). The red kite in Britain is of particular interest because it is a bird which appears to be very close to the limit of its range. Agricultural factors are undoubtedly a partial control and, as will become clear, the relatively small numbers make possible a close comparison of environmental factors with breeding success.

The British colony of kites, which consists only of a total population of about ninety birds (twenty-eight breeding pairs and about thirty non-breeding birds), has been the subject of detailed ecological study which has been prompted by efforts to aid the birds' conservation. Although, as already noted, it is clearly living near the limit of its present range, there has been a slow climb in numbers since 1951 (Fig. 10.9) when there were thought to be only thirteen pairs with nests (Brown 1976), to 1981 when there were twenty-eight pairs with nests, of which twenty-one successfully reared twenty-one young (RSPB 1981). This indicates that conditions are becoming more favourable for the bird although it is still in a precarious position.

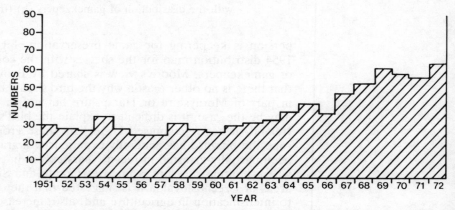

Fig. 10.9   Total numbers of red kite (*Milvus milvus*) in the breeding area in central Wales during 1951–72 (after Davies and Davis 1973).

There are four main landscape components to the kite habitat in Wales (Davies and Davis 1973):

1. Oakwood;
2. Marginal hill-land or ffridd (enclosed hill-ground, subjected to some improvement, used mainly for sheep grazing; generally between 200 and 350 m altitude; large areas have recently been planted for forestry);
3. Mountain sheep walk (mainly above 300 m altitude; now developed into heathland with grasses and areas of heath dominated by ling (*Calluna vulgaris*) and bilberry (*Vaccinium myrtillus*); about a quarter of this area within the kite zone has been planted with conifers);
4. Lowland and valley farmland (mainly below 200–250 m; occupied by enclosed farmland, mainly under grass; reseeding is widely practised, but there is little arable land).

Kites tend to occupy 'home ranges' or areas around nests for several years, despite possible changes in land use (Davies and Davis 1973). Landscape structure (i.e. the dispersion and arrangement of the different habitat types) seems significant in determining the dispersion of nests, not only among kites, but among most other raptors (Newton *et al.* 1981). Dispersed nesting will occur where nest sites are common throughout the feeding area, whilst grouped nests will occur where the sites may be abundant yet concentrated in small parts of the feeding area. In central Wales, the red kite mostly nests in oak (*Quercus petraea*), or, less commonly, beech (*Fagus sylvatica*). Open woodland, mainly of oak, mostly occupies the steep lower valley sides.

Changes in lowland farmland habitat (area 4) in the kite area have generally been very limited, although there has been some recent increase in hedgerow removal. Davies and Davis (1973) point out that since the first requirement of a hunting area appears to be an open landscape, then this development may actually improve the kite's prospects. However, they thought that the considerable increase in coniferous afforestation probably has the converse effect, at least after the conifer cover is about eight years old.

It has been shown that occupation by red kite of territories is relatively greater on land of lower agricultural potential, and nests are built more often (Newton *et al.* 1981). Conversely, Newton *et al.* showed that young were fledged more often on good land than on poor, although, overall, success was no greater on the good land. Decisive factors appear to be the lack of carrion available and greater human disturbance early in the year on good land. In the same study, no correlation could be established between breeding performance and proportion of forest in the nest territory. No evidence was found that afforestation had affected the territorial distribution. Newton *et al.* conclude that, 'the increase in kite numbers in recent years may have been due to less human persecution, an incidental result of changes in land use and in human attitudes' (p. 173).

The conclusions reached regarding the factors affecting kite breeding success should not necessarily be seen as being more generally applicable. In a study of the decline of the raven (*Corvus corax*) in southern Scotland and northern England, Marquiss *et al.* (1978) noted the greatest decline where there was most afforestation and, also, that nest sites were actually deserted when afforestation had taken place. Thus they concluded that coniferous afforestation of former sheep walk was a main cause of diminishing raven populations.

10.8   VERTEBRATES AS PESTS ON FARMLAND

Nowadays, the birds of prey discussed in the previous sections of this chapter are not generally seen by farmers as pests, although they nevertheless still suffer from the attention of gamekeepers and, to a lesser extent, egg collectors and others. Other vertebrates (including some bird species) are more clearly in competition with agricultural activity, the most obvious cases being the rabbit, the fox and, more recently, the badger. In the European Community, carnivores such as the lynx (*Lynx lynx*), pine martin (*Martes martes*), genet (*Genetta genetta*) and wild cat (*Felix sylvestris*) are also listed as endangered species, largely due to the fact that they are seen as pests and are thus subjected to excessive hunting (Nature Conservancy Council 1982).

The rabbit (*Oryctolagus cuniculus*) was introduced to Britain in the twelfth century. Sheail (1971) suggests that the development of the rabbit as a pest may be associated with the planting of enclosure hedges in the eighteenth and nineteenth centuries. In any case by the 1930s there were up to 100 million rabbits in Britain and they were causing substantial damage to arable crops.

In 1953 the virus *Myxomatosis cuniculi* was introduced. The disease is usually fatal and since it causes a discharge is easily transmitted from animal to animal. This led to a vast decrease in the rabbit population in the mid and late 1950s. However, any rabbits that do survive after having contracted the disease are usually immune to further infection. As has already been noted with reference to the stoat and the buzzard, the reduction in rabbit numbers had a substantial effect on those predators which relied heavily on it for food. At the present time, the disease tends to become re-established when rabbit populations reach a significant level. The reduction in rabbit numbers has also had an important effect on the flora of some grassland areas, notably chalk grassland, where sheep numbers had earlier been reduced, as will be discussed in Chapter 11.

It has already been noted that various methods of control of the fox (*Vulpes vulpes*) in Scotland do not appear to limit the size of the population as a whole (Kolb and Hewson 1980). In England, areas where fox-hunting takes place tend to have more foxes than where the animal is controlled in other ways. More recently the urban fox has become well established. Thus, should agricultural or other interests wish to undertake more rigorous control (which might be necessary, for example, should a rabies outbreak occur) then the urban areas will provide a reservoir of animals which will probably make it impossible to reduce the population substantially.

Recently a more controversial issue in parts of England has been the attempted eradication of the badger (*Meles meles*) usually by gassing the setts. Muirhead *et al.* (1974) suggested that badgers are important in transmitting bovine tuberculosis between and to cattle. They suggested that badgers act as a reservoir of the disease. Tuberculosis bacilli were first isolated from a badger carcase from a Gloucestershire farm in 1971 and since that time many infected badgers have been found. Small numbers of infected foxes, rats and moles have also been discovered but it is not thought that these come into contact with cattle. Kruuk *et al.* (1979) suggest that contact takes place between badgers and cattle especially in pastures where badgers may forage within a few feet of grazing cattle and may defaecate along field boundaries. Since the main food of the badger is the earthworm (*Lumbricus terrestris*), Kruuk *et al.* suggest that cattle pastures could be made unattractive to badgers by the application of pesticides such as carbamyl, benomyl and chlordane which would substantially reduce the earthworm population. It is believed that this might also benefit the grassland (Henderson and Clements 1977). Nevertheless, it may be the case that widespread application of pesticides in this way, besides being very expensive, would have extensive, and as yet unassessed, ecological side effects.

The relationship between badgers and bovine tuberculosis is, however, still contested (see, for example, Mabey 1980) and there is strenuous public objection to the gassing of setts. The gassing is, at present, taking place in limited areas of south-west England and the aim is to eradicate diseased animals rather than the population as a whole, although this objective inevitably means that some healthy animals are destroyed (MAFF 1979b; Zuckerman 1980). At the time of writing the policy is again under review following the discovery that the gassing process was very ineffective. Drabble (1981) makes the significant point that by destroying even those animals with mild doses of the disease one is eradicating the animals that have developed disease immunity. Thus the scientific basis of the programme is still open to question, unless one is working towards total badger eradication. It should also be pointed out that the badger suffers from persecution by human activity (e.g. sett digging) particularly around large urban areas.

There are, of course, other vertebrates which are treated as pests on agricultural land, including the brown rat (*Rattus norvegicus*), the grey squirrel (*Neosciurus carolinensis*), the coypu (*Myocaster coypus*) and the mole (*Talpa europaea*). All these are controlled with varying degrees of success, although from an ecological point of view none of them seems under particular threat. Nevertheless, numbers of rats, in particular, seem to have declined quite sharply in recent years. Some smaller birds have built up large populations and are seen as agricultural pests. An example is the starling (*Sturnus vulgaris*) although the feeding habits of this species indicate that the species is by no means entirely deleterious to agricultural activity. Tinbergen and Drent (1980) describe a study in which individual nesting birds were observed on farmland on the Dutch island of Schiermonnikoog. In this case the two main sources of food were the leatherjacket, (the grub of *Tipula paludosa*) and the caterpillar, *Cerapteryx graminis*. However, starlings do cause damage, Feare (1980) suggesting that the three main problems concern ripening cherries, germinating cereals and foods presented to cattle in winter. Part of the last problem is thought to be fairly recent and resulting from changes in agricultural practice. Cattle tend now more often to be housed indoors in winter and feed for them is more concentrated and accessible to the birds.

10.9 CONCLUSIONS

In this chapter we have been concerned with the effect of agricultural practices on various individual plant and animal species. Inevitably the discussion has been selective, mainly because, as was pointed out earlier, research effort has been rather variable. There has been particular concentration on the effects of pesticides and, in terms of the species considered, there has been some emphasis on the effects on birds of prey, the latter having been the subject of intense research effort over the last few years.

Much of the research – particularly on birds of prey – developed as a result of pressures by those concerned with wildlife conservation. This aspect will be considered more fully in the final chapter in this book. However, this also means that research has tended to concentrate on those species which are more obvious to the layman, although, of course, as soon as the research field develops in a serious way then the links with other parts of the ecological system become more explicit. In this context it is interesting to note that research at the Institute of Terrestrial Ecology, Monk's Wood, Huntingdon, England on pesticides led to a whole new area of study – the hedgerow (Pollard *et al.* 1974). As an ecological entity the hedgerow had been virtually ignored until this time. This particular feature will be more fully considered in the next chapter but it is important here in the way it demonstrates how one field of study has led to the development of others.

There are two main threads that can be drawn from the research described in this chapter. The first is that since chemical pesticides are extremely complex and variable compounds, they act on wildlife in a variety of different ways. Plants such as arable weeds will inevitably be reduced in number: that, after all, is the purpose of the herbicide application. There is still some debate as to whether certain plants are liable to suffer extinction, but recent work indicates that this will indeed be the case if herbicides continue to be used in the present volumes. Some types of pesticide, particularly the organochlorines, have been shown to be especially damaging to wildlife. This is because they are both chemically persistent and also capable of being absorbed in the body fat of vertebrates.

A second main conclusion is that pesticides are only the most recent of a variety of human pressures on wild plants and animals. (This is not to say that pesticides are of limited significance: their introduction has clearly had very considerable and wide-ranging effects.) The studies of birds of prey, in particular, demonstrate that these animals have been suffering from (essentially non-agricultural) human pressure, in the form of gamekeepers, egg collectors and others, for many years. Much of this pressure continues despite legislation designed to prevent it. Nevertheless, the scale and nature of organochlorine pesticide applications seemed to be, in several cases, decisive. The almost catastrophic decline in population of several species in the late 1950s and early 1960s demonstrates this. Very many species, notwithstanding the recent recoveries from the early 1960s, seem to be hovering dangerously close to serious decline.

Another main agricultural influence on plants and animals has been through changes in habitat. The effect of marginal changes, as demonstrated with the studies of the red kite, seems to be minimal. Clearly, at some stage significant change in habitat will cause the population of an organism to decline sharply. This aspect will be considered more fully in the next two chapters.

11   THE IMPACT ON ECOLOGICAL SYSTEMS: I THE LOWLANDS

11.1   INTRODUCTION

In the previous chapter the effect of modern farming on individual species was considered: in this and the following two chapters the various components that go together to make up the agricultural landscape will be examined. In lowland areas the rural landscape is – ecologically speaking – made up of many small facets or individual habitats, such as fields, hedges, woods, verges, ditches and ponds. Each of these habitats provides different ecological conditions, although many organisms depend on combinations of habitats for survival.

Modern farming has had a substantial effect on all these habitats, and the data shown in Fig. 11.1 demonstrate the scale of change that can take place, simply using the relatively crude measure of number of species. Figure 11.1 compares the numbers of different types of animals found in various equivalent habitats in unmodernised and modernised farmland. Although any such summary is a generalisation the data on which this figure is based are drawn from a representative variety of British sources (Nature Conservancy Council 1977). To refer to 'unmodernised' and 'modernised' farmland draws attention to the spectrum which extends from natural ecosystems at one extreme to the most intensively farmed land at the other (Fig. 11.2). Within modern farmland there is a range of options available to the farmer and depending upon which option he selects (in terms of type of crop, size of field, use of crop protection agents, etc.), there will be a greater or lesser proliferation of other (non-crop) plants and animals. Many farmers are conscious of the wildlife impact of their activity and use this as one criterion upon which to base the selection from available options. Inevitably this particular criterion will often come relatively low on the list of priorities. For many other farmers it is considered fleetingly, if at all.

Many farmers are also conscious of the importance of habitat diversity as an aid to wildlife conservation. The importance to birds and small mammals of the interrelationships between different habitats was demonstrated in a Canadian study by Wegner and Merriam (1979). They followed the movements of birds and small mammals between woods and adjacent farmland in an area 16 km south of Ottawa. Surrounding the woodland was a network of fencerows. These differ from British hedgerows in that the latter are usually planted and tend to be more species-

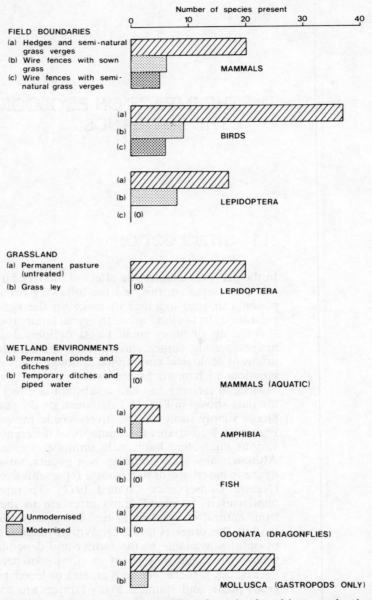

Fig. 11.1 Comparison of numbers of species found in unmodernised and modernised farm habitats (data from Nature Conservancy Council 1977).

rich. Fencerows consist of the vegetation that has grown up around fence lines, the actual fences often having rotted away. Wegner and Merriam showed that more species of birds and mammals moved more frequently between the wood and the fencerow than directly into the farmland. There was concentrated activity of birds and small mammals in the fencerow 'habitat corridors' leading out into the agricultural land. Poorly developed fencerow vegetation restricted foraging by wood nesters into fields. Thus, from the point of view of the birds and small mammals in

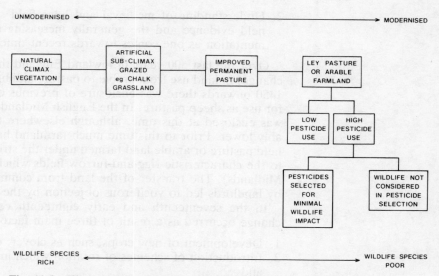

Fig. 11.2   The spectrum from unmodernised to modernised farmland.

this study, the three different habitats (woodland, fencerow and field) are linked together. The fencerows link the nesting and roosting woodland area with the foraging field area. This principle has wide-ranging significance. In many cases ecosystem stability depends upon the effectiveness of the links between its different components. In agricultural landscapes the components frequently consist of fields and areas of woodland, and the links are provided by hedgerows.

11.2   THE HISTORY OF FIELDS

When the open landscape is partitioned into fields, their shape and size are dependent on a range of factors. These include:

1. The limit of ownership or tenure;
2. The limit of relatively homogeneous soil or slope characteristics;
3. The type of use or crop intended; and
4. The method of cultivation (if any).

Subsequent to the initial laying out of the field pattern there may be amalgamations, modifications and internal subdivisions. In Britain, and in most of western Europe, the land has a long history of cultivation; consequently the field pattern is complex. Further, the length of time land has been under cultivation is extremely variable, as is the length of time different agricultural systems have been practised.

Some understanding of the history of fields is important to an understanding of the development of ecological environments. The information that provides a knowledge of field development is divided by Taylor (1975) into two types:

1. Knowledge of fields of prehistoric, Roman and Dark Age times is dependent entirely on remaining physical evidence;

2. Understanding of medieval and later field systems depends both on field evidence and the generally increasing amount of written documentation as one comes towards recent times.

Over the last 500 years in lowland England there have been successive changes in land use from arable to pasture and back again to arable. From 1450 onwards there was enclosure of previous open arable fields, mostly for use as sheep pasture. In the English Midlands about 9 per cent of land was enclosed at this time, although elsewhere the proportion was probably lower. Prior to this time much farmland had either been commonly held pasture or arable land farmed under the strip system (which gave rise to the characteristic rigg-and-furrow fields which are still common in the Midlands). The transfer of the land from common holdings to enclosure by landlords led to vociferous objection by the displaced tenants.

In the seventeenth and early eighteenth centuries more substantial change occurred as a result of three main factors:

1. Development of new crops, such as clover, carrots and turnips;
2. Introduction of schemes of rotation which included a grassland phase; and
3. Development of new implements.

Many of these changes have been described earlier (Ch. 2) and they all led to pressure to enclose previous common land. They also led to various forms of land improvement such as the drainage of wetlands, and to the enclosure of heath and lower moorland areas.

It was at this time that irrigation began to be practised. This was particularly concentrated on the fields alongside streams in chalk valleys in southern England. Small water channels were dug in efforts to induce early grass growth for sheep grazing in spring. The remains of these water-meadows, which gave rise to a distinctive flora, are apparent in many parts of Wiltshire, Hampshire and surrounding counties. Since the practical use of these irrigation meadows has now ceased, conservation of the characteristic flora is proving of some difficulty.

After 1750 Acts of Parliament were passed which resulted in accelerated enclosure of common land, and a large proportion of the enclosed landscape of lowland British farmland was created during the latter part of the eighteenth and the early nineteenth centuries. However, enclosure did not always result in a change of land use. This was particularly so in the upland areas where several hundred miles of stone walls were constructed. (It should be added that some marginal hill-land was ploughed up during the Napoleonic Wars although it was allowed to revert as soon as the pressure for food production declined.)

From 1750 onwards about five thousand individual Acts of Parliament were passed dealing with seven million acres of land (Pollard *et al.* 1974). Despite this the extent of enclosures which actually resulted from the Acts was very variable (Table 11.1). The table shows that parliamentary enclosure was perhaps much less important in some areas than has often been thought. Pollard *et al.* (1974) suggest that a major alternative reason for enclosure was the purchase of all common rights by one individual.

The permanent enclosure of fields led to an increase in underdrainage. Back-filled ditching and mole drainage (Ch. 4) began to be introduced in the seventeenth and eighteenth centuries and became widespread in the nineteenth century.

To summarise, it is clear that the history of fields is an extremely com-

Table 11.1 Examples of areas of land enclosed by parliamentary enclosures Acts for various English counties

County	ha	% of county
Northamptonshire	133,970	51.5
Huntingdonshire	43,711	46.5
Rutland	19,111	46.5
Bedfordshire	55,568	46.0
Oxfordshire	86,113	45.6
Yorkshire, East Riding	121,728	40.1
Leicestershire	81,088	38.2
Lincolnshire	200,903	29.3
Somerset	14,980	3.5
Essex	8,972	2.2
Shropshire	1,137	0.3

Source: Data from Pollard *et al.* 1974

plex subject. In this section we have merely been able to outline some of the main factors which have influenced developments, mainly as they apply to lowland Britain. The date of delineation of field areas varies greatly from area to area and the use to which fields have been put also changes over time and from place to place. Thus the variability of the ecological history of fields becomes apparent.

Therefore, if one considers any individual field in a specific area it is clear that it may have undergone at least two or three changes of use in the period since enclosure. The period of time since enclosure also varies (within certain broad limits) and so there is an almost infinite combination of historical factors influencing the present ecological structure of fields.

11.3 ECOLOGY OF LOWLAND GRASSLAND

11.3.1 Introduction

The agricultural significance of grasslands has already been discussed (Ch. 7). The purpose of the present section is to summarise the more important ecological features of lowland grassland. (The agricultural-ecological relations of upland grassland are discussed in the next chapter.)

In Chapter 7 five types of grassland were delineated from an agricultural point of view:

1. Natural pasture or range;
2. Permanent pasture;
3. Rotation pasture or leys;
4. Temporary pasture; and
5. Supplemental or complemental pasture.

From an ecological standpoint the first four groups represent a scale from minimal to maximal disturbance. (The final group is essentially a management category with no particular ecological significance.) The various degrees of agricultural modification referred to in this classification involve change away from the natural flora, depending on the extent and frequency of cultivation. Thus in the case of rotational or ley pastures, cultivation is usually so frequent that the only 'natural' flora consists of a small number of weed species more characteristic of arable land. Ecologically speaking, therefore, ley pasture is in many ways more akin to an arable crop than to natural pasture. Often non-native grasses, such as Italian ryegrass form the pasture. The agricultural purpose is to maximise grassland productivity and, as Mellanby (1981) points out, some leys dominantly composed of ryegrass are ready for grazing within six weeks of sowing. The animal wildlife within such crops is very limited, although it must be admitted that research on insect wildlife (apart from obvious pests) has been minimal. A further feature of rotational pastures is the intensive use of herbicides and fertilisers, which tend to have an adverse effect on natural wildlife. Some vertebrates, particularly rabbits, may forage on ley pasture although their base is in other adjacent habitats such as hedgerows or nearby fields of permanent pasture.

Notwithstanding the radical change which cultivation invokes, ley pastures will, if left, revert fairly rapidly to a form relatively close to the natural type. Indeed, the re-invasion of ley pasture by weed species is one of the main factors resulting in the 'lean years' during which productivity characteristically declines (Ch. 7). The classic example of this re-invasion is the 'old-field succession' quoted by Odum (1973) from the Piedmont region of the south-eastern United States. Abandoned fields were seen to progress over a period of about 150 years through a sequence of crabgrass, horseweed and aster-broomsedge grassland to grass-shrub, pine forest and ultimately oak-hickory woodland. A study of the reversion of a sown ley over chalk at Strawberry Down, Kent, has been described by Wells (1967). Prior to ploughing there had been thirty-eight grass species. The area was ploughed and sown with a mixture of Italian and perennial ryegrass, cocksfoot and red and white clovers. The Italian ryegrass disappeared by the end of the second season and for the remaining eight years over which records were kept most of the sward was made up of perennial ryegrass, cocksfoot and white clover. Later in the period the cocksfoot and white clover declined markedly. By the second season there were many weeds present which are characteristic of arable land, such as chickweed (*Stellaria media*), together with some more characteristic of grassland, such as dandelion (*Taraxacum officinale*). At the end of the study period there was a total of forty-eight invading species, although there were several species characteristic of the former downland which had not reappeared by the end of the study period.

As this study indicates, the composition and ecological status of grasslands is highly sensitive to management practices. The effect varies, however, in relation to the original nature of the grassland community and its edaphic environment (see, for example, French 1979).

In Britain natural lowland grasslands have generally been described in terms of three main groups (Duffey *et al.* 1974) depending on base status or soil type:

1. Calcareous (base rich);
2. Neutral; or
3. Acidic (base deficient).

11.3.2 Calcareous grassland

Calcareous grasslands are found over brown earth or rendzina soils which have a pH value greater than 7.0 and contain free calcium carbonate. The most common rock formations giving rise to such soils in Britain are Chalk, Jurassic limestone and Carboniferous limestone. In Britain these rock formations (with the exception of some Carboniferous limestone areas) are mostly in the 'lowlands' to the south and east (Fig. 11.3). The grasslands in these areas are therefore a sub-climax and they would revert to deciduous woodland were they entirely free of grazing interference. Those areas of natural grassland remaining today have become rather

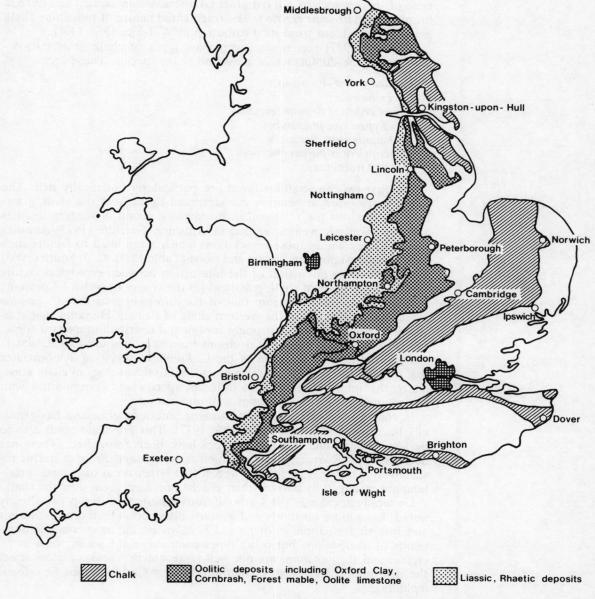

Chalk   Oolitic deposits including Oxford Clay, Cornbrash, Forest mable, Oolite limestone   Liassic, Rhaetic deposits

Fig. 11.3 The distribution of solid Chalk and Jurassic outcrops in England and Wales (after Curtis *et al.* 1976).

more fragmented than the distribution of the rock formations would suggest. This is largely because of the substantial areas of arable cultivation on the calcareous soils in the south and east.

In the late 1960s there were 43,546 ha (107,605 acres) of natural chalk grassland remaining in England, 70 per cent of which was in Wiltshire (Blackwood and Tubbs 1970). Most of these areas were sheep and cattle pastures which were too steep to reclaim. Within Wiltshire, Ministry of Defence land accounted for slightly less than half the total grassland area. Many of these areas were under arable crops early in the twentieth century before being taken over by the military. This is reflected in the presence of species such as tall oat grass (*Arrhenatherum elatius*) and upright brome grass (*Bromus erectus*). The fragmented nature of remaining chalk grassland is evident from its distribution in Wiltshire (Fig. 11.4).

Ratcliffe (1977) recognises seven main types of chalk grassland, depending on the dominance of grass and sedge species. These are:

1. *Festuca ovina–F. rubra*;
2. *Carex humilis*;
3. *Zerna erecta*; (*Bromus erectus*)
4. *Brachypodium pinnatum*;
5. *Arrhenatherum elatius*;
6. *Helictotrichon pubescens*; and
7. Mixed Gramineae.

The first two associations listed are particularly floristically rich. The *F. ovina–F. rubra* association is widespread throughout the chalk grassland areas whilst the *C. humilis* association is found in western districts and is particularly well developed in Wiltshire. Ratcliffe (1977) classifies various facies within these associations which are related to factors such as geographical position, aspect and slope (Table 11.2). C. J. Smith (1980) in an exhaustive discussion of the interaction between ecological factors and different types of chalk grassland (in the course of which he presents a different classification from that of Ratcliffe) suggests that *C. humilis* is a climatic anomaly on the western chalk of Britain. He notes that elsewhere in Europe it has a strongly continental distribution and for some, as yet unexplained, reason it is absent from sites in south-east England.

The debate over the status of the *C. humilis* association demonstrates that despite the very considerable research into the ecology of chalk grasslands, the interrelationships of different species and communities with environmental factors is still not well understood.

In contrast to chalk grassland, Jurassic limestone grassland has generally been less well studied (Ratcliffe 1977). This grassland tends also to be less variable and three main types have been recognised. These are dominated by *F. ovina, Brachypodium pinnatum* and *Bromus erectus* respectively. Duffey *et al.* (1974) suggest that larger areas of Jurassic grassland are managed by burning than are the grassland areas over the chalk.

Generally speaking, the Carboniferous limestone outcrop (as already noted) has a more northerly and westerly distribution than the Cretaceous and Jurassic formation. *F. ovina* and *F. rubra* are the most common dominants of associations but other types common in the south, such as *B. erectus* and *B. pinnatum*, are not well represented. Various factors affect the distribution and floristic composition over Carboniferous limestone, including:

1. The geographic separation from other calcareous grassland;

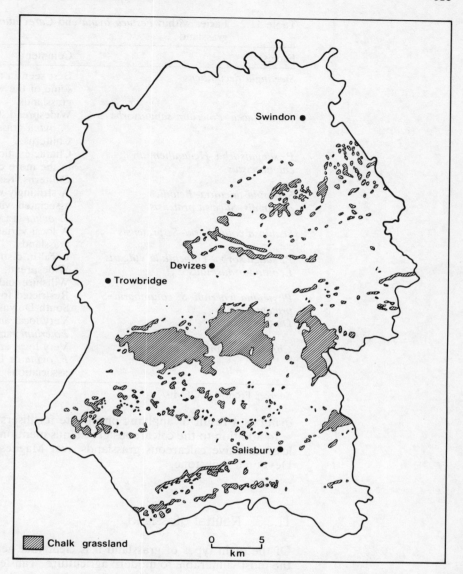

Fig. 11.4   The distribution of chalk grassland in Wiltshire, 1966 (from Duffey *et al.* 1974).

2. The often uneven and rocky nature of the terrain which may cause significant edaphic and microclimatic variations; and
3. The relatively large number of locally abundant species with a locally disjunct distribution.

Many of the Carboniferous limestone grasslands are at a much higher altitude than their Cretaceous and Jurassic counterparts (and, indeed, are often more 'upland' than 'lowland'). For this reason they are seldom under intense pressure from arable activities, which has been the case with much of the chalk grassland. Rorison (1971), however, quotes the example of limestone grasslands now dominated by *B. erectus* and *F.*

Table 11.2   Facies within *Festuca ovina* and *Carex humilis* associations of chalk
grassland

Facies	Comment
Sieglingia decumbens	Best seen in the Isle of Wight and some of the westernmost chalk grasslands.
Carex flacca–Poterium sanguisorba	Widespread, but characteristic of the *F. ovina* grasslands particularly in the Chilterns.
P. sanguisorba–Helianthemum chamaecistus	Characteristic of south-facing slopes, in the more continental areas; e.g. Chilterns, Kent, East Anglia.
Serratula tinctoria–Betonica officinalis–Succisa pratensis	A strikingly western facies, often associated with *C. humilis* but also in *F. ovina* grasslands.
Scabiosa columbaria–S. pratensis	A local variant of the western chalk grasslands.
P. sanguisorba–Filipendula vulgaris	Local in distribution.
Leontodon hispidus	Widespread, but best developed in Wiltshire and Dorset.
Phyteuma tenerum–S. columbaria–S. pratensis	Restricted to Hampshire and the South Downs.
Daucus carota	Very local and associated with the *Poterium* facies.
Chrysanthemum leucanthemum	Very local and always associated with *F. ovina* or *C. humilis* main associations.

Source: From Ratcliffe 1977

ovina as a result of applying phosphate fertilisers.

In addition to the calcareous grasslands mentioned so far, there are also less extensive calcareous grasslands over Magnesian limestone and over Devonian limestone.

11.3.3   Neutral grassland

Of the main types of grassland it is clear that neutral grassland is by far the most vulnerable to modern agriculture. Tansley (1935) defined neutral grassland as 'semi-natural grasslands whose soil is not markedly alkaline nor very acid, mostly developed on the clays and loams'. Ratcliffe (1977) includes also some communities transitional to other vegetation formations. It should immediately be clear that the soils described by Tansley are those which can be agriculturally highly productive. The clays have been particularly vulnerable in recent years to modern drainage developments, whilst the loams have always been among the potentially most valuable soils.

As Duffey *et al*. (1974) point out, neutral grasslands have been extensively studied on the European mainland, largely because of their agricultural importance. The floristic composition and ecological relationships of neutral grasslands are complex and are only briefly summarised in Table 11.3. Floristic composition depends on four main factors:

1. Surface water regime. Ecological structure depends on the ability of

species to withstand inundation, on the one hand, and desiccation, on the other. Ratcliffe (1977) cites two examples: firstly, great burnet (*Sanguisorba officinalis*) which is common on flood meadows, but not able to withstand long inundations such as occur on the washes; secondly, the cowslip (*Primula veris*) which occurs in many permanent grasslands but is absent from flood meadows.

The characteristic species of neutral grassland fall into groups which are able to withstand progressively longer periods under water (Duffey *et al.* 1974) (Table 11.3). The first two of these groups are grasslands, but the latter three are essentially fen-type vegetation.

Table 11.3   Characteristic species groups of neutral grassland, arranged according to soil moisture status

Dry	1.	*Festuca rubra, Anthoxanthum odoratum, Primula veris, Orchis morio.*
Moist	2.	*Holcus lanatus, Alopecurus pratensis, Poa trivialis, Sanguisorba officinalis, Cardamine pratensis.*
	3.	*Carex panicea, C. disticha, Juncus articulatis, Dactylorchis incarnata, Lychnis flos-cuculi.*
Wet	4.	*C. acutiformis, J. effusus, Caltha palustris, Galium palustre.*
	5.	*Phragmites communis* or *Glyceria maxima.*

Source: Derived from Duffey *et al.* (1974), but including minor modifications by Ratcliffe (1977)

2. Soil type. Soils influence neutral grassland composition in a number of ways. The most important soil factors are:
   (a) soil structure and soil texture, through their influence on soil drainage;
   (b) soil organic matter content;
   (c) soil base status.
3. Geographical location. This factor subsumes both climatic influences and geological distribution.
4. Management. The two main management operations on neutral pastures are grazing and hay-cutting. Animals may be turned out to graze either for the whole growing season or for limited periods. Early spring grazing followed by hay-cutting is fairly common in the water meadows (group 6 in Table 11.4) and the northern hay meadows (group 10). These combinations of grazing activity will affect the floristic composition through factors such as palatability, resistance to grazing and ability to colonise a sward opened up by grazing (Ratcliffe 1977).

If animals are not turned out until late in the season a grassland dominated by the agriculturally attractive grasses such as *Lolium perenne* and *Trifolium repens* will result. This type of grassland is less ecologically diverse.

As Ratcliffe points out, grazed grasslands are an important breeding habitat for many species of birds including waders, some ducks and various passerines. The taller vegetation which is standing if hay-cutting is practised is less favourable to these types of birds but others, such as the partridge (*Perdix perdix*), corn bunting (*Emberiza calandra*) and corncrake (*Crex crex*) value this type of habitat. Indeed, the substantial reduction in these three species in recent years can partly be accounted for as a result of the decline in traditional hay-making.

Table 11.4   Classification of neutral grasslands

Group	Examples of constant distinctive species	Geographical and other features
1. Base-rich marshes	*Calthu palustris* *Juncus articulatus*	Often associated with calcareous spring-lines; more common in southern Britain: very limited extent. Grass sometimes high. Grades into rich fen.
2. Base-poor marshes	*Carex rostrata* *Juncus acutifloris*	More common in north. Often associated with non-calcareous rivers. Often found as a minor component associated with groups 1 and 7. Grades into poor fen.
3. Sedge-rich meadows	*Carex nigra* *Carex panicea* *Molinia caerulea*	Dwarf-grass/sedge community. Often found as a minor component associated with groups 1 and 7. Soil base status lies in the middle range between acidic and basic.
4. Tall grass washlands	*Glyceria maxima* *Phalaris arundinacea*	Washlands are a specialised local meadow system associated with flood-relief schemes in the Fenlands.
5. Washlands and wet alluvial meadows	*Alopecurus geniculatus* *Carex hirta*	Group 4 – Distinctive but relatively infrequent group, found in areas subject to prolonged winter flooding. Botanically poor. Group 5 – Botanically richer than group 4.
6. Water-meadows	*Cardamine pratensis* *Festuca arundinacea*	Specialised meadows, man-made between c. 1700 and 1850. Particularly common in the Chalk Valleys of southern England.
7. Alluvial meadows	*Alopecurus pratensis* *Filipendula ulmaria*	Carry vegetation which will develop on alluvial clay-loam subject to inundation by running water. Grass-dominated. Geographically southern and eastern, e.g. in Upper Thames Valley. Species rich sward. Occasional flooding.
8. Calcareous clay pastures	*Briza media* *Senecio erucifolius* *Primula veris*	Geographically a southern group; flooding absent, commonly on heavier boulder clay soils. (Presence of cowslips (*P. veris*) particularly characteristic).
9. Calcareous loam pastures	*Briza media* *Carex flacca* *Poterium sanguisorba*	Only fragments persist, easily worked free draining soils (lighter, sandier soils than group 8).
10. Northern hay meadows	*Bromus mollis* *Geum rivale* *Trisetum flavescans*	Occurs on Carboniferous limestone or calcareous sandstone between 200 and 300 m in northern England and central Wales.
11. Northern grazed meadows	*Briza media* *Carex panicea*	Associated with spring and flush conditions on upland and hill valley pastures of the Carboniferous formations.

The following groups are widespread and often extensive. However, they have a low nature conservation value. They are abundant throughout Britain and, given a suitable soil type, Ratcliffe considers that it is possible to create these groups from one or other of the eleven main groups by appropriate management.

12. Ordinary wet meadows	*Deschampsia cespitosa* *Juncus inflexus*	Occurs on heavy soils with impeded surface drainage.
13. Ordinary damp meadows	*Juncus effusus* *Poa trivialis*	On peaty and/or lighter soils with high water table but not impeded drainage.
14. Ordinary dry meadows	*Cynosurus cristatus* *Lolium perenne*	Most widespread form of neutral grassland, commonly develops when sown grassland is allowed to revert.

Agricultural management	Particular ecological significance for other species	Example sites
Generally unmanaged, though some summer grazing by cattle.	—	Brastey, Glos (Springs rising from Jurassic limestone) Calceby Beck, Lincs
As group 1.	Often very important for wild-fowl, especially geese and ducks.	Oykeil Marshes, Ross-Sutherland Ken-Dee Marshes, Kircudbright
Summer grazing (these are sometimes disused water meadows, e.g. Bransbury).	—	Bransbury Common, Hants Wintringham Marsh, E. Yorks
Summer grazing. Occasional cutting for hay.	Outstanding ornithological importance; very large winter wild-fowl population (especially ducks, swans) and also largest single concentration of rare summer breeders (e.g. black-tailed godwit, ruff, black tern).	Ouse Washes, Cambs. Baston Fen, Lincs
Summer grazing, occasional cutting for hay, but can be easily 'improved' by drainage as grassland or for arable.		
Management very variable, only very rarely in use as working water-meadows. Some areas intensively managed with herbicides and fertilisers, thus reducing botanical diversity and interest.	Provide interesting aquatic habitat where water courses still exist.	Mostly in Avon, Valley of Wilts and Hants, especially Britford, Downton, Wilts
Management constant over long periods: hay-making and/or grazing. Many now reclaimed for agriculture, thus no longer biologically valuable.	In some areas support redshanks, common snipe and other breeding populations.	Cricklade, Wilts Sibson-Castor Meadows, Hunts
Varies from hay-cutting to summer grazing. Groups 8 and 9 have now become rare since they are particularly susceptible to ploughing and reseeding.		Clattinger Farm, Oaksey, Wilts Upwood Meadows, Hunts Moor Closes, Ancaster, Lincs
Grazed by cattle during autumn and winter. Many areas now improved and thus botanically interesting examples are rare.	—	Upper Teesdale Meadows, Durham–Yorkshire
Grazed by sheep and cattle.	—	Upper Teesdale Meadows, Durham-Yorkshire
Lightly grazed by cattle in summer, resulting in tussocky vegetation	Larger examples provide feeding grounds for waders.	(Widespread)
	—	(Widespread)
Grazing	—	Fattening pastures of north Northumberland, Leicestershire and Romney Marsh

At the time of writing, the last substantial habitat of the corncrake, the coastal machair grassland of the outer Hebrides, is under threat from grassland improvement schemes (Cadbury 1980).

More drastic changes have been brought about by the now wide-spread application of artificial fertilisers, which is often preceded by ploughing and sowing of commercial seed mixtures. Herbicides are also frequently applied. Such management, although having clear agricultural advantages, results in grasslands of minimal ecological interest. Many grasslands have also been ploughed up and converted to arable land.

Using these various factors, Ratcliffe (1977) has developed, for Britain, the neutral grassland classification of D. A. Wells (Duffey et al. 1974). This, together with various other information, is summarised in Table 11.4. Groups 1, 2 and 3 of Ratcliffe's classification can be grouped as the Carex–Juncus communities of marshes and silty peats. These are not, Ratcliffe suggests, grasslands in the literal sense but are included because they often pass into one or other of the true grassland types on adjacent areas where the water table is lower and also because they tend to be managed in the same way.

Neutral grasslands sensu stricto are included in groups 4 to 14 in Table 11.4, although the last three groups are considered to have a low nature conservation value and are also fairly easily 'reconstructed'.

The vulnerability of the neutral grasslands is well illustrated by the following extract from Ratcliffe (1977: 194):

Since this review began (1965–77), two grade 1 neutral grassland sites have been totally destroyed by ploughing; at Worlaby and Corby, in Lincolnshire. Ploughing has also destroyed part of Sibson meadows and part of the Upwood site has been treated by herbicides. Drainage schemes are taking place in the Derwent valley which will make possible agricultural improvements by ploughing etc. A major drainage scheme is proposed for the Avon valley south of Salisbury which will place in jeopardy not only the Britford water meadows but also much of the ecological interest outside the water meadows. Recently the farmer at Pen yr Hen Allt applied artificial nitrogenous fertilizers for the first time and is likely to continue this substitute for farmyard manure to the detriment of the botanical interest of the sward.

Most of the grade 1 and 2 neutral grassland sites are being farmed by the older generation of farmers. A change of ownership or tenancy inevitably means destruction of this habitat and it can be predicted with great confidence that very few of these sites will remain by 1990, unless strenuous efforts are made to safeguard them. As a class, this is the most threatened of all British habitats with high nature conservation interest, but it is also perhaps the most neglected in terms of actual conservation.

11.3.4 Lowland acid grasslands and heaths

Some areas of sandy soils in lowland England are covered with what Tansley (1953) described as 'grass heath'. These are grasslands which occur on the same soil as Calluna heath. The largest remaining area of lowland acid grasslands in England is the commons of Surrey, Berkshire, Hampshire and Dorset, together with parts of the Breckland area of East

Anglia. In all these cases the continued existence of the grassland communities, which are dominated by *Festuca ovina*, *Deschampsia flexuosa* and *Agrostis tenuis* set in a matrix of lichens and mosses (Duffey *et al.* 1974), depends upon continued grazing. When grazing ceases the vegetation will be invaded by dwarf shrubs. Conversely grazing will convert heath into grassland. The heath is dominated by *Calluna vulgaris* although *Erica cinerea* is also present. Gorse (*Ulex* spp.) is also common.

Some of the soils in the areas mentioned contain appreciable quantities of calcareous material and the grassland communities are characteristic of the calcareous types rather than the acidic. This is the case in Breckland, the grasslands of which have been intensively studied by A. S. Watt (1936–1940 and 1981a, b). The Breckland soils are a mixture of chalky till and sands overlying chalk. Acidity ranges from a pH of 3.6 (strongly acidic) to 8.0 (highly calcareous) and substantial variation may be found over very small distances, of even a few centimetres (Corbett 1973). The plant communities change according to these soil variations and Ratcliffe (1977) points out that the characteristic grasslands of the open brecks probably originated through cultivation. A further crucial biotic influence on the grasslands has been the rabbit. The influence of the rabbit on floristic composition has been examined more recently by Watt (1981a, b).

The land use of the Brecklands is of interest in that much of the land is now marginal to agriculture. In recent years the use of fertilisers has enabled cultivation of areas of soils that were previously too acidic. The area of semi-natural habitat has decreased in the last 50 years by approximately 87 per cent, from about 60,000 ha to about 8,000 ha today. Of the present area, 6,070 ha form the military Stanford Practical Training Area and the remaining 2,000 ha or so are fragmented areas of grassland (Ratcliffe 1977). In the mid-1930s large areas of Breckland were planted to commercial forestry and there are, today, 18,600 ha of forests, mainly comprising Scots and Corsican pine. The remaining losses of semi-natural habitat (33,330 ha) have been mostly to agriculture, with some urban developments, especially around Thetford (Fig. 11.5).

The extent of pressure on the heaths and grass heaths in Dorset is shown in Fig. 11.6. This series of maps illustrates the reduction in the extent of the Dorset heathlands (defined by Webb and Haskins (1980) as associations of *Calluna vulgaris*) from 1759, when there were about 40,000 ha, to 1978. By the latter date only about 6,000 ha remained. Figure 11.7 shows the decline in this area and it is clear that if the current rate of decline continues then this form of vegetation will disappear from this area at the start of the twenty-first century.

Webb and Haskins (1980) also point out that grazing on these Dorset heathlands continued into the eighteenth century but declined thereafter and had more or less ceased altogether by the end of the nineteenth century. There is now so little grazing that burning is unregulated. The diminution up to 1896 was mainly the result of reclamation to agriculture but after that time there was considerable loss to urban development (especially of Bournemouth) which continues up to the present time. In addition there were losses to forestry and to mineral extraction.

Losses of semi-natural acid and calcareous grassland are usually easier to quantify than are those of the neutral grasslands because the former tend to be in geographically contiguous areas.

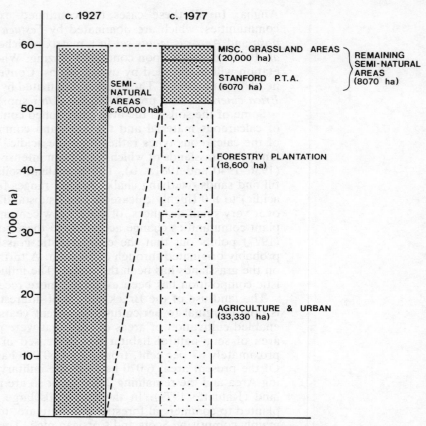

Fig. 11.5   Reduction in the area of semi-natural habitat in the East Anglian Breckland (data from Ratcliffe 1977).

11.3.5   Fauna of lowland grassland

Habitat modification, of course, not only affects the floristic composition but also has an impact on the faunal populations which the vegetation supports. The effect of grassland management and cultivation of pasture on earthworm populations and other soil organisms has already been discussed in Chapters 7 and 8. This section is, therefore concerned with organisms which spend most of their life cycle above the soil surface.

Invertebrates

Probably the best known invertebrates of the lowland grasslands are the Lepidoptera. The numbers of species of butterflies and moths recorded for the various lowland grassland and heath types in Britain are summarised in Table 11.5. The figures demonstrate that calcareous grassland and also the Breckland heaths (which are in part calcareous) are particularly species-rich. However, this table takes no account of the relative abundance or scarcity of particular species. Some species, such as, for example, the large blue butterfly (*Maculinea arion*), formerly found on

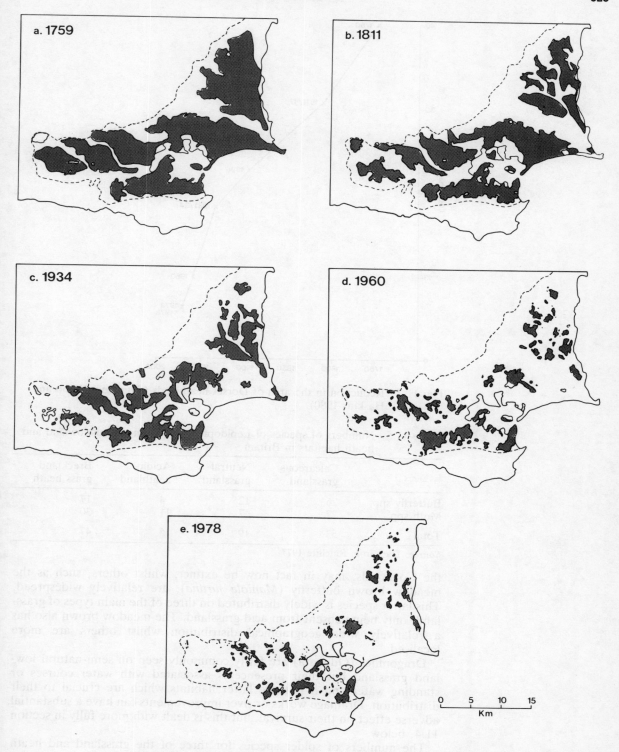

Fig. 11.6 Extent of heathland in Poole Basin, Dorset, 1759–1978 (after Webb and Haskins 1980).

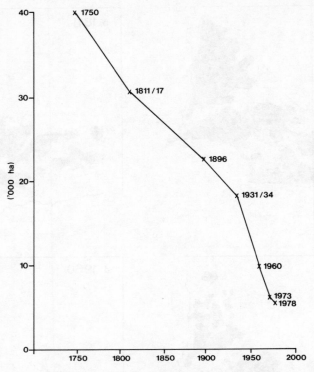

Fig. 11.7   Reduction in the area of Dorset heathlands (data from Webb and Haskins 1980).

Table 11.5   Numbers of species of Lepidoptera known in various lowland and heath habitats in Britain

	Calcareous grassland	Neutral grassland	Acidic heathland	Breckland grass heath
Butterfly spp.	26	12	4	17
Moth spp.	27	7	11	30
Total	53	19	15	47

Source: Data from Ratcliffe 1977

the Cotswolds, may in fact now be extinct, whilst others, such as the meadow brown butterfly (*Maniola jurtina*), are relatively widespread. This latter species is widely distributed on three of the main types of grassland, only being absent from acid grassland. The meadow brown also has a relatively wide geographical distribution whilst others are more localised.

Dragonflies (Odonata) are fairly commonly seen on semi-natural lowland grassland but their presence is associated with water courses or standing water and it is these latter habitats which are crucial to their distribution. Drainage works or river improvements can have a substantial adverse effect on their survival, but this is dealt with more fully in section 11.4, below.

The numbers of spider species for three of the grassland and heath types are summarised in Table 11.6. As with the information given for the Lepidoptera, this table masks the scarcity and abundance of particular

Table 11.6   Numbers of species of spider known for three lowland grass and heath habitats in Britain

	Calcareous grassland	Acidic heathland	Breckland grass heath
Numbers of species known	46	63	26

Source: Data from Ratcliffe 1977

species, many of which are very local in their distribution.

Very little is known of the interrelationship of invertebrate communities and, indeed, as was noted in Chapter 10, some groups are very under-recorded. It follows that there is little knowledge also of the effect of management practices on the invertebrate fauna. In one of the relatively few recent studies of the effect on invertebrates of grassland management practice, Morris (1979), working on a site at the Caistor Hanglands NNR, has described the almost catastrophic impact of cutting of grassland on the Heteroptera, compared with grazing.

Working near Annisdale, New South Wales, Australia, King and Hutchinson (1980) showed that superphosphate application to grassland led to a fourfold increase in the number of Collembola and Acarina. However, increasing the stocking rate usually resulted in a decrease in numbers. As we noted in Chapter 7, grassland management also has a fundamental effect on earthworm numbers.

Birds

The number of bird species that actually breed in lowland grassland is relatively few, although there are some notable examples. The stone cur-lew (*Burhinus oedicnemus*) requires the grassland to be well grazed by rabbits and then nests directly on the soil surface; amongst the flint pebbles it is particularly well camouflaged. Formerly common in Breckland and on part of the chalk downland, numbers have declined dramatically in recent years as a result both of reclamation and of the general increase in vegetation caused by the absence of the rabbit after myxomatosis. The survival of the stone curlew in Britain is now greatly at risk (Ratcliffe 1977). Other birds which nest in similar habitats include the ringed plover (*Charadrius hiaticula*) which has almost disappeared from its former Breckland habitats and the wheatear (*Oenanthe oenanthe*) which has declined markedly.

Fuller (1982) suggests that water level is the primary influence on breeding bird species composition on neutral grasslands. Thus wildfowl and waders will breed most frequently where there is nearby standing water or a high water table, respectively.

The acid heaths in Dorset form the habitat of the Dartford warbler (*Sylvia undata*). The case for conservation of these heaths is made by Bibby (1978), and it is clear that numbers of this species also have severely diminished in the last few years as a result of habitat change. The skylark (*Alauda arvensis*) and the lapwing (*Vanellus vanellus*) are much more common species of grassland, particularly on neutral grassland. Nevertheless, the lapwing also has declined somewhat as a breeding species in recent years. The damper forms of neutral grassland are the nesting places of a wider range of species including snipe (*Gallinago gallinago*), redhawk (*Tringa totanus*), yellow wagtail (*Motacilla flava*), and reed bunting (*Emberiza schoeniclus*), none of which can be described as common species, although some are widespread. The very scarce black-tailed godwit (*Limosa limosa*) and the ruff (*Philomachus pugnax*) both breed in the Ouse Washes of East Anglia (Sharrock 1976). The curlew (*Numenius arquata*) breeds in a variety of habitats including most forms of heath and grassland.

The number of species breeding on calcareous grassland is less, as indeed is the density of breeding birds. Table 11.7 summarises the density of birds for various habitats and shows that even in 1941 before the recent intensification of agricultural activities, the density of birds on chalk downland during the breeding season was much lower than on heaths or, in the summer, in woodland or on arable land. In their study Colquhoun and Morley (1941) counted only ten different species on the Berkshire down in the breeding season. These include the skylark, meadow pipit (*Anthus partensis*), partridge (*Perdix perdix*) and lapwing. The partridge, as was noted in Chapter 10, has subsequently become much less common, both on the downland and elsewhere.

Table 11.7   Density of birds in different types of habitat

	Bird population per hundred ha		
	Winter	Summer	
		Pre-breeding	Breeding
Chalk downland, Berkshire (Colquhoun & Morley 1941)	34	86	79
Heaths and woodland (Lack 1935)	—	—	165
Oak woodland (Murton 1971)			1235–1360
Farmland, Cambridgeshire, predominantly arable (Murton 1971)	1148	—	748

Source: Data from Duffey *et al.* 1974

Whilst the number of birds that actually breed on grassland is relatively low, the number of species that forage on grassland is considerably higher. Most of the large number of species that nest in woods and hedgerows forage in adjacent meadows, where these are available. Where feeding areas are limited in any way then, in general terms, the population levels tend to be lower.

The intensive use of grassland for winter feeding by birds can pose particular problems. In the last few years the number of wigeon (*Anas penelope*) wintering on the Ouse Washes has increased almost threefold (Owen and Thomas 1979) to numbers as high as 42,500 (Prater 1981). Although the total population of wigeon in Britain and north-west Europe appears stable, birds seem to be resorting increasingly to inland habitats and salt-marshes as feeding habitats. As Owen and Thomas suspect, this may lead to increasing conflicts with farmers as the birds resort to arable land.

Grazing geese are also a problem on some coastal grasslands and the effect of certain insecticides on geese has already been noted (section 10.5). Patton and Frame (1981) have described a study of three sites in Scotland – two on the Isle of Islay, Inner Hebrides, grazed by barnacle geese (*Branta leucopsis*) and near Castle Douglas, Kirkudbright, grazed by greylag geese (*Anser anser*). The amount of grass consumed by the geese was carefully estimated and it was considered that on Islay there was a maximum loss of about eighty cow grazing days per hectare, and at Castle Douglas a maximum loss of about ninety cow grazing days per hectare. A study of the gizzard contents of twenty geese indicated a preference for sown ryegrass species (*Lolium* spp.) as opposed to indigenous grass species. Patton and Frame concluded that compensation to farmers is the only effective resolution of the problem if the habitats are to be preserved.

Mammals and reptiles

The significance of the most important mammal inhabitant of lowland grasslands and heaths – the rabbit – has already been discussed (section 10.8). Rabbits have had most success in re-establishing themselves where the substrate is both calcareous and sandy: the calcareous base-rich soils provide sufficient fodder and the sandy nature of the soil provides good dry burrows. Thus, as Ratcliffe (1977) points out, they have been particularly successful in the East Anglian Breckland. As has already been mentioned, the rabbits themselves exert a considerable influence on the flora. In addition, reduction in rabbit numbers will, for example, cause an increase in those small herbivorous rodents that need cover, such as the harvest mouse (*Micromys minutus*).

Species which prey on the rabbit, such as the fox (*Vulpes vulpes*), the stoat (*Mustela erminea*) and the weasel (*M. nivalis*), tend to flourish in proportion to the size of the rabbit population. The relationship between prey and predator populations is not a simple one and the work of Jefferies and Pendlebury (1968) on gamekeeper 'vermin bags' (referred to earlier in section 10.6) showed that the weasel population tended to be stable in relation to the decline in rabbits, whilst the stoat population fell. This presumably reflected the fact that weasels could more easily survive on the various rodents which increased as the vegetation 'recovered' from intensive rabbit grazing.

Observation suggests that the hare (*Lepus capensis*) is somewhat in decline in recent years and once again, this may be due to changes in agricultural practice (see section 10.6). It is likely that traditional annual hay-cutting will favour the hare population, compared with two or three mechanised silage cuts. High densities of hares are associated with chalk and limestone areas (Ratcliffe 1977). Broekhuizen (1982) quotes data for a number of countries in Europe showing consistent trends in hare populations, and suggests that the changes are due to both land use effects and natural population dynamics (Fig. 11.8). Ratcliffe also suggests that the decline in the bat population, at least in southern England, may be due to the conversion of grassland and heathland into intensively managed farmland. Changing agricultural practices and effects of pesticides on food supplies are also cited as a threat to bat populations in Europe (Nature Conservancy Council 1982).

All six species of British reptile – the slow-worm (*Angius fragilis*); the sand lizard (*Lacerta angilis*); the common lizard (*L. vivipara*); the grass snake (*Natrix natrix*); the smooth snake (*Coronella austriaca*) and the adder (*Vipera berus*) – have declined, according to Ratcliffe, through destruction of their heathland habitats. Similarly the common frog (*Rana temporaria*) and common toad (*Bufo bufo*) have also declined. These are discussed more fully in Chapter 13.

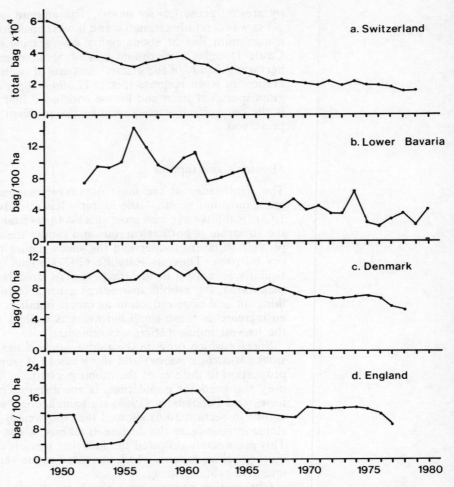

Fig. 11.8   Changes in hare populations in European countries as recorded by data on hare-bags reported by gamekeepers (from Broekhuizen 1982).

11.4   ARABLE LAND

The agricultural system of arable farming has already been discussed in Chapter 8, and many of the direct ecological effects of arable techniques – particularly the application of fertilisers – on individual species were examined in Chapter 10. The purpose of the present section is mainly to summarise these various effects in the context of the arable ecological system as a whole. There has been relatively little published work on the 'natural' ecology of arable fields – if, indeed, that particular phrase has any very precise meaning in view of the essentially artificial nature of the arable field – and most of the following information derives, unless otherwise stated, from points made by Ratcliffe (1977).

The presence of wild plants or weeds in arable fields indicates to a greater or lesser extent, depending on their number, a failure in agricultural terms. The effect of herbicides on weed species diversity and population has already been discussed (section 10.3). It was noted there that some weed species such as *Bromus erectus* are threatened by changes in cropping practice and herbicide usage. The secondary effect of these reductions on invertebrate animal life is not very well understood. A small number of Lepidoptera are restricted to one or two plant species, an example being the moth *Endothenia ericetana*, which is restricted to the field woundwort (*Stachys arvensis*). About twenty-five species of Lepidoptera and twenty-five species of Coleoptera are restricted to weed species as food plants.

As was noted in the introduction to this chapter, most birds use fields in conjunction with other nearby habitats, particularly hedges (section 11.5, below). This also applies to small mammals, although, in the latter case, some species appear to be able to survive in the relative isolation of arable fields. Jefferies *et al.* (1973) studied the population of three species of small mammal, the long-tailed field mouse (*Apodemus sylvaticus*), the bank vole (*Clethrionomys glareolus*) and the short-tailed vole (*Microtus agrestis*), in a field near Ramsey, Huntingdonshire. The field was a fen-type arable field 225 m by 55 m in size (1.9 ha or 4.7 acres) bounded by water-filled dykes 3 m wide and 2 m deep. The number of mammals present was sampled by means of Longworth traps before and after the field was drilled with cama winter wheat treated with dieldrin/mercury seed-dressing. A number of long-tailed mice were trapped by Jefferies *et al.* well away from the field edge. They concluded that mice occupied both the field edge, moving out diurnally to forage, and also more central field locations.

Some birds have become adapted to nesting on arable fields. Examples cited by Ratcliffe (1977) are the lapwing (*Vanellus vanellus*), oystercatcher (*Haematopus ostralegus*), common curlew (*Numenius arquata*), stone curlew (*Burhinus oedicnemus*), skylark (*Alauda arvensis*) and reed bunting (*Emberiza schoeniclus*), although in none of these cases, with the possible exception of the lapwing, could arable land be said to form a very successful breeding habitat. Indeed one is tempted to speculate that the selection of such a habitat, with the very critical timing of cultivation, could well have accelerated the rate of decline of some of these species. The population reduction in the grey partridge (*Perdix perdix*), probably partly resulting from the decline in insect food, has already been noted (section 10.6).

11.5   FIELD MARGINS AND BOUNDARIES

As was mentioned in the introduction to this chapter, much of the animal wildlife depends on the juxtaposition of field and the less disturbed habitat at the field edge. Most lowland cultivated fields have a limited area of 'headland' or uncultivated area around the margin which will be referred to briefly in section 11.5.2 below. By far the most important habitat at the field margins is, however, the hedgerow and in recent years the ecology of the hedgerow has been the subject of considerable study.

11.5.1 Hedgerows

The most comprehensive study of British hedges is that of Pollard *et al.* (1974). As that study indicates there is no doubt that there has been a substantial reduction in hedgerow lengths in recent decades. Hooper (1979), for examples, estimated that between 1945 and 1970 hedgerows were destroyed at a rate of about 8,000 km y^{-1}, giving an overall loss of 25 per cent of the total hedgerow length in Britain over the 25-year period. The rate of removal shows a marked regional variation, however, and is greatest in eastern England where there is a greater concentration on arable farming. The rate of removal in one part of eastern England is summarised in Table 11.8. Ratcliffe (1977) calculates that in eastern England typical areas now have between 8 and 13 km of hedgerow per km^2, whilst in lowland areas of western Britain the figure is between 24 and 29 km per km^2. It has also been claimed by Hooper (1970) that rates of removal in eastern England are about ten times the national average, whereas on mixed farms in the Midlands annual losses are only half the national average.

Table 11.8   Rate of removal of hedgerows in eastern England, as estimated from aerial photographs.

Date	Rate of removal (m y^{-1})
1946–54	800
1954–62	2400
1962–66	3500
1966–70	2000

Source: Data from Pollard *et al.* 1974

The fundamental reasons for hedgerow removal are economic. Although it has been argued that, in terms of instalment and maintenance costs, hedgerows have an advantage over fences, there are distinct cost benefits from hedgerow removal in agricultural systems. Ratcliffe (1977) believes that the cost of maintenance has probably been decisive in accelerating hedgerow removal, in that even where they are required for stock control, they still require regular trimming. Moreover, increases in field size allow considerable time-savings in cultivation and harvesting and give greater flexibility in field management. Edwards (1970), for example, calculated the effects of increased field size on time spent turning machinery at row ends (assuming a speed of 3.2 km h^{-1} and a turning time of 36 seconds). An increase in field length from about 110 m to 800 m reduced the turning time by about 80 per cent. Davies and Dunford (1962) estimated that because of these savings the removal of 1½ miles of hedges on a Devon farm led to a reduction of 33 per cent in the time it took to harvest an acre of corn. Similarly, Green (1981) calculated that a change from a row length of 200 m to 500 m, which in a square field would mean an enlargement from 3.25 ha to 20 ha, could save 15 per cent in work time. Estimates of the combined effects of larger field size and use of wider implements are also shown in Fig. 11.9.

Small gains in the area of land available for cultivation are also produced by hedgerow removal. Green (1981) suggests that with a 2 ha field, 2.6 per cent of the land is under hedgerows (assuming an average width of 2 m). A survey in south Yorkshire indicated that gains of from 210 m^2 km^{-2} to almost 1,020 m^2 km^{-2} were made between 1967 and 1978

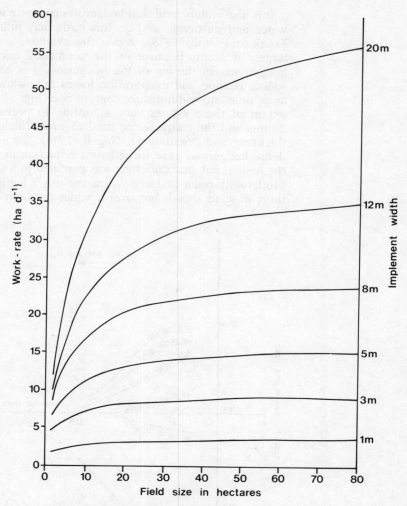

Fig. 11.9 The combined effects of field size and implement width on tillage work-rates (after Sturrock and Cathie 1980).

as a result of hedgerow removal; this represents an increase of 0.02 to 0.1 per cent in cultivable area (Briggs, unpublished data).

In addition to these direct cost advantages of hedgerow removal, there are also a number of more indirect savings. Hedges harbour pests, diseases and weeds. Aphids, pests and certain invertebrates, such as flea beetles and weevils, can pass part of their life cycle in hedgerows. Other pests, such as the wheat bulb fly (*Leptohylemyia coarctata*) and the cabbage aphid (*Brevocoryne brassicae*) generally complete their whole life cycle in the host crop but may use hedgerow plants as alternative hosts. Conversely, Pollard *et al*. (1974) point out that the numbers of pests that actually reside in hedges are relatively insignificant in terms of their population as a whole. Furthermore, if the hedge acts as a reservoir for the pest it will also act as a reservoir for the predators (e.g. ladybirds and hoverflies). Consequently, though it may be assumed that hedgerow removal may reduce the incidence of pest attack in crops, Pollard *et al*. (1974) conclude that there is no firm evidence to support this.

It is also widely held that hedgerows compete with crop plants for light, water and nutrients, and on this basis may influence crop growth (e.g. Frank and Willis 1978). Again the available evidence is contradictory, largely it seems because of the conflicting microclimatic influences of hedgerows. In the lee of the hedgerow, it is generally found that wind speeds decline and evaporation losses fall, while humidity, temperature amplitude and soil moisture content rise (Fig. 11.10). The magnitude and extent of these effects vary according to prevailing microclimatic conditions and the character of the hedgerow (including its height, density, thickness and orientation). Nageli (1946), for example, found that very dense hedgerows gave the greatest reduction in wind speed in the lee of the hedge, but that the effect was confined to a small distance downwind. Moderately open hedgerows, on the other hand, produced small reductions in wind speed, but over a wider area. In general, the effect upon

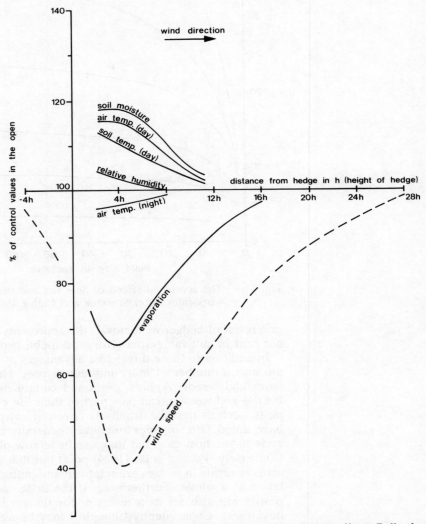

Fig. 11.10 Microclimatic variations in the lee of a hedgerow (from Pollard *et al*. 1974).

wind speed can be detected for a distance up to thirty times the hedgerow height (30 h) downwind. In contrast, the effects on evaporation and other microclimatic conditions tend to be restricted to a downwind distance of 1–16 h (Fig. 11.10).

These microclimatic effects have a number of implications. By affecting soil moisture availability and rates of evapotranspiration they may influence crop yield, and Pollard *et al*. (1974) suggest that in areas where moisture in not limiting yields may be depressed immediately downwind of the hedge and increased in the zone beyond (Fig. 11.11). Few studies have detected any consistant effect on yield, however, in part because of the influence of other factors, such as the tendency for soil compaction in the headland area and for differences in seed rates due to manoeuvering of machinery during sowing. Shepherd (1970) reported that the effect on yield was limited to a distance of between 8 and 12 h, and Green (1981) points out that this means that a field of about 50 by 50 m would be needed to exploit fully this advantage. In addition, hedgerows provide shelter for livestock. Few data are available to indicate whether this has any significant effect on animal health or productivity, but it is likely that survival rates of field-born lambs, for example, may be increased where shelter is available.

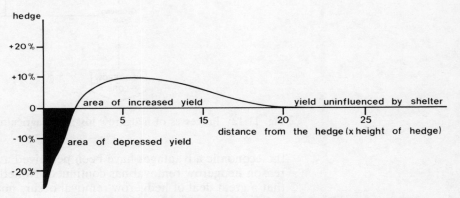

Fig. 11.11 Variations in crop yield in the lee of a hedgerow (from Pollard *et al*. 1974).

More certain is the influence of hedgerows on soil erosion. In general, the reduction in wind speed and increase in soil moisture downwind of hedgerows results in diminished wind erosivity and an increased resistance of the soil to erosion. As a consequence, wind erosion losses are often much reduced for distances of 10 to 30 h downwind. Where hedgerows have been removed in areas susceptible to erosion, soil loss seems to increase. Spence (1955) and Sneesby (1966), for example, cited hedgerow removal as a cause of wind erosion in the Fenlands of East Anglia, while Radley and Simms (1967) attribute wind erosion in east Yorkshire to loss of hedgerow shelter. Similarly, in a study in south Yorkshire, Briggs and France (1982) noted the importance of hedgerows in restricting soil loss, and in a separate survey it was calculated that hedgerow removal between 1967 and 1978 has increased the potential for wind erosion in parts of the county by up to 20 per cent.

Overall it is therefore clear that there are both advantages and disadvantages to agriculture in hedgerow removal. On balance, however,

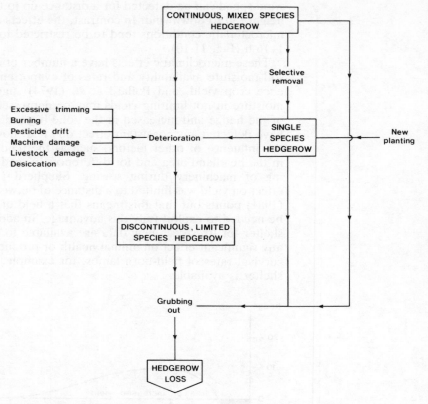

Fig. 11.12   Processes of hedgerow loss from agricultural land.

the economic advantages have been perceived as the greater, and for this reason hedgerow removal has continued. Nonetheless, it is also apparent that a great deal of hedgerow removal occurs not as a result of deliberate policy, but as a side effect of other management practices. In south Yorkshire, for example, it was apparent that hedges were lost by a process of progressive degradation rather than by active 'grubbing out'. Comparisons of hedgerow status and distribution in 1967 and 1978 showed that a tendency existed for continuous hedgerows containing trees to decline to continuous hedges without trees, either by selective removal of trees or due to mechanical trimming. At the same time many continuous hedgerows became fragmented, while hedgerow fragments were themselves lost probably by removal. Overall, therefore, hedgerow loss can be viewed as a step-by-step process as indicated in Fig. 11.12. Deterioration of hedgerows in this process is a result of many factors, including excessive trimming and damage by machinery, fire (during stubble burning) and herbicides.

The ecological implications of hedgerow removal and degradation vary with age, history, conditions and location of the hedgerow. In Britain, hedgerows are generally of two main types: remnant hedgerows which have either been created by leaving lines of trees or bushes during clearance of the surrounding land (asserts), or which have grown up in unused areas between fields; and planted hedgerows laid mainly during enclosure. The former tend to be richer in species, while the latter are often dominated by a single species; hawthorn, for example, was widely planted to provide 'quicksets' during the later periods of enclosure. In addition, there is some evidence that species diversity increases with increasing age due to the invasion of hedgerows by new species (Pollard *et al.* 1974). Great regional variations in hedgerow composition nevertheless occur, as a result both of regional differences in the history of hedgerow management and of differences in environmental conditions. For these reasons, it is difficult to quantify the effects of hedgerow removal.

Ratcliffe (1977) suggests that, in Britain, about 500 vascular plant species occur in hedges, although about half of these are more common in woodland and grassland, and the other half will grow in these situations. Three important genera – *Ulmus, Rosa* and *Rubus* – have their main habitat in hedges, however, and *Ulmus* has, of course, been decimated in recent years by Dutch elm disease. Significant losses of *Ulmus* in particular may therefore occur as a result of hedgerow removal, and the fact that it is a basic food for about 100 species of insect must also raise the question about the survival of some individual species. Hooper (1979) claims that in total about twenty plant species would be seriously affected by hedgerow removal, while a further thirty to forty would be markedly affected, possibly leading to their extinction in some counties.

The consequence of hedgerow removal for faunal populations is more significant. No individual bird species is totally confined to hedgerows as a breeding habit, and for most birds, hedges are secondary to woodland habitats. Nevertheless, very large populations of birds actually breed in hedges and reduction in total hedgerow length will possibly cause a decline in most populations especially where alternative breeding grounds are not locally available or are already occupied. Breeding is also likely to be affected by hedgerow degradation. Loss of trees and dissection of hedgerows by excessive trimming, especially with mechanical flayers, and by fire damage, produces a less stable and less diverse habitat within which relatively few breeding sites occur. The associated loss of species due to deliberate spraying and accidental herbicide drift may also reduce insect numbers and further reduce the habitat potential of hedgerows. Such changes are probably unimportant in areas of rich, alternative habitats, but in many intensive arable areas woodland and scrub is scarce, and hedgerows represent a last refuge for faunal populations; here, therefore, the impacts are likely to be severe.

11.5.2 Headlands

The areas around arable fields are known as headlands. These vary in width depending on the location and particular local circumstances. Where land is in continuous arable cropping the headland may continue relatively undisturbed for several years, although there is a tendency to spray them with herbicide. The general development of machinery – particularly with regard to manoeuvrability – has meant a reduction in size of headlands. Equally, the decline in hedgerow length has meant a corresponding decline in headlands (although in some areas it is notable that a headland remains where a hedge has been grubbed out: this is usually because of a significant change in ground level, the headland having the advantage of being narrower than the hedge and needing less maintenance).

Most headlands consist of an assemblage of weed plants, which are able to develop in the absence of competition from the crop. Plants such as *Poa annua, Lolium perenne* and *Plantago major* are common. Ratcliffe (1977) suggests that although the flora has some ecological interest in supporting a varient of the field flora merging into the boundary flora, the fauna is not known to have any special interest.

11.6   LOWLAND CASE-STUDIES

The complex environmental changes that occur as a result of fundamental developments in agricultural land use can be illustrated by regional examples. These examples help to demonstrate both the extent to which environmental conditions are affected, and the feedback relationships which exist between these conditions and agricultural management decisions. They also help to emphasise the way in which environmental conditions reflect the cumulative impact of agricultural activities, often over many centuries.

11.6.1   The Fenlands of eastern England

The Fenlands include the largest extent of grade 1 agricultural land of any region in Britain. It is also an area which has been drastically modified by agricultural practices. The location and extent of the Fenlands is shown in Fig. 11.13. Broadly speaking, they can be divided into areas of silt soil (the skirtlands) and areas where the mineral substrate is covered with a layer of peat. Most of the peat accumulated as reedswamp during the gradual marine regression which affected the area during the Flandrian period. By Roman times these peats were of the order of 4 to 5 m or more in depth in many parts of the area. The surface comprised a mosaic of mixed oak woodland, alder and willow scrub, interspersed with brackish and fresh water channels.

Reclamation of the Fens began in Roman times and, according to Astbury (1957), several main elements of the modern river system were initiated during this period. Until the Middle Ages, however, land use was restricted to wild-fowling, reed-cutting and local summer grazing for cattle. The main impetus for reclamation came in the seventeenth century. In the early years of that century, attempts were made to shorten the course of the River Ouse between Ely and Littleport as a means of improving drainage and increasing the amount of summer grazing. After £1,000 had been spent on the scheme, however, the work was stopped because of opposition 'by some of the better sort about Ely', apparently due more to 'mislike of the persons that pursued it, than of the work itself' (Wells 1830, quoted in Seale 1975). During subsequent years conditions in the Great Level deteriorated and attempts to improve drainage were thwarted by the local fenmen who saw their traditional livelihood threatened, and the Civil War led to the abandonment of many projects. In 1649, however, the Earl of Bedford invited the Dutch engineer, Cornelius Vermuyden, to England to undertake reclamation operations.

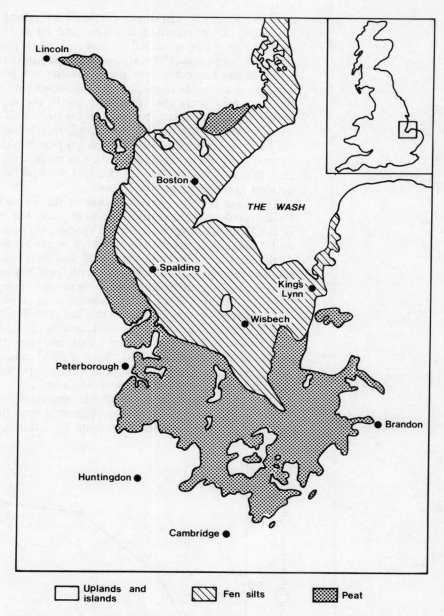

Fig. 11.13   Generalised map of fenland peats and silts: the Bedford Levels
(after Curtis *et al*. 1976).

Under his direction the Great Level was divided into three parts – the
North, the Middle and the South Levels – and the New Bedford River
was cut. By 1652 the main work had been completed and the framework
of the present Fenland drainage had been established.

Initially, the reclaimed land was used solely for grazing, but progressively, as drainage improved and the peat dried out, more arable crops
were grown and the typical rotation in the area developed into an eleven-
year sequence containing six or seven years of wheat and three years of

oats (Seale 1975). Land was prepared by paring and burning, the turf being cut off, overturned in ridges and burned. During the eighteenth century the water table fell as drainage work continued, albeit with intervals during periods of war, and wind-pumps had to be used to raise the water from the ditches into the rivers. By the last years of this century, wind-pumps were being replaced by steam-driven engines, but throughout the next fifty years the continued fall in the water table increased the height of lift necessary to raise the water. This meant that the scoop-wheels used to lift the water became increasingly cumbersome, and in 1851 they were replaced by centrifugal pumps. At the same time, further extension of the drainage network was made with the construction of the Eau Brink Cut near King's Lynn and the Middle Level Cut between the Sixteen Foot River and the Ouse.

This long history of reclamation of the Fenlands began ultimately to have a profound effect upon the peat soils. For the most part, the peats of the Fenlands consisted of pure reedswamp material, rich in *Phragmites* and *Sphagnum*, and with a very high water retaining capacity. Drainage of the peat led to irreversible drying and shrinkage, and in the new, aerobic environment which was created the peat began to oxidise. As a result, peat wastage occurred and the Fenland surface was gradually lowered. In 1848, at Holme Moss in Cambridgeshire, a stake was inserted flush with the peat surface, and since then this has provided a reference point against which to assess the degree of wastage (Fig. 11.14). By 1950, the peat surface had been lowered by over 4.0 m, and since then further wastage has occurred (Hutchinson 1980). Seale (1975) calculated that wastage was taking place at a rate of about 1.8 cm y^{-1} between 1934 and 1960, and he estimated that between 1965 and 1985 the area of peatland greater than 60 cm in depth would decrease from 160,000 ha to about 5,000 ha. Summers (1976) has suggested that three-quarters of the remaining peatland would degenerate into skirtland over the next thirty

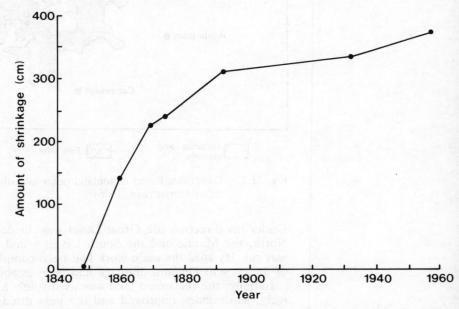

Fig. 11.14   Shrinkage of the peat surface at Holme Moss, Cambridgeshire (data from Russell 1971b).

years. Richardson and Smith (1977) measured peat wastage relative to paving stones which had been buried at 34 sites in 1941 by the then Great Ouse River Board. The rate of wastage varied from 0.27 to 3.09 cm y^{-1}, but was generally less in the period 1955–71 than from 1941–55. The average rate of wastage, assessed from 8 sites for which continuous records were available, was 1.37 cm y^{-1}.

The loss of peat as a consequence of drying and oxidation has had a number of implications. Lowering of the peat surface beneath the water level in the main dykes and rivers has necessitated improved pumping and the construction of sluices to control flooding. Even so, floods have occurred on a number of occasions, most notably in 1795 and, more recently, in 1939 and 1947. Cultivation has been hampered by the exposure of bog oaks from the lower layers of the peat and the uncovering of tile drain systems. Uneven shrinkage has also resulted in the development of an irregular surface topography, whilst buildings and bridges have been undermined (see, for example, Astbury 1957). At the same time, increased cultivation is encouraging wind erosion (Sneesby 1966). More importantly, the underlying estuarine and lacustrine clays and silts are now being uncovered and, where peat is shallow, these relatively sterile materials are being brought to the surface by ploughing. Where ploughing of the clays is careful, beneficial results can be obtained, for the clay can be mixed with the peat to produce a stable and fertile soil. In many cases, however, mixing is not carried out efficiently and, according to Mason (1975), is too expensive for general use. Summers (1976) takes the pessimistic view that the struggle to conserve the Fenland peats has already been lost, though in a number of areas efforts to prevent further wastage are being made by maintaining a high water table throughout the winter and only allowing the water table to fall at the start of the growing season.

The degeneration of the Fenland peats is clearly a serious agricultural problem, for it involves the loss of as much as 150,000 ha of prime agricultural land over a period of 40–50 years. This is equivalent to about 20 per cent of the land loss to urbanisation over the the same period. Fenland degradation, however, also represents a serious ecological problem. Very few areas of undrained Fen woodland now remain, one of the main exceptions being the Wickham Fen which is now protected as a nature reserve. As a consequence, the wildlife potential of the Fens is being much reduced, while the visual character of the landscape is changing markedly. The dykes that remain tend now to be cleared using herbicides and machines, rather than by hand, whilst pollarded willows have been removed from many areas because they obstruct mechanical clearance of the dykes.

The amount of wildlife that can be supported in the Fenland, if farms are carefully managed, is surprisingly high. At Church Farm, Little Ouse, particular efforts have been made to reconcile modern agricultural techniques with wildlife conservation. This 164 ha (406 acre) farm is almost entirely on peat soils and is farmed mainly as arable, including 73 ha (179 acres) of wheat, 20 ha (70 acres) of sugar beet and 20 ha (70 acres) of onions. Rather exceptionally for this area the farm has been planted with various trees and hedges. Thirty-eight species of bird were reported as 'breeding' or 'probably breeding' whilst a further twenty-two species visited (Herring 1977). In addition, attempts are made to conserve the habitat potential of the dykes and ditches by clearing the ditch sides on alternate sides each year.

Westmacott and Worthington (1974) believe that this sort of approach

can have a substantial conservation impact. They draw attention to the
fact that some farmers have surreptitiously taken public road verges into
cultivation, thus effectively precluding extensive public tree planting, even
if that were financially feasible. The water authority does not allow tree
planting within thirty feet of river banks so further planting along these
water courses is not possible. Nevertheless, as Church Farm demonstrates, there are areas where it is possible to plant valuable wildlife
habitats.

11.6.2   Chalk downland

The ecological basis of calcareous grassland was discussed in section
11.3.2 (p. 312) and the general distribution of the chalk in England given
in Fig. 11.3 (p. 313). The open landscape of the chalklands has changed
considerably in use over the last few hundred years. Although there was
a noticeable shift from the historic sheep pasture to increased arable usage
in the first part of the nineteenth century, much arable land later reverted
to pasture (C. J. Smith 1980). In tracing the complicated land use history,
Smith points out that it was not until the late 1930s that arable use of the
chalkland really began to develop. With the onset of the Second World
War and, later, the availability of artificial fertilisers, the soils over the
chalk began to support the continuous arable cropping for which they are
agriculturally best known today. In many areas there is continuous cropping of barley (Jarvis 1973). Nevertheless, there is some variability in land
use which partly results from the presence of various superficial geological
deposits and consequent soil complexity (Curtis *et al*. 1976).

One area of chalk downland was studied by Westmacott and Worthington (1974). This lay a few miles north-east of Dorchester. In the area they
examined there was a mixed pattern of farming with dairying taking place
in the valley bottoms. Holdings consisted of specialist arable farms, as
well as specialist dairy farms and mixed farms. Westmacott and Worthington note that, compared with some of their other study areas, relatively
few hedges had been removed. To a large extent, this was due to the fact
that, on the chalk downland at least, the extent of hedgerows had traditionally been limited, and field sizes have always been large. Westmacott and Worthington go on to suggest that 'most would agree that
hedgerow removal has not produced a poorer landscape'. Whilst aes-
thetically this might be so, in ecological terms hedgerow removal inevitably tends to reduce the population densities, if not actual numbers of
wildlife species. As noted in section 11.5, this is likely to be particularly
significant in areas in which hedgerows and woodlands are, to start with,
of limited extent.

As Westmacott and Worthington indicate, the major conservation issue
on chalk downland is the retention of the remaining areas of natural or
semi-natural grassland. They suggest that one way of promoting their conservation in a modern agricultural landscape is to ensure that as far as
possible banks and verges are retained. Roads and track verges are also
important in this respect.

Westmacott and Worthington's proposal is not altogether satisfactory
from an ecological viewpoint. Their changes are those which, they say,
one might expect a landowner keen on game shooting to make. The
hypothesis is that there is also a change from livestock farming to arable

cropping. Many of the hedgerows are removed but there is an increase in tree cover on some of the slopes. Most of the grassland slopes have had some sort of cover planted on them and – although Westmacott and Worthington do not say so – this would greatly reduce the already small area of traditional chalk grassland. They point out that in their survey of farmers, many would wish to plant all slopes with shrubs or trees, especially where they have no stock available for grazing or where the slopes are too steep to mow.

It may be that the only way in which sizeable areas of traditional chalk grassland can be maintained in southern England is by subsidy to the farmer. At Butser Hill, Hampshire, an area of downland has been taken into ownership by the county council. In order to maintain the grassland, local farmers are allowed to graze sheep and cattle under very restricted arrangements (Lowday and Wells 1977). Thus the tenancy agreement is subject to various conditions, including the following

1. Minimum stock numbers and areas not to be winter grazed;
2. Free public access at all times;
3. Application of fertilisers prohibited;
4. Application of herbicides prohibited;
5. Various compartments to be systematically grazed by arrangement with the county recreation officer. At the conclusion of grazing the compartment shall be chain-harrowed;
6. No winter feeding is to be allowed on unploughed downland turf;
7. One month's notice may be given by the county council.

Lowday and Wells conclude that management and maintenance costs of the grassland exceed the income and it is clear that a farmer using land at an unsubsidised rent level would today find these restrictions uneconomic. At Butser Hill this arrangement has, however, continued successfully for some years but it does demonstrate that traditional chalk downland can only be maintained today by subsidy.

12 THE IMPACT ON ECOLOGICAL SYSTEMS: II THE UPLANDS

12.1 INTRODUCTION

Although farming is less intensive in the uplands than in the lowlands, the impact of agriculture on upland ecosystems has nevertheless been significant. This is partly because the ecosystems within the uplands are inherently more susceptible to change, often because they occur under less favourable climatic conditions and because the soils are less chemically stable. Thus even quite small changes in the intensity or nature of management in these areas may trigger off fundamental adjustments in the ecosystems.

In order to assess agricultural impacts on ecological systems in these areas it is necessary to examine briefly the main features of the uplands. Throughout most of the temperate world, these areas consist of natural or semi-natural pasture, heath or forest. With the exception of the forest, the uplands are generally used for extensive agriculture, mainly grazing of sheep, cattle or goats. Increasingly, however, agriculture in the uplands is becoming more intensive and is encroaching on the few areas of previously natural vegetation. As a consequence, there is a tendency for the area of all three vegetation types to decline as they are replaced by more productive, artificial pasture.

In Britain, upland areas are dominated by rough pasture and heathland or moorland. The distribution of these is shown in Fig. 12.1. Ratcliffe (1977) suggests that the main diagnostic feature of the upland habitats is their low temperature. Thus the British upland flora shows a similarity to that of Scandinavian regions, although rather less to Alpine Europe. In very general terms British upland habitats represent a more maritime phase of their Scandinavian counterparts. Again broadly speaking, grasslands tend to predominate on the western and more basic soils, whilst heather moor (or heath) is more common in the east and on more acid soils.

Ratcliffe points out that the distinction between 'lowland' and 'upland' is difficult to define. He proposes that a useful working distinction is that in many hill areas (for example, central Wales) the upper limit of enclosed land gives a useful indication of the extent of the 'lowland' area. However, in the context of the present discussion a major problem is that this definition is dependent on a man-made modification and one of the key issues in areas marginal to the upland – such as the lower slopes of Ex-

Fig. 12.1 Distribution of rough pasture and moorland in Britain (from Fuller 1982).

moor – is the extent of moorland enclosure. There is clearly some danger of a circular argument developing; namely, that if land has been enclosed into more intensive agricultural use then it is no longer upland. In ecological terms such an argument makes little sense. Ratcliffe recognises this difficulty particularly in the context of southern Britain, but also points out that it is not possible to adopt an altitudinal distinction. This is because natural vegetation zones vary according to regional differences in climate, which themselves may bear only a limited relationship to altitude.

In the case of the heathlands the distinction between 'lowland' and 'upland' is perhaps a little easier. Ratcliffe suggests that the presence of the red grouse (*Lagopus lagopus scoticus*) is the best single diagnostic feature. The use of this bird for such diagnosis might seem surprising since success of the grouse population depends to a certain degree on human management. It is therefore worth examining the factors governing grouse distribution in a little more detail.

Figure 12.2 shows the distribution of red grouse in the British Isles for the period 1968–72. The British red grouse is regarded as one race of the polytypic willow grouse *Lagopus lagopus*. Outside the British Isles the habitat extends broadly across the higher latitudes of the west Palaearctic (Fig. 12.3). In the breeding season it is concentrated wherever there are ample supplies of heather (*Calluna* spp.), berry-bearing small shrubs, such as bilberry (*Vaccinium* spp.), or other food plants (Cramp and Simmons 1980).

There have been a number of attempts at artificial introduction of the species including into areas in Norway, Sweden, Denmark, West Germany and Poland (Cramp and Simmons 1980). In Britain there were attempts to introduce the bird to Exmoor in the 1820s which failed (Sharrock 1976). Later attempts in 1915–16, both on Exmoor and Dartmoor were successful (see Fig. 12.2). The population density depends on the nutrient quality of the food plants, which is itself a reflection of the base status of the substrate. Thus numbers are highest over base-rich soils containing diorite or epidiorite (Jenkins and Watson 1967).

Up to about 1940 numbers in Britain had been high but after this time they showed a marked decline and this decline prompted considerable research into grouse management. The main reason for the decline was seen to be a decrease in the heather quality. The balance between different ages of heather vegetation is critical with old deep heather being needed for sheltering and nesting although young heather shoots are needed for feeding. The latter are encouraged through careful rotational burning (see also section 7.2.1) and also the application of fertilisers or even drainage where there is excessive waterlogging (Sharrock 1976). There is a danger that over-burning of moorland, particularly where there is also large herbivore grazing, can result in conversion to *Festuca–Agrostis* grassland.

12.2   LAND RECLAMATION FOR AGRICULTURE

One of the most pervasive changes in the uplands has been the reclamation of land for agriculture. As will be seen later, reclamation does not always involve the primary cultivation of land previously unused for agri-

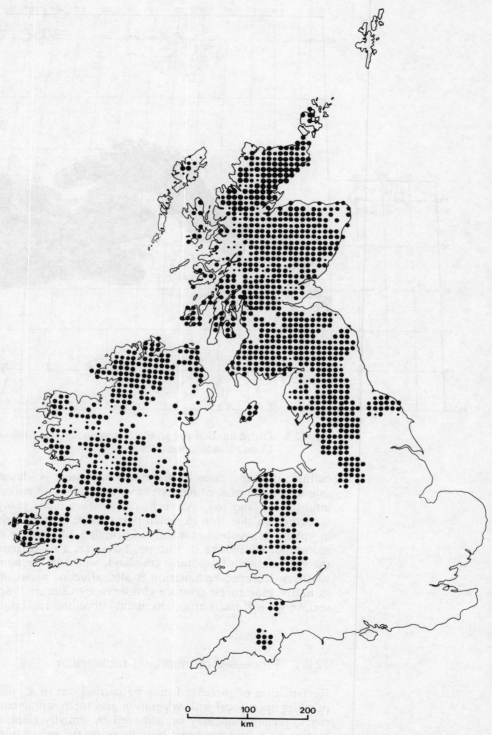

Fig. 12.2 Distribution of red grouse (*Lagopus lagopus scoticus*) in the British Isles (from Sharrock 1976).

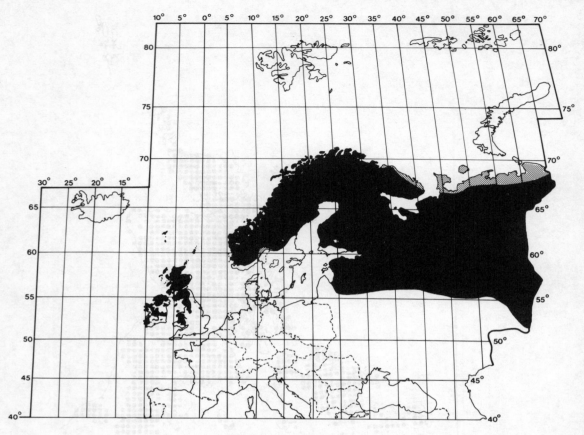

Fig. 12.3   Distribution of red grouse (*Lagopus lagopus scoticus*) in Europe (from Cramp and Simmons 1980).

culture; in many cases it may more accurately be described as a wholesale upgrading of existing, extensively used land to make it fit for a higher intensity of land use. As such, reclamation takes many different forms according to the area in which it is carried out. In Britain, reclamation in upland areas consists mainly of bringing into cultivation heathland and moorland. Elsewhere in Europe, however, a more significant process is the reclamation of montane grassland, whilst in the more arid regions of southern Europe, reclamation is also affecting areas of maquis, as well as native pine forest (Nature Conservancy Council 1982). In the present section we will focus attention mainly upon the reclamation of moorland.

12.2.1   Processes of moorland reclamation

Reclamation of moorland may be carried out in a number of ways, depending upon local soil, vegetation and topographic conditions. In many cases, reclamation may be achieved by initially clearing the heath vegetation (e.g. by paring and burning or by the use of herbicides) and then by applying lime and, where appropriate, phosphate and nitrogen fertilisers. This allows agriculturally preferred species such as ryegrass (*Lolium* spp.), meadow grass (*Poa* spp.) and clover (*Trifolium repens*) to become established. Reseeding is not necessary in these cases, though herbicides may be necessary to control bracken invasion. Reclamation by these processes is carried out extensively in the North York Moors and Dartmoor.

Where the land is poorly drained, or where the soil is intensely podsolised, more drastic reclamation measures may be necessary. Deep ploughing (to a depth of 40–50 cm or more) may be carried out to break up iron-pans and to mix the surface mor humus with the underlying mineral soil. In addition, land drainage may be installed, boulders may be cleared, and herbicides may be applied to prevent regeneration of bracken, *Calluna* and *Juncus*. Reseeding is then carried out and the land limed and fertilised.

12.2.2   Extent of moorland reclamation in Britain

In Britain, some 6.4 million ha of land are classified as moorland and heathland. The majority of this is in the uplands of Scotland, Wales, the Pennines, North York Moors, Dartmoor and Exmoor. All this area has been subject to reclamation in recent decades, though perhaps nowhere has the question of reclamation been more fiercely debated than on Exmoor. Here, over 4,000 ha of heathland were reclaimed between 1947 and 1979, representing over 17 per cent of the total area, and an annual rate of loss of 160 ha of moorland (Porchester 1977). Blacksell and Gilg (1981) also note that a survey by the Ministry of Agriculture showed that 64 per cent of the remaining moorland within the Critical Amenities Areas defined by the Exmoor National Park were potentially reclaimable. From a field survey, they show, further, that reclamation in Exmoor has, in recent years, been concentrated mainly in the moorland fringe, in areas which have in the past fluctuated between farmland and heath.

This pattern of change is repeated in many other moorland areas, albeit to a smaller extent. Parry *et al.* (1982a), for example, show that a total of 6,050 ha of moorland was reclaimed for agriculture in the North York Moors between 1950 and 1979. Of this, about 3,588 ha involved primary reclamation of open moorland, while the remainder was secondary reclamation of land which had, in the past, been farmland but which had subsequently reverted to moor. Most of this secondary reclamation was of land which had reverted during the period 1904 to 1950 – most of it apparently during the 1930s (Table 12.1); but in 1979 about a third of the land which was abandoned during that period was still under rough pas-

Table 12.1   Reclamation of moorland for agriculture in the North York Moors

Date of secondary reclamation	Total area (ha)	% of secondary reclamation abandoned in each period				
		1853–95	1895–1904	1904–50	1950–63	1963–74
1974–1979	1462	8.0	4.5	59.8	11.8	15.9
1963–1974	1635	9.1	3.4	69.0	18.6	—
1950–1963	4131	11.8	2.9	85.3	—	—
1904–1950	597	80.9	19.1	—	—	—
1895–1904	251	100.0	—	—	—	—

Source: From Parry *et al.* 1982a

ture or moorland. Parry *et al.* suggest that it is this area which would, in future, be the prime target for further reclamation.

Similarly, on the Brecon Beacons in Wales, 4,542 ha were reclaimed between 1948 and 1975, of which 2,738 ha involved secondary reclamation, most of it representing land which had reverted between 1909 and 1948 (Parry *et al.* 1982b). Again, a third of the land abandoned during this period remained as moorland in 1975, and seemed likely areas for future reclamation. On Dartmoor, 93 per cent of the reclamation between 1971 and 1979 has been secondary; about a half of this affected land which had reverted prior to 1885, and about a third of this had in turn been abandoned before 1800 (Parry *et al.* 1982c).

Some indication of the scale and speed with which agricultural operations in the uplands can influence the pre-existing vegetation system is given in a detailed study of part of Dartmoor (Countryside Commission 1978). The unenclosed areas of Dartmoor centre on two extensive areas of blanket bog. Dissecting the edges of the blanket bog areas are small valley marshes. Surrounding this complex are areas of heather and heather-derived vegetation. Much of this latter area has been invaded by bracken and gorse.

In the Countryside Commission Report (1978) it is suggested that if grazing and burning were increased to a maximum on Dartmoor, then about half the remaining heather areas would change to grassland in the next twenty-five years. The blanket bog would probably change to a purple moor grass dominated grassland and similar changes would take place in the valley marshes. The changes in the upland margins might be effected within about 50 years, whilst at higher altitudes changes might take of the order of 100 years. Thus under this kind of regime Dartmoor would eventually become entirely grassland dominated.

At the other end of the management scale, if all grazing were to cease then there would be a tendency to invasion by woodland and scrub. The Report envisages that eventually, under these circumstances, much of Dartmoor would be covered by woodland and scrub vegetation although the process might take over 100 years to complete.

12.2.3 Ecological effects of reclamation

The ecological effects of reclamation may be far-reaching. Reclamation fundamentally changes the character and composition of the upland vegetation, *Calluna, Vaccinium* and *Pteridium* species, for example, being replaced by agricultural grasses and clover. This change also affects animal species which are dependent upon the heathland vegetation for food, cover or nesting grounds. In Britain, there are few species whose survival is threatened by current reclamation on a national scale, but locally it seems likely that the loss of upland heath may lead to loss of habitats for species such as the raven (*Corvus corax*). In addition, as we noted in section 10.7, upland heath is an important habitat for the red kite, whose distribution is already greatly restricted. Clearly, reclamation of moorlands on which this bird depends could threaten the species.

Nevertheless, moorland reclamation does not always result in reductions in the diversity or abundance of the fauna. Reviewing recent work, for example, Stewart and Lance (1983) note that drainage, fertilisation and fencing of peat moorland at Glenamay in Ireland resulted in a tenfold increase in numbers of wood mice (*Apodemus sylvaticus*) and the establishment of pygmy shrew (*Sorex minutus*) populations. Similarly, at Eskdalemuir in Dumfriesshire, drained moorland plots showed increased numbers of field voles (*Microtus agrestis*) which, in turn, led to rises in the populations of foxes (*Vulpes vulpes*), short-eared owls (*Asio flammeus*), long-eared owls (*A. otus*) and kestrels (*Falco tinnunculus*). Many studies have also reported large increases in insect populations following reclamation.

It is also important to stress that the main areas of reclamation are still at the moorland fringe, and in many cases agriculture is merely regaining land that it once held. Ecologically, these fringe areas are probably less significant than the open moors, for rarely has regeneration been complete, and in many cases the original *Calluna* heath has not been re-established following abandonment; instead, reversion has often been to bracken, gorse (*Ulex* spp.) and coarse grasses such as *Nardus stricta* and *Holcus mollis*. Indeed, as the presence of old field systems and remnants of lazy-beds within the moorland fringe indicates, agriculture has not yet totally recovered the land it once claimed in some of these areas. On the other hand, the very fact that this land has not reverted to true moorland demonstrates that the process of reclamation is rarely wholly reversible. Even quite short periods of disturbance by cultivation, herbicide usage and burning may cause long-lasting changes in the ecosystem. Moreover, as the moorland fringe retreats, so the buffer zone between the area of intensive agriculture and the open moorland becomes narrower, bringing more areas within reach of the secondary effects of agriculture such as damage by herbicide or fertiliser drift, and desiccation due to land drainage.

12.3 CHANGES IN MANAGEMENT PRACTICES

Although the conversion of moorland to pasture or other intensive uses is generally referred to as reclamation, this definition is not strictly accurate. Most areas of moorland are already used, at least extensively, for grazing, and reclamation commonly represents no more than a gross improvement of the land for farming. Even without such deliberate conversions of heathland, however, upland areas are subject to numerous management practices which influence their ecology. These include burning, fertiliser usage, pesticide usage and grazing control.

12.3.1 Burning

As noted in Chapter 7, heather burning is carried out both to control the age structure of the heather and to provide open ground for grouse. This dual purpose reflects in part the dual use of the heather moorlands for sheep grazing and as grouse reserves. Over the years, these two uses of the moorlands have come into conflict and as a result there have at times been attempts to modify burning practice.

Burning originated as a widespread practice in the eighteenth century with the expansion of sheep rearing in the uplands. It was readily appreciated that sheep performed better on land which had been burned to encourage heather regeneration, and over time burning frequency increased in an attempt to maintain a consistent supply of young heather. During the mid-nineteenth century, however, the use of the heathlands for gameshooting became widely established, partly in response to the increased accessibility of the moorlands. Initially, sheep rearing and grouse rearing on the moors appeared compatible, but in the 1920s grouse populations dwindled and game-bags declined. This represented a significant loss of both sport and income to the landowners in these areas, and it initiated a major clash between the grazing and sporting interests. In many areas, attempts were made to restrict the frequency of heather burning, or to eliminate it completely, in the belief that this would provide more cover for grouse and allow populations to build up. The policy met with little success, however, and in the 1940s the number of grouse fell drastically. In order to determine the cause of this decline, a Royal Commission was established under Lord Lovett. Its findings were that the main reason for declining populations was a virus disease which was spreading through many heathland areas. The Commission demonstrated, further, that far from reducing the incidence of this disease, lack of heather burning was in fact increasing the problem by allowing more extensive contacts between grouse populations. In addition, it was shown that young grouse benefited both from the availability of open ground following hatching, and from the supply of young heather shoots on which they could graze. Consequently, the mutual benefit of heather burning – to both sheep and grouse – was established.

Since then, further research has demonstrated the factors which determine the efficacy of heather burning. Miller and Miles (1970), for example, have shown that the regenerative capacity of heather reaches a maximum at an age of six to eight years, then declines to a minimum by twenty-six to thirty years. This is partly because, after burning, regeneration takes place largely from shoots on the lower part of the stem. As the heather plant ages, these sites become covered by new wood so that vegetative regeneration is inhibited. In addition, as the heather community gets older, individual plants die and the density of stems per unit area declines. Miller and Miles (1970) quote a reduction in stem density from 1,108 m^{-2} at an age of 6–8 years, to 184 m^{-2} at an age of 34–37 years; the density of potential sites for shoot development may be assumed to decline accordingly.

Nonetheless, regeneration does not take place only from stems. Germination of seeds also provides a means of regeneration, and this, too, is greatly affected by burning practice. Gimingham (1972) refers to studies showing that excessive soil temperatures during heather burning (i.e. 200 °C or more) reduce the viability of buried seeds. Lower temperatures (40–120 °C), however, appear to increase germination rates.

It is also apparent that the nutritional value of the heather is greatest during the early years of growth, but declines beyond about ten years as the heather enters its building stage. Thus, modern practice is to burn heather every seven to fifteen years. Generally, small plots of 0.5 to 2.0 ha are burned to ensure good control of the fire and to provide a local range of heather conditions. In recent years, however, there is some evidence that the interval between burning has decreased whilst the area burned has increased. Today there are relatively few areas of unburned heathland in Britain, the main examples being some of the high moors in the North York Moors and in parts of Scotland. Thus, major ecological changes have occurred in many moorland areas.

These changes are a function of the successional pattern of regeneration following burning, which depends in turn upon both the soil conditions and the nature of the burn. Where the temperatures during the burn are controlled (i.e. about 400 °C), and on dry soils, *Calluna* regeneration is relatively rapid, though various species gain at least temporary dominance during the early years. Algae such as *Zygogonium ericetorum*, for example, frequently form an extensive mat on humose surfaces. Lichen also invade the bare surface of many burned peatlands, and *Lecidea* species may form a thin film which reduces infiltration of rainfall and encourages runoff and soil erosion. These pioneer species are soon ousted, however, by other colonists such as *Cladonia* spp. and mosses (e.g. *Polytrichum*). Subsequently grasses including *Agrostis* spp. and *Festuca* spp., together with herbs such as *Galium saxatile*, *Potentilla erecta* and *Veronica officinalis*, invade. Indeed, by repeated burning on dry soils it is possible to replace heather by these grass and herb species.

As this indicates, burning frequency and intensity influence the successional development of the heathland. In areas where burning is controlled and rotated on a cycle of about fifteen years, a considerable diversity of species may be found, reflecting the different successional stages represented by the various plots. Lack of burning, on the other hand, tends to result in a mature heather community in which other species are scarce, whilst overburning may cause fundamental changes in vegetation composition, often with the loss of heather species. The most notable and widespread effect of over-burning is bracken invasion, a problem which will be discussed in section 12.4.2. On wet sites, however, species such as *Erica tetralix*, *Juncus squarrosus*, *Empetrum nigrum*, *Nardus stricta* and *Molinia caerulea* may have a competitive advantage which inhibits *Calluna* regeneration and results in deterioration of the heath community.

12.3.2 Fertiliser usage

Lack of nutrient availability is one of the main factors inhibiting the use of upland areas for agriculture, and as noted in Chapter 7 the productivity of many upland areas can be greatly increased by improvements in nutrient supply. One way in which this can be achieved is through the use of fertilisers, particularly lime, nitrogen and phosphate. Except where the aim is to reclaim heathland, fertilisers are rarely used in heather moorland. In areas of rough or improved pasture, however, fertiliser application provides an economically viable means of controlling sward composition and herbage production.

Some of the ecological effects of fertilising and liming upland pasture have been mentioned in Chapter 7. Typically, these practices result in an increase of preferred species such as *Poa pratensis*, *Lolium perenne* and *Dactylis glomerata* at the expense of *Agrostis* spp., *Nardus stricta* and *Festuca* spp. (e.g. Elliott *et al.* 1974; Sandford 1979). More generally, Wells (1981) quotes the study by Perring and Farrell (1977) listing 321 flowering plants considered to be under threat in Britain. Of these, he notes, 117 are associated with grasslands and these are all likely to be threatened by fertiliser application. Similarly, he comments that a base-rich meadow may support over 100 species of flowering plants and grasses, 'which, by the continual application of herbicides and/or fertilisers, can be reduced to a few species within a few years'. This problem is seen most acutely, perhaps, in areas which support relatively rare native assemblages, such as the Sugar Limestone in Northumberland. Meadows in this area contain a wide range of alpine species, apparently representing a Late Glacial relict flora. Intensification of agriculture in Teesdale is threatening these communities.

12.3.3   Grazing intensity

The intensity of grazing exerts a major control on ecological conditions in the uplands. Grazing animals affect plant vigour and thus competitive processes by defoliation, by trampling, and by their influence on nutrient cycling. These effects show marked spatial disparities, however, due to the behavioural patterns of the grazing animals (section 7.3.3). Animal behaviour, in turn, is influenced by palatability of the sward, so that intricate feedback mechanisms operate between grazing patterns and vegetation composition.

On the whole, sheep are more selective in their grazing behaviour than cattle, and thus the effects of sheep grazing upon vegetation conditions are spatially more variable (e.g. Hunter 1964; Dale and Hughes 1978). Sheep select herbage on the basis of many factors, including the nutrient content, abundance, availability and structure of the material on offer. These conditions vary during the growing season so that the pattern of grazing and its effects on vegetation change through the year. Eadie (1967) discussed the consequences of this for vegetation composition. He claimed that during the period of active pasture growth, sufficient quantities of the most favoured species are available to meet herbage requirements. Thus grazing is markedly selective, and the ungrazed species mature and reseed, reducing the quality of the remaining herbage. During the winter, sward deterioration continues as over-grazing of the limited supply of favoured species takes place, whilst senescence of the mature, ungrazed species results in retention of large amounts of dead plant materials which are carried over to the next spring, inhibiting subsequent regrowth. In addition, in the over-grazed areas, trampling and soil erosion may prevent regeneration.

As noted in section 7.3.3, the problems of sward selectivity are most acute where grazing intensities are low, and thus prolonged undergrazing of upland pastures tends to result in loss of nutritious species such as *Agrostis* spp. and *Festuca* spp., whilst hardier species including *Nardus stricta*, *Molinia caerulea* and *Holcus mollis* increase in abundance. Species diversity also tends to decline for these hardier grasses suppress many of the herbaceous plants which could otherwise survive in the close-cropped, nutrient-rich sward.

The extreme effects of reductions in grazing intensity are found where animals are excluded from upland pasture. Rawes (1981) studied changes in vegetation composition at the Moor House National Nature Reserve in Cumbria, England. After twenty-four years of protection from sheep grazing, the decline in species numbers in experimental plots was still continuing. Figure 12.4 shows one sample area and the changes that have taken place between 1956 and 1979. At first sight these may appear to be

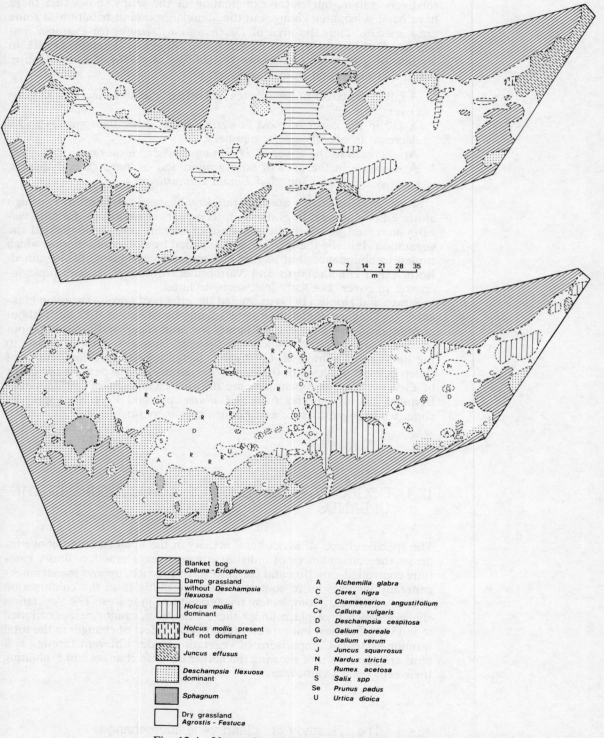

Blanket bog
Calluna - Eriophorum

Damp grassland
without *Deschampsia
flexuosa*

Holcus mollis
dominant

Holcus mollis present
but not dominant

Juncus effusus

Deschampsia flexuosa
dominant

Sphagnum

Dry grassland
Agrostis - Festuca

A   *Alchemilla glabra*
C   *Carex nigra*
Ca *Chamaenerion angustifolium*
Cv *Calluna vulgaris*
D   *Deschampsia cespitosa*
G   *Galium boreale*
Gv *Galium verum*
J   *Juncus squarrosus*
N   *Nardus stricta*
R   *Rumex acetosa*
S   *Salix spp*
Se *Prunus padus*
U   *Urtica dioica*

Fig. 12.4   Vegetation of Green Hole, Moor House National Nature Reserve,
Cumbria, England. Top: 1956; bottom: 1979. (From Rawes 1981.)

relatively minor, but closer examination of the maps shows that there have been substantial changes in the abundance and distribution of some grass species. Thus the area of *Deschampsia flexuosa*, for example, has increased, as has the area of *Holcus mollis*. Similarly, Ulmanis (1982), in a study of a long-term exclosure (established in 1956) at Moor House, noted that protection from grazing led to:

1. A fall in the cover of bryophytes and lichens due to an increase in grass cover;
2. A fall in the cover of most flowering plants other than grasses and a decrease in overall species numbers;
3. An increase in the cover of *H. lanatus* and *D. flexuosa*;
4. A decrease in the area of bare ground and of 'bare ground species' such as *Euphrasia* spp. and *Linum cartharticum*.

Some of these effects are illustrated by changes observed in two high altitude Pennine peat bogs during fifteen years of exclosure. *Calluna vulgaris* increased markedly from a state of scarcity until it dominated the vegetation. Initially the increase was matched by *Empetrum nigrum* which tripled in abundance, but as ling became more abundant, this declined. Both *Rubus chamaemorus* and *Narthecium ossifragum* showed rapid increases in cover, but leafy liverworts declined.

Rawes and Hobbs (1979) examined the effects of grazing sheep on blanket bog. They concluded that light grazing by sheep, at 0.37 ha, without burning produces a good food supply for both sheep and grouse without adversely affecting the blanket bog system. They also drew attention to the finding of Grant *et al*. (1976) that increasing the stocking density of sheep beyond the level of 0.55 sheep per ha is likely to depress the cover of *Calluna*. Similar conclusions are drawn by Gimingham (1972); under frequent grazing, species such as *Calluna* spp. and *Vaccinium* spp. decline, whilst *Agrostis* spp. and *Festuca* spp. increase.

12.4　ECOLOGICAL EFFECTS OF AGRICULTURE IN THE UPLANDS

The specific effects of agricultural activity in the uplands noted above indicate the consequences of individual management practices under carefully controlled experimental conditions. In practice, upland management generally involves more complex effects resulting from the combination of many different, complementary agricultural practices. The vegetation changes which take place under these conditions cannot be apportioned to individual causes, but represent the overall effect of changes in the total farming system. Comparisons of vegetation under different farming systems are one means of showing the nature of these changes and evaluating their ecological significance.

12.4.1　The ITE study of upland vegetation change

One of the most thorough investigations of changes in upland vegetation is that which has been carried out by the Institute of Terrestrial Ecology (Ball *et al*. 1981a, b). Using indicator species analysis, they analysed 938 sites in 12 upland areas in Britain and identified 4 major vegetation types. These broadly reflect a gradation in agricultural land use from semi-natural upland heathland to relatively intensively managed improved pasture.

Ball *et al*. (1981a) describe the four vegetation types as follows:

1. Shrubby heaths. These represent the least agriculturally affected areas. They are characterised by the predominance of dwarf shrubs such as *Calluna vulgaris* (ling), *Vaccinium myrtillus* (bilberry), *Erica cinerea* (bell heather), *E. tetralix* (cross-leaved heather) and *Empetrum nigrum* (crowberry), together with *Ulex* spp. (gorse), *Carex* sedges, rushes and various grasses (e.g. *Molinia caerulea, Deschampsia flexuosa* and *Agrostis tenuis*). Wetter sites, which also tend to be the least managed, include *Eriophorum vaginitum* (cotton grass) *E. angustifolium* (common cotton grass) and *Sphagnum* spp. (bog moss). Soils are generally strongly acid, with a pH below 4.2.

2. Grassy heaths. These represent the slightly more intensively managed heathlands. Native grasses such as *Nardus stricta* (mat grass), *Deschampsia flexuosa* (wavy hair grass) and *Molinia caerulea* (purple moor grass) are dominant, along with *Galium saxatile* (heath bedstraw) and *Vaccinium myrtillus. Pteridium aquilinum* is locally abundant, whilst sedges, cotton grass and rushes occur in wetter sites. Soil pH is variable, but more than half the sites investigated had a pH below 4.7.

3. Rough pastures. These result either from the partial improvement and reclamation of the moorlands, or from the deterioration of abandoned reclaimed pasture. A wide range of native species are found together with some introduced species. In the least intensively managed sites, *Festuca ovina* (sheep's fescue) and *Agrostis tenuis* (common bent) are abundant where soils are relatively freely draining, but in wetter sites *Juncus* spp. (rush), *Deschampsia caespitosa* (tufted hair grass), *A. canina* (brown bent), *A. stolonifera* (creeping bent) and *Nardus stricta* are common. As the intensity of management increases, species diversity increases and species such as *Holcus lanatus* and various herbs become abundant. Soils in the rough pastures are moderately acid, with a pH between 4.7 and 6.0.

4. Improved pastures. These show a high proportion of introduced species such as *Lolium perenne* (ryegrass), *Dactylis glomerata* (cocksfoot) and *Trifolium repens* (white clover). As the level of management falls, less nutritious species such as *Agrostis tenuis* and *Holcus lanatus* (Yorkshire fog) increase in abundance, whilst bracken occurs as an invading species in the least intensively used improved pasture. In wetter sites, rushes and thistles are common. Soils are generally moderately to slightly acidic, with a pH above 5.2 and often above 6.0.

Examples of species composition for each of these vegetation types are shown for an area near Glascwm in Powys, Wales, in Tables 12.2 and 12.3. This area has been subject to progressive agricultural intensification in recent years, with improvement of existing pasture and reclamation of the surrounding moorland where common land constraints permit. As these data indicate, each of the sites contains a fairly distinctive species assemblage. To some extent the differences between the sites may result from inherent differences in environmental conditions within each vegetation type, but they also reflect the impact of agricultural management. As intensity of use increases, native species such as *Calluna vulgaris*,

Table 12.2 Plant species recorded as present in sample 5000 m² quadrats from Glascwm, Powys (recorded 1977)

Plant species	Vegetation group and class				
	Shrubby heath – Vaccinium/Calluna	Grassy heath – Festuca/Vaccinium		Rough pasture – Festuca/Agrostis	Improved pasture – Herb-rich Lolium
		(high bracken variant)	(moderate bracken variant)		
Achillea millefolium					X
Agrostis canina/stolonifera				X	X
A. tenuis				X	X
Aira praecox		X			
Bellis perennis					X
Bromus mollis					X
Calluna vulgaris	X	X			
Campanula rotundifolia			X		
Carex panicea		X			
Cerastium holosteoides				X	X
Cirsium arvense					X
C. vulgare					X
Conopodium majus				X	X
Cynosurus cristatus				X	X
Dactylis glomerata					X
Deschampsia flexuosa	X	X	X	X	
Digitalis purpurea		X	X		
Empetrum nigrum	X				
Erica tetralix	X				
Festuca ovina		X	X	X	
F. rubra			X	X	X
Galium saxatile		X	X	X	
Holcus lanatus			X	X	X
Juncus conglomeratus				X	
J. effusus				X	X
J. squarrosus		X	X	X	
Leontodon spp.				X	X
Lolium perenne					X

Table 13.1. Species composition of vegetation types at different... Pasture... percentage cover (mean of 5, replicates)

Species	Shallow Grass heath	Heath moderate bracken	Rough pasture	Improved pasture
Lotus corniculatus				
Luzula campestris/multiflora	X X X	X X X	X X	X X
Nardus stricta				
Phleum pratense				
Plantago major				
Poa annua				
P. pratensis				
Potentilla erecta	X X	X X X	X X X	X X
Prunella vulgaris				
Pteridium aquilinum				
Rumex acetosa	X			
R. acetosella				
R. obtusifolius				
Sagina procumbens				
Sieglingia decumbens	X	X X	X	X
Stellaria graminea				
S. media				
Trifolium repens	X		X	X
Ulex europeaus/gallii				
Urtica dioica		X		X
Vaccinium myrtillus				
Veronica chamaedrys				
V. serpyllifolia				
Cladonia arbuscula/impexa	X X	X	X X	
C. pyxidata				
Dicranella spp.				
Dicranum spp.	X	X X		
Hypnum cupressiforme			X	X
Polytrichum spp.	X	X		
Rhytidiadelphus squarrosus			X	
Sphagnum spp.	X	X		

Species present. X = species present... field records made in 1977 by D. F. Helliwell and colleagues of the Institute of Terrestrial Ecology... vegetation change in upland landscapes (Ball et al. 1981a, b, 1982).

Vegetation... light and Clauge's doubt... and introduced species such as Lolium perenne and Trifolium repens and the pasture weeds e.g. Bellis perennis and Poa annua, increase. It is clear also that an overall increase in... increase in plant species diversity and in species constancy. Nevertheless it is apparent that whilst species diversity at a single site increases with reclamation and greater improvement, ulti- mately the total range of species will decline if agricultural intensification results from the loss of heathland vegetation which...

Source: From D. F. Ball, pers. comm. Data are derived from field records made in 1977 by D. R. Helliwell and colleagues of the Institute of Terrestrial Ecology as part of a DoE-contracted research programme on vegetation change in upland landscapes (Ball et al. 1981a, b, 1982).

Table 12.3   Species composition of vegetation types at Glascwm, Powys

Species	Percentage cover (mean of 5 quadrats/site)				
	Shrubby heath	Grassy heath		Rough pasture	Improved pasture
		High bracken	moderate bracken		
Calluna vulgaris	87	—	—	—	—
Vaccinium myrtillus	3	—	20	—	—
Deschampsia flexuosa	3	—	—	—	—
Cladonia spp.	4	—	—	—	—
Pteridium aquilinum	—	66	34	—	—
Galium saxatile	—	10	2	—	—
Potentilla erecta	—	8	—	—	—
Agrostis canina/stolonifera	—	6	—	—	—
A. tenuis	—	4	—	—	5
Nardus stricta	—	—	12	—	—
Holcus lanatus	—	—	10	—	5
Juncus effusus	—	—	8	—	—
Lolium perenne	—	—	—	19	50
Festuca ovina	—	—	—	16	—
Cynosurus cristatus	—	—	—	15	—
Trifolium repens	—	—	—	14	15
Dactylis glomerata	—	—	—	10	—
Poa pratensis	—	—	—	8	—
Cirsium vulgare	—	—	—	3	2
Rumex spp.	—	—	—	2	2
Cerastium holosteoides	—	—	—	—	10
Phleum pratense	—	—	—	—	3
Bellis perennis	—	—	—	—	2
Litter	—	—	14	—	—
Bare ground	—	—	—	7	—
No. of species recorded	15	18	20	32	27

Source: From Ball, D. F., pers. comm. Data are derived from field records made in 1977 by D. R. Helliwell and colleagues of the Institute of Terrestrial Ecology as part of a DoE-contracted research programme on vegetation change in upland landscapes (Ball *et al.* 1981a, b, 1982).

Vaccinium myrtillus and *Galium saxatile* decline, and introduced species such as *Lolium perenne* and *Trifolium repens* and the pasture weeds (e.g. *Bellis perennis* and *Rumex obtusifolius*) increase. It is clear also that improvement results in an increase in plant species diversity and in species constancy. Nevertheless, it is apparent that whilst species diversity at a single site increases with reclamation and pasture improvement, ultimately the total range of species will decline if agricultural intensification results in the loss of heathland vegetation types.

12.4.2   The bracken problem

One of the most far-reaching ecological changes which has occurred in many upland areas of Britain has been the extension of bracken (*Pteridium aquilinum*) into grassland and heathland sites. The problem of bracken infestation is not confined to Britain – it is found as widely as the USA, Australia, New Zealand, Canada, the USSR and even Hawaii. Taylor (1980), however, calculated that the area of bracken in Wales almost doubled between 1936 and 1966, from 62,000 ha to 118,000 ha – an annual rate of expansion of about 2,072 ha. Similarly, Taylor quotes an estimate of an 8 per cent increase in the bracken area in Dartmoor between 1969 and 1976, from 5,880 ha to 6,370 ha, and a further estimate that between 1966 and 1980 the area of bracken on Eglwysg Mountain in Clwyd, Wales, rose from 77 ha to 96 ha. On the basis of these and other data, Taylor calculated an annual rate of bracken expansion for the UK as a whole of 10,360 ha, approximately equivalent to the rate of land loss to either urbanisation or afforestation (Best 1981).

This rapid extension of the bracken area has considerable significance both ecologically and agriculturally. Bracken is a relatively poor food source for ruminants, being eaten only when young by sheep. Moreover, it may be toxic when eaten in large quantities: 'bright blindness' in sheep and 'bracken staggers' in horses are common results of bracken poisoning (MAFF 1974), while Evans (1976) has demonstrated that long-term consumption of bracken may cause cancer, particularly in calves. Indeed, carcinogens retained in milk and other dairy products may be passed to humans and it has been suggested that this pathway may contribute to locally high incidences of gastric cancer mortality in the Lleyn Peninsula in Wales (Taylor 1980). Ecologically, bracken is also of limited value. Jeffreys (1917) estimated that light intensities below the litter of bracken fronds were reduced to one fortieth of those in the open, indicating the intense shading the plants cause. As a result, few other species are able to survive and bracken is able to compete successfully with most other plants, including *Calluna vulgaris*. For this reason bracken tends to be associated with a limited range of shade-tolerant species. Gliessman (1976) also suggests that toxins released by the plant reduce the ability of other species to survive in association with bracken. Bracken invasion therefore leads to a reduction in both the species diversity and biomass production of the ground flora, an effect illustrated by the data in Fig. 12.5. Nevertheless, there is some evidence to suggest that the development of an almost monospecies cover of bracken is a phase in the evolution of the community rather than a climax situation, and over time most bracken communities become more open, with the development of at least a partial ground cover of species such as *Festuca ovina* (sheep's fescue), *Agrostis tenuis* (bent) and *Anthoxanthum odoratum* (sweet vernal grass). Moreover, it has also been argued that bracken may improve moorland soils by reversing podsolisation and encouraging structural development (Tivy 1973); the association of bracken with these better soils can also be interpreted, however, as an indication that the species preferentially invades more fertile areas. Certainly it is commonly found in abandoned fields and on better drained slopes, as well as along old field boundaries in moorland areas, and Page (1976) considered the success of the species was due to its ability to colonise previously cultivated sites.

As this indicates, the problem of bracken expansion can often be related to changes in agricultural practice: in particular, the abandonment of marginal cultivated land or the reduction of grazing pressures in these areas. In addition, expansion has undoubtedly occurred because of a decline in the use of bracken as a bedding material for both livestock and labourers, and as thatching for farm buildings. Bracken was similarly used as a fuel and, because of its high potassium content, was exploited for

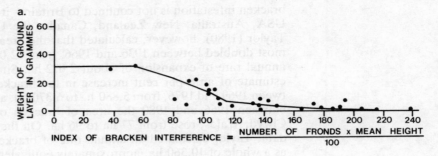

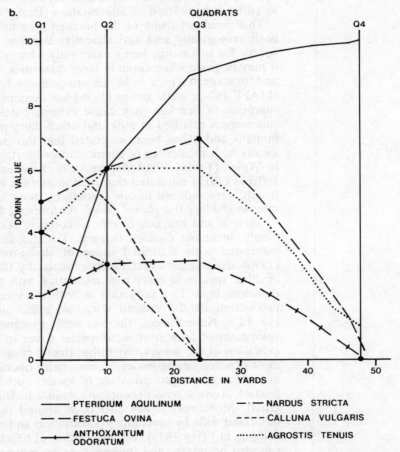

Fig. 12.5   Effects of bracken upon (a) ground layer biomass and (b) species diversity (from Tivy 1973).

glass and soap, while it has also been employed as a substitute for hops in brewing and as a base for bread (Rymer 1976). Cutting of bracken for these functions in the past helped keep the fern in check, but over the last century this practice has declined as a result of depopulation in these areas and modernisation of farming systems. Conversion of the moorland from cattle to sheep also diminished the traditional control of the plant, for cattle, while not eating bracken to any extent, did reduce its competiveness by trampling. At the same time, with the development of moorland sheep farming, the practice of heather burning grew up. Excessive burning has often destroyed the heather stock, allowing bracken to become established.

Today control of bracken is largely by the use of chemical herbicides, notably compounds of asulam. While these seem to be capable of reversing the process of bracken invasion, at least with repeated treatments, it also has implications for other components of the heathland ecosystem. Martin (1976) suggests that, in the short term, eradication of bracken may result in invasion by other weed species such as nettles, foxgloves and thistles, while Davies *et al.* (1979) noted that asulam also reduced the abundance of grass species such as *Agrostis tenuis, A. canina* and *Poa pratensis*. On the other hand, the fine-leaved grasses, *Festuca ovina* and *Deschampsia flexuosa* proved to be more resistant, and recovery of the grass sward to something like its original cover was achieved within three years. Cadbury (1976) also showed that *Ulex* spp., *Rumex acetosella* and grasses of the *Poa, Holcus* and *Agrostis* genera are susceptible to spraying with asulam. Since aerial spraying is normally practised in moorland areas because of inaccessibility, these effects may be serious due to herbicide drift into neighbouring woodland and grassland. Nonetheless, as yet the problems of spraying are small, for according to Drummond and Soper (1978) only about 3200 ha of bracken are treated each year. To date, however, little research has been carried out on the ecological effects of asulam and there is clearly a need for careful monitoring of sites treated with the herbicide to ensure that fundamental damage to associated communities does not occur.

12.5 EXMOOR: A MANAGEMENT CASE-STUDY

Exmoor is an upland area of south-west England predominantly formed of Devonian grits and sandstones rising in altitude to 522 m. Much of the moorland fringes have been enclosed at one time or another and the land-use history is very complex, some areas having reverted from earlier improvement. (The land-use history is detailed by Orwin and Sellick 1970 and summarised by Curtis *et al.* 1976.) The central upland area (The Chains) is covered by blanket mire and is of relatively low ecological interest (Ratcliffe 1977). The surrounding open moorland is dominated by *Calluna, Vaccinium* and *Ulex gallii* (gorse). There is a range of heath vegetation with a gradual transition from lowland to upland. Adjacent to the heathland are steep-sided valleys with valley sides covered in oakwoods, the latter of very high conservation interest.

Exmoor forms one of ten national parks in England and Wales set up under the National Parks and Access to the Countryside Act in 1949. The 1949 Act defined the purpose of the National Park Authority as:

(a) preserving and enhancing the natural beauty of the area specified;
(b) promoting its enjoyment by the public.

The natural beauty of Exmoor is such that, as the *Exmoor National Park Plan* states (Exmoor National Park Committee 1977a) there is very little need for enhancement. The main difficulties that have arisen have been associated with the preservation of the moor.

The conflict between agriculture and conservation interests in the Exmoor area first came to a head in 1962. In April of that year an area of heathland on the north side of the park, near Countisbury, North Devon, was fenced off by the owner prior to ploughing up and reseeding. The purpose of this agricultural operation, which was supported by a substantial grant from the Ministry of Agriculture, was to enable an increase in stocking rates, either of sheep or beef cattle. (It is important to appreciate that in England and Wales, the conferring of national park status does not imply any change in land ownership; a very high proportion of land in national parks remains in private ownership.)

The Park Authority made considerable, yet unsuccessful attempts to stop agricultural improvement of this particular land (which are summarised in Exmoor National Park Committee (1977b)). Similar difficulties arose elsewhere on Exmoor such that it was estimated that 9,500 acres (3,846 ha) out of a total of 59,000 acres (23,886 ha) were ploughed up between 1947 and 1976. In 1977 the Park Authority designated 40,000 acres (16,194 ha) as a 'critical amenity area'. Of this, the Authority considered that some 12,560 acres (5,085 ha) were capable of agricultural improvement (Exmoor National Park Committee 1977b). In other words, about 30 per cent of the most important amenity area was capable of being substantially ecologically modified, although it should be pointed out that some of the area was already in public ownership and, therefore, to a considerable extent, protected.

An emotive aspect of this issue, already referred to, is the fact that ploughing up moorland is eligible for grant aid. (A similar conflict exists in the areas of wetland referred to in the next chapter.) Newby (1979) describes the accusations and counter-accusations that were made in the late 1970s when the Park Authority was accused by amenity groups and conservationists of being biased in favour of agriculture. On the other hand, the National Farmers Union and the Country Landowners Association considered that the Park Authority was unnecessarily influenced by amenity interests. The Countryside Commission (the central government agency concerned with national parks) became involved and matters culminated in the setting up of an inquiry under Lord Porchester. Porchester produced a report which recognised that voluntary agreements between the Park Authority and individual farmers had proved largely ineffective in conserving the moorland from agricultural improvement (Porchester 1977).

Porchester recommended that a new survey should be undertaken to re-define the critical amenity areas so that within the areas a 'final commitment' could be made to conservation, the presumption being that in these areas agricultural improvement would not be allowed. The financial and legal implications of this development are substantial and somewhat outside the scope of the present volume. Suffice it to say that, at the time of writing, it is by no means clear whether the National Park Authority has sufficient resources to be able to resist ploughing of all the designated areas.

The rationale behind the new 'Porchester maps' which delineate the significant areas is explained by Curtis and Walker (1982). Porchester recommended that two maps should be drawn. The first of these ('Porchester map one') identifies and defines all areas which are predominantly moor and heath. Porchester's suggested guidelines make interesting reading in the light of the difficulties of defining upland areas, referred to in section

12.1 of this chapter. He suggested, *inter alia*, that, in constructing map one:

(a) The Authority should not be too theoretical in drawing up criteria;
(b) The principal factor would be that of vegetation;
(c) The Authority should look at the area as an informed layman would and should identify those tracts of land which would commonly be regarded as moorland or heath;
(d) The Authority should not be too influenced by the existence of various forms of enclosure since much of the land on Exmoor which most people would rightly regard as moorland has, at some stage, been enclosed.

Map one forms the basis of voluntary arrangements whereby farmers inform the Authority of any improvement proposals for the defined area. Map two defines those particular tracts (within those defined on map one) which the National Park Committee wish to see preserved for all time. It is clear that the delineation of map two is a more subjective exercise than that of map one.

The criteria used in determining whether to include land on map two were:

(a) Visual considerations of vegetation and relief: pattern, colour and texture;
(b) Enclosures and openness;
(c) Extent: real and apparent including considerations of remoteness and wildness;
(d) Views and edges;
(e) Public access, rights of way and road access;
(f) Landmarks and landscape features.

Curtis and Walker (1982) describe the application of these criteria in detail using landscape evaluation techniques. After modification following consultation with outside bodies, landowners and the general public, the National Park Committee finally adopted 38,610 acres (16,036 ha) as having map two status. (It will be recalled that 40,000 acres (16,194 ha) had previously been designated as the extent of the former critical amenity area, although it must be stressed that the map two areas do not necessarily coincide with the former CAA.)

In September 1980 the National Park Committee resolved that, in relation to map two:

(a) There should be the strongest possible presumption in favour of conservation of natural flora, fauna and landscape;
(b) There should be the strongest possible presumption in favour of traditional rough grazing systems compatible with conservation; and
(c) There should be periodic reviews of the map two moorland areas concerned.

Guidelines for management agreements were also concluded at the same time and it now remains to be seen how effective this new, more rigorous delineation of important areas will be in arresting the process of agricultural improvement. Indications are that the combination of the Porchester maps and clear management agreement guidelines are much more effective that the previous system. Nevertheless, as already noted, sufficiency of financial resources is, in the final analysis, a critical factor.

12.6   CONCLUSIONS

Uplands account for some 40 per cent of the land surface in Britain. As Wells (1981) notes, therefore, these areas are of considerable ecological significance because of their extent alone. On the whole, wildlife in these areas is of restricted diversity and abundance, yet the value of the uplands as refuges for many species and as the habitat for certain rare species should not be overlooked. Moreover, much of the ecological and aesthetic value of upland communities lies in their extensiveness and intactness. Fragmentation of these areas by reclamation and improvement for agriculture must inevitably reduce their value in these terms.

At present, the main land conversions are in the fringe areas, which are largely lands that have reverted in the past from previous periods of agricultural activity. Increasingly, however, the pressure upon more ancient upland communities is growing. In Europe, this pressure is encouraged by the Hill Farming and Favoured Areas policies of the European community, which are facilitating intensification of upland farming. The threats, however, do not only come from agriculture. In Britain, afforestation has affected large areas of uplands (see, for example, Parks 1980), whilst in alpine areas of Europe tourism is disturbing both the ecological and cultural stability and integrity of the land (e.g. Danz and Henz 1981). Climatic changes and long-distance industrial pollution of the atmosphere are also affecting sensitive upland ecosystems (e.g. Drabløs and Tollan 1980). It is apparent, therefore, that agriculture is but one contributory factor to the ecological changes that are occurring in temperate upland areas. Nevertheless, it remains an important factor, and one which needs to be closely monitored and, if necessary controlled, if unwanted ecological effects are to be avoided.

13 THE IMPACT ON ECOLOGICAL SYSTEMS: III WETLANDS

13.1 INTRODUCTION

Wetland habitats are among the most vulnerable to agriculture. They comprise a range of different environments including running streams, relatively still-water lakes and ponds, land which is only seasonally flooded (e.g. salt-marshes, water-meadows) and wet terrestrial environments such as peat-bogs and river-banks. Almost all these areas (which occur variously in uplands and lowlands) are being reclaimed for agriculture by drainage and infilling. Because of their often small size and relative proximity to farmland, they are also susceptible to the side effects of agricultural practices, such as pollution by agricultural chemicals and the lowering of the water table by agricultural drainage. At the same time, many wetland habitats are ecologically highly important in that they harbour rare and diverse communities of plants and animals. Not surprisingly, therefore, the threat to wetlands posed by agriculture is of widespread concern.

In this chapter we consider the ecological impact of modern agriculture on wetlands; hydrological aspects are dealt with in Chapter 9.

13.2 AQUATIC ENVIRONMENTS

True freshwater aquatic environments can, for convenience, be divided into running-water (streams) and still-water bodies (lakes and ponds). In both these types of environment, the suitability of the habitat for wildlife depends to a great extent upon water quality (see Ch. 9). In general, water bodies with low pH are nutrient deficient and relatively unproductive (oligotrophic), whereas more alkaline waters with higher nutrient content (eutrophic) support a larger organic population. Nonetheless, the introduction of excessive levels of nutrients may cause eutrophication of the water, leading ultimately to a lack of oxygen and a severe decline in organic activity. One of the main sources of these nutrients is agricultural runoff containing fertiliser and livestock wastes. In addition, accumulation of pesticide residues in aquatic systems may reduce their ecological viability, whilst increased turbidity, reduced sunlight penetration and rapid aggradation of bed sediments may also affect animal and plant populations in these habitats. In the case of running-water bodies, agriculture may have further impacts though its influence on discharge and flow velocity (Ch. 9). Thus, the main threats to these habitats and the species they contain are by:

1. Chemical pollution (e.g. by fertilisers, animal wastes, pesticides);
2. Sediment pollution due to soil erosion;
3. Flow modifications due to changes in runoff rates (e.g. as a result of land drainage, canalisation or damming).

13.2.1 Chemical pollution

As we saw in Chapter 9, agriculture has in many areas been causing increasing chemical pollution of streams and lakes due to runoff of farm wastes. Nitrates from inorganic fertilisers, and both nitrates and phosphates from animal slurries have been particularly implicated in pollution of water bodies, and increasing concern has been expressed about the effects upon aquatic ecosystems. The impact of these pollutants is most marked in ponds and lakes, where the processes of dilution, decomposition and dispersal are restricted by the slow turnover rate of the water. As a result a progressive accumulation of nutrients tends to occur within the system. Typically this leads to increased organic activity and, most notably, a rapid rise in the populations of algae. Initially, other organisms at higher trophic levels follow suit in response to the increased food availability, but the end result is that the expansion in faunal populations leads to a depletion of the oxygen reserves within the system and a sudden and often catastrophic collapse of the organic communities.

Possibly the classic example of this process is Lake Erie in the USA (Beeton 1971). Here, between 1900 and 1962, concentrations of chloride rose by over 300 per cent from 7 p.p.m. to 23 p.p.m., while the sulphate content increased from about 13 p.p.m. to 24 p.p.m. As a result fish populations were all but eradicated. Apart from the ecological effects, this had severe impacts upon the fishing industry. The catch of lake herring, for example, declined from 2.6 million kg in 1925 to 3,200 kg in 1962. Similarly, by 1962 the recorded take of blue pike had fallen to only 450 kg, about 0.003 per cent of its pre-1958 level.

Elsewhere in the temperate world the effects of eutrophication are generally less severe, although comparable in nature. In Scotland, Lund (1971) cites the eutrophication of Loch Leven. Quoting early descriptions, he notes that in 1905 species of *Chara, Nitella* and *Tolypella*, were abundant, and that in a bay at the east end of the lake *Elodea canadensis* was so abundant that, in summer, 'it is very difficult to row a boat through them'. By the late 1960s, the *Chara* had declined considerably and there were few macrophytes, but phytoplankton had increased markedly. Lund notes also that the mayflies *Caenis horaria* and *Cloeon simile* have become extinct in the area, dragonflies are absent and the numbers of caddis flies have decreased. In view of the progressive increase in fertiliser usage in the surrounding area, it seems likely that these changes are a result of eutrophication by fertiliser wastes.

Similar effects have been recorded in Malham Tarn, a eutrophic upland lake situated on the Carboniferous Limestone plateau of the Pennines in North Yorkshire. Its ecological significance derives not only from its unusual position, but also from its biological diversity: apart from a rich flora of *Chara*, other macrophytes (e.g. *Potamogeton lucens*), and planktonic and benthic algae, it is one of the few British lakes containing *Cladophora* balls. Lund (1971) comments that because of its shallowness the lake would be highly susceptible to inputs of fertilisers from the surrounding grassland.

In much of upland Britain, pressures upon lake ecosystems are limited due to the low levels of fertilisers used. Lowland lakes, however, are subject to much higher fertiliser inputs. One of the most critical areas is the Broadlands of Norfolk and Suffolk (Fig. 13.1). This area consists of a series of lakes connected by slow-flowing rivers. It is now acknowledged that they were man-made, probably being flooded medieval peat workings (Ellis 1965). The flora and fauna of the area are of considerably interest and the Broads also support a substantial tourist industry in the form of boat-hire operations. For some years there has been considerable concern at the increase of eutrophication caused by an excess of nitrate and phosphates. The source of these nutrients is not only runoff from farmland but also from sewage outfall and from the action of motor cruisers in churning up the peaty lake-bed deposits.

Philips (1977) examined the mineral nutrient levels in three Norfolk broads differing in trophic status, and produced an annual mineral content budget for one of them. He considers that because the source of enrichment is input sediments one possible solution may be diversification of input.

The Broads suffer not only from increases in nutrient levels but also in some cases from rapid and drastic changes in pH level. In November 1970 the acidity of Calthorpe Broad suddenly changed from near neutrality to pH 3 (Gosling and Baker 1980). This had drastic results: the entire fish population together with freshwater mussels and most aquatic macrophytes died. There was a substantial invasion by the acidophilous alga, *Tribonerna uriuus*. The pH level later returned to normal but Gosling and Baker report similar fluctuations in each year from 1971 to 1979. The problem was eventually traced to improvements in drainage in the surrounding agricultural land which had been completed shortly before the first fluctuation. During the winter a perched water table remained in the area surrounding the Broads, artificially sustained by dams. However, in the summer, under the new regime, increased evaporation together with seepage caused the soil to dry out and the soil profile, which was rich in pyrites, was subjected to alternate wetting and drying. This apparently resulted in oxidation of pyrite and the production of sulphuric acid. The solution to the problem was treatment by lime and this was successful from 1978 onwards in maintaining neutrality in the Broad. Various plants and animals re-colonised the Broad as a result.

Eutrophication may have an indirect effect on some plant species. A study has been made of the variation in extent of reedswamp in eighteen broads between 1945 and 1978 by Boorman and Fuller (1981). There was a substantial loss of reedswamp from 121.5 ha in 1946 to 49.2 ha in 1977, most of the change being by reversion to open water. Changes for three sample broads are shown by maps in Fig. 13.2 and for all eighteen broads graphically in Fig. 13.3.

Boorman and Fuller have a number of conclusions on the cause of reedswamp decline, which has dramatically altered not only the Broadland

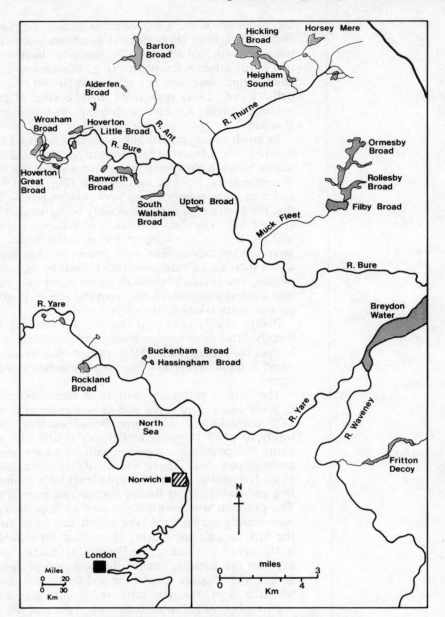

Fig. 13.1   The Norfolk Broads (from Boorman and Fuller 1981).

ecosystem but also the visual amenity. They conclude that, particularly between 1950 and 1963, the coypu (*Myocastor coypus*), the large, introduced, amphibious rodent, could have been primarily responsible for reedswamp decline. However, by raising the rate of sedimentation, eutrophication could have assisted in making the reeds more accessible for grazing. Also Boorman and Fuller point out that under anaerobic water conditions the plants might be more susceptible to shoot damage when the oxygen supply is cut off.

As these examples indicate, lakes and ponds in many areas are threatened by eutrophication by fertiliser and animal wastes. Possibly the most

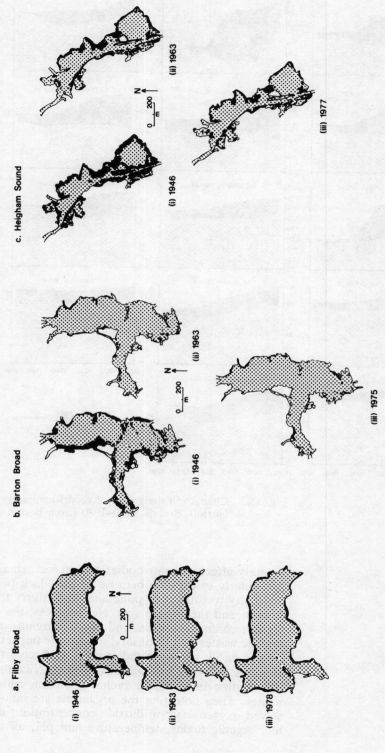

Fig. 13.2 Changes in the distribution of reedswamp (black) and open water (stippled) in the Norfolk Broads, 1946–78 (from Boorman and Fuller 1981).

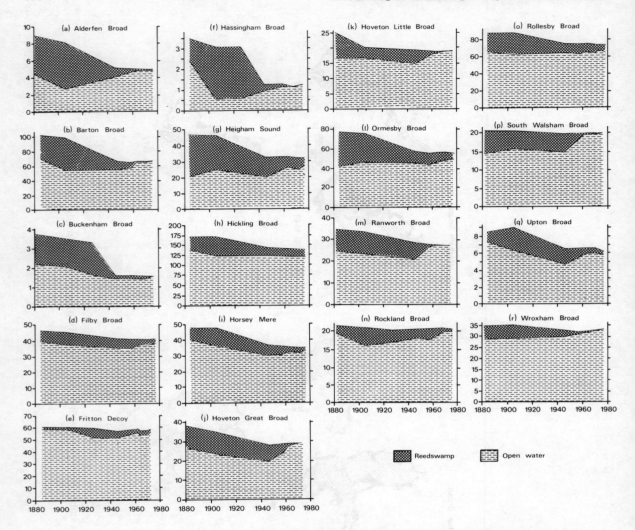

Fig. 13.3   Changes in the extent of reedswamp and open water in eighteen
Norfolk Broads, 1880–1980 (from Boorman and Fuller 1981).

severely affected water bodies are, in fact, small farm ponds. These are
particularly susceptible because of their lack of mixing or outflow, their
proximity to sources of pollutants (e.g. slurry from farmyards and silage
stores), and their small size. Nevertheless, the effects of eutrophication
are not confined to lakes and ponds. Streams, also, may be polluted by
organic wastes and fertilisers, though the impact of agricultural chemicals
cannot easily be separated from the effects of industrial and domestic
effluents. Hawkes and Davies (1971) describe the changes which occur in
the benthic invertebrate fauna of stream riffles polluted with organic
wastes. They note that the organisms are influenced by changes in dissolved oxygen, carbon dioxide concentration, hardness, suspended matter, organic toxins, temperature and pH, all of which are affected by

pollution. From studies in the River Cole south-west of Birmingham, they conclude that groups such as the Plecoptera and most of the Ephemeroptera and Trichoptera are particularly sensitive to organic enrichment and are suppressed or eliminated by pollution. Conversely, groups such as the Enchytraeidae, Hirudinea and Chironomidae tend to increase in numbers and dominate the faunas. This reflects the synecological effects of organic enrichment: the decomposer organisms expand relative to the producers and consumers. Since most riffle communities are naturally dominated by producers and consumers, this represents a major change in faunal composition.

So far we have been concerned mainly with eutrophication effects caused (at least partly) by fertiliser application. A further chemical problem in aquatic habitat is caused by the use of aquatic herbicides. These are employed mainly to keep drainage channels clear and use is still increasing though not as much as in recent years (Scorgie 1980). An assessment of the use of aquatic herbicides is given by Newbould (1979).

The secondary effects of aquatic herbicides are in general terms predictable. Scorgie (1980) studied the effect of the application of cyanotryn to part of a drainage channel. The dominant plants, which were mainly *Myriophyllum spicatum* (spiked water-milfoil) and *Vaucharia* spp., were completely eradicated from the treated section after twelve weeks. Although the total number of invertebrates did not change substantially, considerable numbers of species declined, and the invertebrate colony became one associated with detritus. Plants began to colonise the treated section about a year after application followed by the gradual regeneration of the invertebrate colony. It should, of course, be recognised that in agricultural conditions applications are likely to be repeated on an annual basis which is likely, in turn, to lead to an elimination of certain invertebrates.

The effects of chemical pollution of aquatic habitats upon both plant and animal species are far-reaching. In Europe many aquatic organisms are now considered to be in danger of extinction largely (or at least partly) as a result of pollution (Nature Conservancy Council 1982). Three species of plant, in particular, appear to be threatened: *Aldrovanda vesiculosa*, *Coleanthus subtilis*, and *Myosotis rehsteineri*. The last of these occurs along lake margins in southern Germany and is believed to exist in Italy, although records here have not been confirmed. In Germany, a major cause of decline is eutrophication, whilst around Lake Lugano the species has become extinct as a result of pollution. *A. vesiculosa* is similarly restricted in its distribution. It is a submerged carnivorous aquatic found in still, eutrophic waters and occurs in Germany around the shores of Lake Constance, in a number of localities in France (e.g. in the Gironde, the Bouches-de-Rhone and the eastern Pyrennees) and in Italy. Its population appears to be declining rapidly and it is now believed to be extinct in the Siechenweither reserve near Meersburg, where it was once prolific. In Italy it now seems to be confined to northern areas, although it was once widespread. In many cases, this decline seems to be related to increased pollution due, among other things, to intensification of agriculture (Nature Conservancy Council 1982).

Many freshwater fish are also threatened by aquatic pollution. The Nature Conservancy Council (1982), while stressing the lack of information on many other fish populations, identified four species which are considered to be endangered, and a further nine which are vulnerable.

In several cases, these populations are under direct stress due to pollution by agricultural chemicals. Similarly, eutrophication threatens a number of reptiles and amphibians. *Bombina bombina* (the fire-bellied toad) is an aquatic species whose survival in Europe is now endangered in part because of pollution (Nature Conservancy Council 1982) (see also section 13.2.3 below). The natterjack toad (*Bufo calamita*) has been severely reduced in numbers in Mediterranean regions because of pesticide pollution (Petit and Knoepffler 1959). In southern France and Italy aquatic pollution is also contributing to the decline in the European pond tortoise (*Emys obicularis*), most particularly along the Mediterranean coast (Honneger 1978).

13.2.2   Sediment pollution

The ecological effects of sediment pollution on aquatic ecosystems have not been well documented. Siltation is part of the natural cycle of evolution of lakes and ponds and accounts for the characteristic seral succession seen in many aquatic environments. As such it contributes to the ecological diversity of such sites, for it helps produce a range of habitats from open water to lake-side marsh (Fig. 13.4). Accelerated sedimentation, however, undoubtedly impoverishes the biological status of aquatic ecosystems. Bottom-dwelling plants may be buried, while the increased turbidity inhibits both the growth of submerged floating species and the activity of the benthic fauna. Ultimately, of course, continued sediment inputs result in infilling of the lake and elimination of the original aquatic habitat.

Sediment pollution is most acute in areas where agricultural practices encourage excessive soil erosion. Typically, these conditions are associated with intensive arable cultivation of unstable soils. Erosion is also a problem in upland environments, however, where slope angles are often high and rainfall intensities sufficient to cause runoff. Vegetation clearance, either by cultivation or by burning (e.g. Imeson 1971; Kinako and Gimingham 1980), thus leads to severe erosion. As has been noted, few studies have considered the effects of these processes on the ecology of nearby aquatic environments, but research is clearly required.

13.2.3   Flow modification

As noted in Chapter 9, streamflow may be affected by agricultural practices such as land drainage which influence runoff rates. In addition, as part of more general soil conservation and catchment management schemes, streams may be deliberately modified: river meanders may be straightened, channels may be dredged eliminating pools and shallows, river banks may be raised and graded and the water level itself may be permanently changed. Thus the river-side vegetation may be substantially altered and bank-side vegetation may be entirely removed. Similarly, ponds and ditches are often deliberately infilled for cultivation or to remove obstacles to farming activities.

These kinds of changes have been most carefully studied in the context

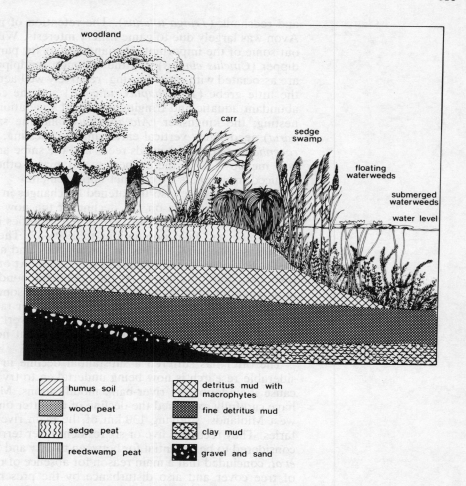

woodland

carr

sedge
swamp

floating
waterweeds

submerged
waterweeds

water level

	humus soil		detritus mud with macrophytes
	wood peat		fine detritus mud
	sedge peat		clay mud
	reedswamp peat		gravel and sand

Fig. 13.4   Successional sequence in an enclosed lake (from Open University 1972).

of the higher order animals and in particular birds and also the otter (*Lutra lutra*). E. Williams (1980) reported two sets of comparisons of breeding bird densities. On the River Wye in Gwent, on a heavily managed section there were eleven species of birds breeding at a density of less than five pairs per kilometre. A comparable but relatively unmanaged section held twenty-five breeding species at a density of greater than fourteen pairs per kilometre. The difference was thought to be due to dredging and to removal of bank-side vegetation.

In a similar comparison between the Wiltshire/Hampshire Avon and the Dorset Stour, Williams showed that on a section of the well managed, freely flowing Stour only two pairs of waders (both lapwings, *Vanellus vanellus*) were noted. On a similar length of the Avon, where water levels are kept high and the surrounding meadows are poorly drained, there were more than 180 pairs, including lapwings, snipe (*Gallinago gallinago*) and redshank (*Tringa totanus*). The retention of natural habitats on the Avon was largely due to game fishery interests. Williams goes on to point out some of the important riparian habitats for particular birds. Thus the dipper (*Cinclus cinclus*) and the common sandpiper (*Actitis hypoleucos*) are associated with rapid flowing, relatively well aerated water with riffles; the little grebe (*Tachybaptus ruficollis*) and the coot (*Fulica atra*) need abundant aquatic and fringing emergent vegetation for both feeding and nesting; the kingfisher (*Alcedo atthis*) and the sandmartin (*Riparia riparia*) need sandy vertical earthbanks for nesting, the reed warbler (*Acrocephalus scirpaceus*) needs reeds and the snipe and redshank both need wet meadows. All these species and many others are displaced and reduced in numbers by channel improvements.

Other species are also threatened by changes in aquatic environments. Moore (1980) reports on the decline of the now rare damselfly (*Lestes dryas*). None were found in 1978 in ten localities in south and east England which had held populations in 1940–47. The main habitat for this species is ponds and ditches on flat, alluvial land near the sea, and in the past it tended to breed on small farm ponds. At only two of the revisited sites was there no change. At the others the pond had either been filled in or was considerably modified. Moreover, at some sites the surrounding marsh had been drained or, at least, the water table had been lowered. Moore noted that if the site was allowed to revert to its original state the related dragonfly *L. sponsa* often returned but not the rarer species *L. dryas*.

The otter has suffered a substantial decline in recent years and considerable research is now being undertaken to try to find the cause. One cause is undoubtedly river-bank modifications. MacDonald *et al.* (1978) have carefully examined the decline of the otter on the River Teme in the west Midlands, studying 130 km of the main river and 213 km of tributaries. They recorded five or six resident otter territories out of what they considered to be potential for between twenty and forty pairs. MacDonald *et al.* concluded that a main reason for absence of residents was a shortage of tree cover and also disturbance by the presence of anglers. Jenkins (1980) similarly confirms the importance of cover and seclusion after studying otters in north-east Scotland.

In Europe, infilling and physical modification of aquatic environments appears to be threatening a number of species. The Nature Conservancy Council (1982) notes that seventy-four species of bird were listed as endangered by the European Council Directive on the conservation of wild birds (Official Journal of the European Communities 1979), and to these it adds a further thirty-eight species. A large proportion of these species are wetland species such as ducks, geese, swans, waders, terns, rails and warblers. The ICBP (1981), for example, list important bird-breeding areas in the European Community. Most of these sites are wetlands (Table 13.1) and many have international importance. Distribution maps of these sites for Britain are shown in Fig. 13.5. Many of these wetlands are aquatic sites and are thus subject to threat from drainage and infilling.

In the same way, agricultural reclamation of aquatic sites poses a serious threat to many reptiles and amphibians. Within the European Community the most critically endangered species is the fire-bellied toad (*Bombina bombina*). As noted earlier, water pollution has contributed to the decline in this species, but it is also affected by severe pressure from reclamation. It is a diurnal, highly aquatic toad which prefers medium to

Table 13.1 Important bird-breeding areas in the European Community

	Ireland	UK	Denmark	Germany	Netherlands	Belgium	France	Italy	EEC*
Wetlands: (i) total	35	99	89	87	60	13	56	41	465
(ii) of international importance	32	91	83	72	52	8	40	21	384
Non-wetlands important for non-breeding waterfowl	3	7							10
Areas important for seabirds	16	60	5	1			15	10	107
Areas important for terrestrial species		15		4	2		22	69	112
Total	54	181	94	92	62	13	93	120	694

* Fifteen areas extend across national boundaries

Source: From ICBP 1981

large areas of partially vegetated, unpolluted still water in open lowlands. It occurs in Denmark in shallow ponds near the coast, in a few scattered localities in the eastern islands of Fyn and Sjaelland. It is also found in a few sites in the extreme north-east of Germany. Over the last 150 years the population has declined persistently: of 68 sites reported in Germany before 1960, only 14 still harboured the species in 1976 (Müller 1976). In Denmark there are only an estimated 15 localities where the toad still occurs (Carstensen 1976).

The decline in the populations of *Bombina bombina* seems to be at least partly related to climatic change. In the Fyn area of Denmark, for example, the disappearance of the species from several localities in the early 1960s was correlated with a period of relatively cool, moist summers (Carstensen 1976). More important in many cases, however, has been drainage and infilling of ponds as a result of intensification of agriculture, and Wederkinch (1976) considers that climatic change has merely served to increase the species' sensitivity to human influence.

Although in less immediate danger of extinction, several other species are threatened by loss of habitats due to reclamation or abandonment of aquatic sites. The midwife toad (*Alytes obstetricans*), for example, is threatened in the Netherlands by the infilling of cattle ponds and drainage of marshlands (Bergmans, 1978, Honnegar 1978). In France and Italy, populations of the yellow-bellied toad (*Bombina variegata*) are declining for similar reasons. The painted frog (*Discoglossus pictus*) is threatened in Sicily because its main habitats – pipes and ditches associated with the traditional irrigation systems – are being abandoned as agriculture is modernised (Riggio 1976). Drainage of wetland habitats is also reducing populations of the European pond tortoise (*Emys orbicularis*). This carnivorous species breeds in muddy ponds, sluggish streams, stagnant ponds and associated marshland in lakeland areas. Although widely distributed in France and Italy, it is now extinct as an indigenous species in Germany and survives only locally due to accidental or deliberate introductions (Müller 1976). Throughout its range, however, its natural habi-

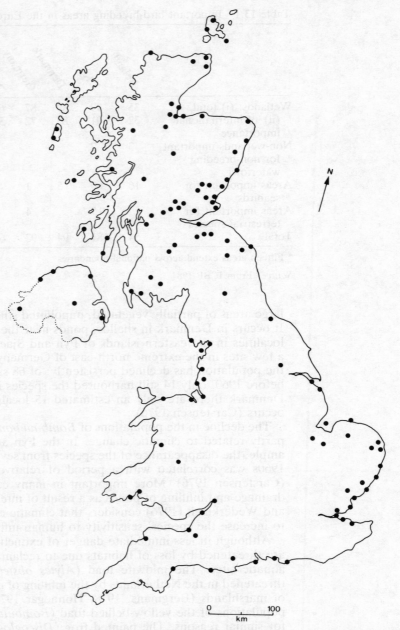

Fig. 13.5   Important wetland bird-breeding sites in the United Kingdom (after
ICBP 1981).

tats are being reclaimed and the Nature Conservancy Council (1982)
recommends that this practice should be curtailed to ensure survival of
the species.

13.3 PEATLANDS AND MARSH

13.3.1 Character and distribution

In many parts of the temperate world, peatlands and marsh provide some of the most important habitats for wildlife. Marshlands consist of a range of different habitats, including coastal salt-marsh, basin- or valley-floor alluvial marshes and swamps, and areas of wetland associated with springs. Peatlands are equally varied and, for convenience, can be divided into three main types:

1. Upland or blanket bog. These are acid ombrotrophic peatlands, dominated by plants with low nutrient requirements, such as *Sphagnum* spp., *Eriophorum vaginitum* and *Calluna vulgaris*, together with *Empetrum nigrum, Vaccinium myrtillus* and *Rubus chamaemorus*. Such bogs are widespread in Canada, the USSR, Scandinavia and Ireland, and in Britain examples occur on the Pennines, Welsh mountains and Scottish highlands. The peats are maintained by rainwater, and plant nutrient cycles operate independently of the groundwater system.
2. Raised bogs. These represent a late stage in the development of lowland fen bogs (see below). They evolve due to the accumulation of peat above the regional water table (or due to a fall in the water table) and thus become dependent upon rainwater for nutrient inputs. In vertical sequence they therefore show a gradation from minerotrophic to ombrotrophic conditions, and the surface vegetation is consequently similar to that found in blanket bogs. In Britain examples include Shapwick Heath in Somerset and Thorne Waste and Hatfield Moss in South Yorkshire. Extensive raised bogs are also found in Ireland and Scandinavia.
3. Fen bogs. These are minerotrophic peatlands formed by the accumulation of vegetable matter in lowland basins supplied with nutrients by groundwaters and runoff from adjacent areas. They tend to be dominated by a wider range of species than the ombrotrophic bogs, including *Scirpus lacustris, Phragmites* spp., *Typha* spp., *Glyceria* spp. and *Carex* spp. Fen bogs and associated fen-carr are widespread (e.g. the Somerset Levels and East Anglian Fens in England; Holland Marsh in Ontario, Canada; and the extensive lowland rivers in Denmark and the Netherlands).

All these wetland types have considerable ecological significance both for the diversity and the rarity of the plant and animal communities they support.

Pressures upon these peatlands and marshlands come from a large number of sources, including peat cutting for industry and agriculture, drainage, pollution, erosion and cultivation. Moore and Bellamy (1974) review the threats to the world's peatlands and note that throughout the world pressures are increasing due to extraction and agricultural intensification. In Finland about 90 per cent of the peatland, which once covered some 11 million ha (40 per cent of the land surface) is drained and planted, much of it for forestry. In Ireland extraction for fuel is the main threat, about a third of the country's energy coming from peat-fired power stations. Increasingly, however, attempts are being made to extend arterial drainage systems in areas of peatland in order to improve their grazing potential and in the future agricultural pressures are likely to grow. Similarly, drainage of peatlands in the Netherlands has affected about 180,000 ha of land; little more than 3,600 ha have been left undisturbed. Hill (1976) quotes estimates indicating that in the Dakotas and Minnesota in the USA, 56,000 ha of wetlands are drained annually.

13.3.2   Agricultural impacts

The effects of agriculture upon peatlands and marshes are considerable. As noted earlier (Table 13.1) wetlands, including peat-bogs, account for many of the most important bird-breeding areas in Europe, and they also support varied and often rare plant, reptile and mammal communities. Drainage, reclamation and cultivation of these areas cause irreversible changes in the ecology of the peatlands, and in many instances pose serious threats to individual species.

The effects of drainage of peatlands upon plant communities have been demonstrated in a number of studies. In Finland, Heikurainen (1972) has developed a peatland classification on the basis of vegetation which distinguishes between drained and undrained sites. Parfenov (1979) recorded changes in floral composition of swamplands in Poland and noted that, in the short term, species such as *Glyceria maxima, G. fluitans* and *Agrostis stolanifera* declined, whilst *Calamagrostis neglecta, Poa pratensis* and *A. canina* increased. In the long term, hydrophilous species tended to be replaced by mesophylous species including *Festuca rubra* and *A. alba*.

In Europe the effects of marshland drainage threaten a number of plant species, most notably *Euphrasia marchessettii* and *Sisymbrium supinum* (Nature Conservancy Council 1982). Additionally, many reptiles and amphibians are endangered because of the loss of these habitats. The western spadefoot toad (*Pelobates cultripes*) is declining rapidly in many Mediterranean areas of France (its sole area of occurrence in the EEC) due to wetland drainage and pesticide pollution (Fretey 1975). The same is true of the more widely occurring common spadefoot (*P. fuscus*) (Bruno *et al.* 1974). The moor frog (*Rana arvalis*) is also threatened in some countries. In Germany, for example, several localities in the north which were known prior to 1960 have not been confirmed since (Muller 1976), and the species has become extinct in several parts of the Rhineland (Sinsch 1978). Even the adder (*Viper berus*) which is widespread in the European Community, is decreasing in numbers in many localities as a result mainly of hunting, but also of marshland reclamation and drainage (e.g. Muller 1976).

Changes in the vegetation and associated invertebrate fauna of drained peatland and marsh may have a major impact upon bird populations. The effect varies according to wetland type, however, for different habitats support different ranges of birds. In general, bird populations are relatively restricted in ombrotrophic peatlands, and the main resident species consist of the greenshank (*Tringa nebularia*) and the crane (*Grus grus*). In addition, migratory species frequently use raised or blanket bog areas as seasonal feeding grounds. In Europe, for example, warblers such as the sedge warbler (*Acrocephalus schoenobaenus*) and the reed warbler (*A. scirpaceus*), are common summer visitors, whilst the Greenland white-fronted goose (*Anser albifrons*) uses ombrotrophic bogs in the winter, where it feeds on the seeds of the white-beaked sedge (*Rhyncospora alba*) (Moore and Bellamy 1974). Minerotrophic bogs, such as the reedswamp and fenlands of Germany and Denmark, support a wide range of waterfowl, including ducks such as the mallard (*Anas platyrhynchos*), teal (*A. crecca*) and wigeon (*A. penelope*). In addition, insectivorous birds (e.g. warblers) and birds of prey, including the marsh harrier (*Circus aeruginosus*), feed off the herbivorous mammals which occupy these wetlands.

By undermining the basis of these ecosystems – namely the vegetation and associated consumer organisms – drainage leads to a loss of many of these bird species. The study of drainage effects on the Somerset Levels by Ferns *et al.* (1979) amply illustrates the consequences. The Somerset and Gwent Levels are particularly important as feeding areas for migrant whimbrels (*Numenius phaeopus*). The whimbrel is a large wader, similar in size and general features to the curlew but which migrates each spring from Africa to its breeding grounds in the subarctic (Iceland, the Faeroes, Scotland, northern Scandinavia and north-west Russia). About 2,000 individuals congregate annually in late April and early May on the Somerset and Gwent Levels.

In their study Ferns *et al.* showed that the whimbrels spent most of the day feeding in small groups on grassy fields, most favouring rough grazing, with large tussocks of *Juncus* spp. and *Glyceria maxima*. Much of the diet consists of wireworms (Elateridae) and caterpillars (Noctridae). At night the birds gather at two main roosts on the shores of the Severn at Stert Island and Collister Pill. Ferns *et al.* suggest that the Levels probably constitute the last feeding grounds for those birds *en route* to Iceland. They argue that because of this species' habit of concentrating at a small number of sites on migration they are particularly vulnerable to changes in land use and that the decline in the amount of damp, permanent pasture on the Levels, as a consequence of improved drainage, represents a particular threat. The reality of this threat is indicated by *The Strutt Report* (MAFF 1970a) which suggested that 53,000 ha (77 per cent of the total area) of the Somerset Levels could be made dry enough to support arable farming by improving the drainage. (A further threat to this species is the possible construction of the Severn Barrage which, at the time of writing, is under detailed consideration.)

In the same way, drainage of wetlands is causing a serious decline in the numbers of some mammal species. Insectivores, including the Pyrenean desman (*Gatemys pyranaicus*), the southern water shrew (*Neomys anomalus*), the beaver (*Castor fiber*) and the European mink (*Mustela lutreola*) are all considered to be endangered as a result of wetland drainage (Nature Conservancy Council 1982). The otter (*Lutra lutra*) which was affected in the 1950s by pesticides, especially dieldrin, is similarly under pressure now from loss of habitats by drainage and reclamation.

13.4 CONCLUSIONS

The examples quoted in this and the previous two chapters indicate the ways in which modern farming practices are affecting various habitats and their associated plant and animal communities. All too often, the full effect of agriculture cannot be evaluated because of lack of quantitative data on biological populations. As the inventories produced by the Nature Conservancy Council (1982) and the ICBP (1981) demonstrate, attempts at gathering information are being made in the European Community and similar studies have been conducted elsewhere (e.g. Ratcliffe 1977; Scott 1980). Nevertheless, there is a clear need for more detailed monitoring and analysis of threatened sites, species and habitats.

Wetlands appear to be among the most vulnerable habitats, for they may act as sinks for pollutants and are subject both to deliberate drainage for reclamation and also to accidental impacts through drainage of surrounding areas. At present, significant conflicts are developing over plans to drain several such areas, including the Somerset Levels in England (a large part of which has recently been declared a site of special scientific interest) and large areas of peat-bog in western Ireland. The effects of such schemes, not just upon the local ecology but upon species and sites of international significance, are likely to be considerable. There is thus a strong case for the more active evaluation of the ecological value of such areas, and for more stringent protection of selected sites.

The effects of agriculture upon other lowland and upland areas are more difficult to specify or to control, if only because the habitats at risk are in many cases geographically less compact. Relatively rarely, in fact, can particular habitats which are critical to the survival of communities or species be identified. Where they can, as in the case of the legume *Astragalus verrucosus* and the crucifer *Brassica macrocarpa* (Nature Conservancy 1982), there is both the necessity and the opportunity for protective measures to be taken. More generally, however, changes in these habitats operate in a dispersed and unco-ordinated fashion across a wide area. As a consequence the effects are often difficult to predict and control is difficult to implement. In addition, because habitats are under the constant influence of a multitude of external effects – not only agriculture, but also industrial and domestic pollution, recreational pressures, fire and climatic changes – it is rarely possible to isolate agricultural practices which threaten these ecosystems. Monitoring of the effects is nevertheless vital, both because of the potential impacts upon communities and individual species, and because of the cumulative effects upon the ecosystems as a whole and upon rural landscapes. These wider impacts will be considered in the final chapter.

14   REFLECTIONS AND CONCLUSIONS

14.1   LARGE-SCALE ENVIRONMENTAL CHANGES

Despite the rationalisation of farms and the changes in farm ownership which have occurred in recent years, agriculture throughout most of the temperate world is still essentially a piecemeal operation. Management decisions are made by individuals, often in response to wholly individual perceptions of needs and consequences. At the local level, therefore, the individual farmer can often gain some understanding of the effects of his decisions and management practices, but rarely can the larger-scale impacts be appreciated with any accuracy. Nor can they easily be predicted.

It is hardly any easier for the consequences of these actions to be analysed scientifically. At the scale of whole ecosystems and landscapes, the analytical techniques available to the ecologist are of limited applicability. Monitoring land use changes with any precision has until the advent of the latest generation of Earth Resource satellites (e.g. Landsat) not been feasible, and macroscale environmental changes have generally been detectable only on the basis of visual observation on the ground. Similarly, the information that is needed on the decisions made by each farmer – and the factors governing those decisions – can rarely be accumulated. Moreover, at the large scale, even more than at the level of the species or habitat, numerous different processes interact. These include both agricultural processes and non-agricultural effects such as urbanisation, industrial activity and pollution, afforestation and recreation. As a result, many of the conclusions drawn about the large-scale impacts of agriculture on the environment are at best qualitative, and many have been couched in somewhat emotive terms (see, for example, Shoard 1980).

As a consequence generalisations about farming decisions and practices and their effects upon the environment need to be made with care. Management decisions – even those that appear consistent in effect – may be made for a multitude of different reasons. The changes that occur in the environment are thus the net effects of many different and often conflicting actions, and general relationships between environmental trends and agricultural practice may hide the true patterns of cause and effect which operate at the detailed level. Moreover, many land use and environmental changes may show parallel time trends which are nevertheless causally unrelated. It is all too easy to make false interpretations, therefore, on the basis of spurious correlations.

One example which illustrates this dilemma is the case of peatland erosion in the Pennines of England. Erosion has been continuing for many decades and has been variously attributed to:

1. Increased grazing intensities and resulting loss of vegetation cover (Evans 1977);
2. Increased frequencies of heather burning (Tallis 1973);
3. Expansion of moorland drainage with consequent desiccation of the peat (Tallis 1964);
4. Climatic changes: notably a trend towards cooler and drier summers which restrict *Calluna* growth (Conway 1954);
5. Natural cycles of peat bog development (Bower 1962).

Evidence can be adduced for all these causes, though causal relationships cannot normally be established in the field, except at a very local level. Thus definitive conclusions about the effects of land use are not justified. Indeed, it seems likely that here as in many environmental systems equifinality operates and similar outcomes may be produced by very different input conditions.

The discussion thus far indicates the difficulties in drawing firm conclusions about large-scale effects of modern agriculture on the environment. Nevertheless, in some cases the direct impact of farming upon large ecosystems or upon whole landscapes can be identified. In the next section we therefore consider the nature of these large-scale changes.

14.2   LAND USE CHANGES

The impacts of agriculture at the large scale are reflected in changes in the extent and distribution of major land use types: areas such as deciduous woodland, moorland, heathland, rough grassland, permanent pasture and arable land. Variations in the extent of these land use types in England and Wales throughout the last eighty years are shown in Table 14.1. As these data indicate, marked fluctuations in the extent of these land use types have occurred, though a number of general trends are visible:

Table 14.1   Changes in the extent of land use types in England and Wales, 1900–71

Land use type	Area (million ha)			
	1901	1931	1951	1971
Arable cropland	4.904	3.878	5.536	5.666
Permanent pasture	6.232	6.354	4.365	3.965
Rough pasture, etc.	1.440	2.151	2.203	1.884
Total agricultural	12.576	12.383	12.104	11.515
Woodland and forest	0.748	0.840	0.971	1.115
Urban	0.674	1.005	1.339	1.646
Other non-agricultural land	1.027	0.800	0.614	0.750
Total non-agricultural	2.449	2.645	2.924	3.511

Source: Data from Best 1981

1. A decline in the total area of agricultural land;
2. An increase in the area of forest and woodland;
3. A decline in the area of permanent grassland;
4. An increase in the area of non-agricultural land.

Some caution is necessary in interpreting these figures, for the impression may be gained that the main process of change have been the conversion of grassland to woodland and urban use. In reality the changes have been far more complex. Throughout the present century, in fact, there has been a gradual loss of deciduous woodland to agriculture, but this has been more than offset by the afforestation of many upland areas for coniferous plantations. Best (1981) shows that the average annual rate of transfer of farmland to forestry in Britain was about 25,000 ha between 1960 and 1975, most of which (about 20,000 ha y^{-1}) took place in Scotland. Between 1972 and 1975, in Scotland alone, 30,000 ha of land was afforested each year. In comparison, losses of farmland to urban uses have in fact been small, averaging no more than 18,000 ha y^{-1} between 1960 and 1975 in Britain as a whole.

Within the category 'agricultural land' further changes have occurred during the present century. As we noted in Chapters 7 and 8, fluctuations in the acreages of pastureland and arable cropland have been marked, especially as a result of the plough-up campaigns of the wars and, conversely, the effects of economic depressions during the 1930s. It is interesting to note, however, that extension of cultivation onto the moorlands, which had occurred widely during the nineteenth century, was reversed during the period from 1920 to 1950. Only in the last thirty years has reclamation of these areas for agriculture been revived on a large scale.

Two further cautionary points must, nevertheless, be emphasised. First, as Best (1981) discusses in detail, the quality of general land use data is often variable and comparisons of results from different surveys may be unreliable. Second, as we have already noted, the net changes reflected in such figures disguise the myriad contrasting events which occur at the local level. It is also clear that agriculture is not the only land use to be affecting the environment, and that forestry and urban uses are also significant.

Fluctuations in the area of agricultural land and in the character of land use thus reflect a variety of different processes. In terms of agricultural impacts on the environment the most important land use changes are:

1. Land reclamation for agriculture;
2. Changes in cropping system (e.g. between arable and grassland);
3. Abandonment of cultivated land.

The effects of these different processes will be considered in the next three sections.

14.2.1   Land reclamation for agriculture

As we saw in the previous chapter, reclamation of land for agriculture represents one of the main mechanisms by which farming affects aquatic habitats. At the larger scale, however, reclamation may have impacts upon many other ecosystems. The main activities involved are:

1. Woodland clearance;
2. Moorland reclamation;
3. Extensive drainage improvements.

Woodland clearance is an ancient practice (see Ch. 2), and many of the features of modern agricultural environments are a product of past clearance activity. Thus, the distribution of many woodlands and coppices, of hedgerows and even individual trees, reflects the cumulative effects of historical land management. Curtis (1956), considering the effects of agriculture in mid-latitude USA, shows the progressive loss of woodland cover in the Cadiz Township of Green County, Wisconsin (Fig. 14.1). Between 1831 and 1882 the total area of woodland declined from 8,800 ha to 2,600 ha, whilst the number of separate wood lots increased from 1 to 70. These small, isolated wood lots had an average area of 37 ha. Over the next 40 years, the remnant patches of woodland were progressively

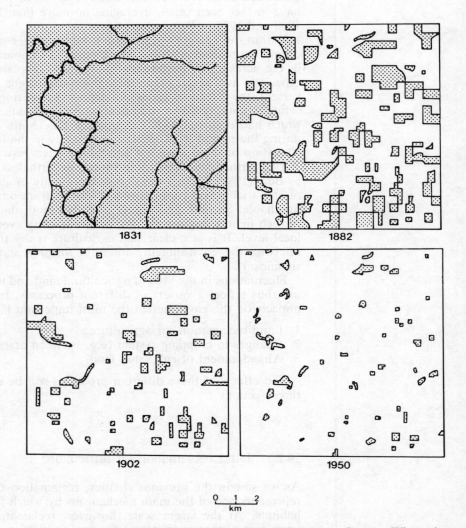

Fig. 14.1 Woodland clearance in Cadiz Township, Green County, Wisconsin, 1831–1950 (from Curtis 1956).

removed, until in 1935 there were only 57 lots with an average area of 7.4 ha. Rackham (1976) notes that in Britain, the rate of woodland removal since 1945 has increased markedly, not only for agriculture but also for forestry plantations. Peterken and Harding (1975), in a study of woodland changes in three areas of Northamptonshire, Cambridgeshire and Lincolnshire, estimate that 42, 36 and 46 per cent of the ancient woodland were removed or fundamentally modified between 1946 and 1974 in these three counties respectively.

Changes in the extent of moorland and heathland as a result of agricultural expansion have similarly increased in recent decades. In Britain and Denmark, many of the moorland areas have considerable potential for intensive grazing, and reclamation is feasible at relatively low cost. The heathland vegetation can be removed by the use of herbicides (such as asulam in the case of bracken) or by ploughing. Lime and NP fertilisers are then applied and a good, mixed species grass sward develops without the need for reseeding. Land drainage is only necessary on wetter soils such as peats or peaty gleyed podzols.

Large areas of moorland and heathland have been reclaimed in this way. Moore (1962) describes the gradual reduction and fragmentation of the heathlands in Dorset in England, and Gimingham (1972) notes that since 1811 the total area of heathland in Dorset and Hampshire has fallen by 67 per cent. Similar transformations of lowland heath took place during the nineteenth and twentieth centuries in Sussex, Surrey, Kent and East Anglia (the Brecklands). Today, one of the most severely threatened areas of heathland in Britain is Exmoor. Curtis *et al.* (1976) review the history of reclamation on the moor and cite data indicating that between 1957 and 1966 up to 3,264 ha (13.75 per cent of the total area) were reclaimed, mostly for agriculture. They also note that these statistics were disputed by the County Landowners Association and National Farmers Union at the time of publication. At present, however, regular monitoring by aerial photography is being planned to help identify the progress and effects of moorland reclamation (Curtis and Walker 1982). Dartmoor and the North York Moors have similarly been subject to considerable reclamation, though the upland moors of Scotland, Wales and the Pennines are less affected due to the more severe climatic and soil conditions which limit their agricultural potential.

Elsewhere in Europe reclamation has been equally extensive. Following the formation of the Danish Heath Society in 1866, large tracts of heathland were reclaimed in Denmark, aided by state subsidies. By 1955, 700,000 ha of heath and bogland had been reclaimed and cultivated and Gimingham (1972) comments that the only surviving heathlands are in Jutland. These are now protected as nature reserves.

We have already examined the effects of drainage on wetland habitats and species, but in some parts of the temperate world extensive drainage schemes have been carried out which have affected large areas of land. The draining of the English Fens was documented by Darby (1969), and was discussed previously (section 11.6.1.), whilst Curtis *et al.* (1976) outline the history of drainage in the Somerset Levels. As we noted in Chapter 4, the rate of land drainage in Britain is increasing annually, and Trafford (1977) estimates that between 1969 and 1976 alone 650,000 ha of land were drained.

The large-scale effect of such drainage is most marked in areas of peatland or semi-natural woodland. Hill (1976), for example, noted that between 1961 and 1969 the area of bottomland (valley-floor) hardwoods in northern Louisiana declined from 1.3 million ha to 1.0 million ha as a result of drainage for soya bean cultivation. Similarly, as we saw in the previous chapter, the great majority of peatlands in Finland and the Netherlands have been drained and cultivated, leading to the virtual destruction in many areas of the original ecosystem.

14.2.2   Changes in cropping system

Changes in crops are an essential and routine feature of agricultural management, and at the local level provide an important means of maintaining soil fertility and controlling pests and diseases. As was noted in Chapter 2, however, recent decades have seen more fundamental changes in cropping systems as mixed farming has given way to greater specialisation in arable or grazing systems. Moreover, there has in Britain and several other European countries been a gradual shift since 1939 from permanent grassland to arable cultivation (see Table 14.1, p. 382). In addition, Lazenby and Down (1982) have pointed out that the acreage of rotational grass, which rose during the Second World War, has subsequently declined from 1.94 million ha to 1.69 ha. Similar changes can be seen throughout Europe (see, for example, CEC 1982, Table 60).

In so far as these changes reflect basic trends in farming methods and cropping systems, they clearly have significance for large-scale ecosystems and landscapes. Conversion of pasture to arable land, for example, is often associated with loss of hedgerows and walls, whilst cultivation of permanent pastures involves a marked change in ecological conditions and the visual character of the landscape.

14.2.3   Abandonment of cultivated land

Although the overall trend in agricultural land use over the present century has been for increased intensification, numerous exceptions have occurred in which farming has retreated from marginal lands. In some cases, particularly in the drier parts of the temperate region, this retreat has been occasioned by climatic changes; or by increased problems of drought and soil erosion (e.g. Saarinen 1966; Knight and Richard 1971). Elsewhere, abandonment of cultivated land has been motivated by changing economic and political conditions. Low market prices and lack of state aid, for example, have contributed to the decline in hill farming in many areas of Europe, a problem which is still a matter of concern within the EEC (e.g. Danz and Henz 1981).

Data on the extent of recent abandonment are scarce, but an indication of the changes which have taken place is provided by Blacksell and Gilg (1981), in a survey of land use adjustments in Devon. The data they present show that between 1950 and 1976 significant abandonment of marginal farmland occurred both on Dartmoor and Exmoor. In the southeast area of Dartmoor, they recorded that 187 ha of land had been converted from farmland to heathland, and a further 31 ha to 'unused land' (e.g. scrub). This represents about 14.5 per cent of the total conversions in the area during this period. In western Exmoor, over the same period, 124 ha changed from agricultural to heathland, whilst 48 ha changed to unused land. Together, these account for about 22 per cent of the total area of conversions between 1950 and 1976 within the study area.

Blacksell and Gilg also quote data from *The Changing Face of Devon*, a survey conducted by the Nature Conservancy Council (1979). Out of a total of almost 52,500 ha of land which experienced a change in use between 1900 and 1978, 5,775 ha reverted from farmland to heathland. In addition, 1,704 ha changed from rough pasture or heathland to scrub. Thus, as in the Dartmoor survey, almost 14.5 per cent of the land changes recorded have involved the abandonment of land used for agriculture. Such changes rarely show up in national or regional trends, however, because they are more than offset by the expansion of agriculture onto the heathlands.

14.3   ECOSYSTEM CHANGES

The changes in agricultural land use discussed in the preceding sections all have marked ecological impacts. Some indication of their significance is given by the data from Perring (1970), quoted in Table 14.2. This shows the causes of plant species decline and extinction in Britain for the period from 1900 to 1970. Clearly, drainage, ploughing and changes in arable cropping practices have had major effects. Similarly, Fig. 14.2 shows examples of the extent of habitat loss for selected areas and periods in Britain. At a more specific level, however, it is difficult to quantify these impacts, for the ecosystems which develop are fundamentally different.

Ideally, the effects of fundamental changes in agricultural land use upon whole ecosystems could be assessed by monitoring changes over time at selected sites. As noted in the introduction to this chapter, this is rarely if ever possible. It is therefore necessary to interpret the effects by comparing ecosystems which are considered to be representative of different stages of agricultural modification, for example neighbouring natural and cultivated areas. Such studies suffer all the problems of counter-factuality: namely, that assumptions have to be made about the original comparability of all the ecosystems and differences between them must be assumed to reflect only differences in agricultural land use. These problems tend to reduce the general reliability of the conclusions. An example of

Table 14.2   Causes of plant species decline in Britain, 1900–70.

	Extinctions	Very rare	Rapid decline	Total	%
Arable change	1	7	3	11	15
Ploughing	0	6	4	10	14
Drainage	4	5	7	16	22
Habitat removal	4	0	0	4	5
Collecting	2	5	0	7	9
Forestry	1	2	3	6	8
No management	0	1	0	1	1
Natural causes	8	8	3	19	26
Total	20	34	20	74	100

Source: From Perring 1970

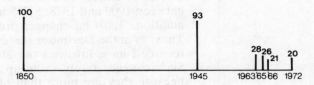

a. **HEDGES (length)**
   - Hunts.

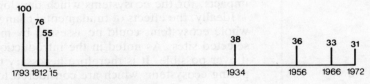

b. **HEATHLANDS (area)**
   - Dorset.

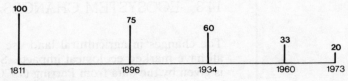

c. **CHALK DOWNLANDS (area)**
   - Dorset

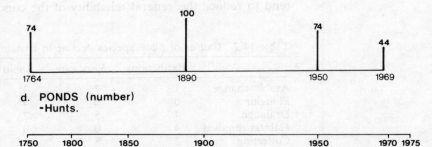

d. **PONDS (number)**
   -Hunts.

Fig. 14.2   Examples of habitat loss in lowland England, 1750–1975 (from Nature Conservancy Council 1977).

such a study is the upland vegetation study of the Institute of Terrestrial Ecology (Ball *et al*. 1981a, b,) referred to in section 12.2.

Comparisons of this nature clearly provide only a partial picture of large-scale ecological impacts of agriculture. For a more general understanding of these impacts it would be useful to have a more universal index of ecological value by which to assess different areas. Much attention has in fact been given to deriving such indices as a basis for conservation and ecological planning (see, for example, Margules and Usher 1981). Many of the techniques, however, are essentially subjective in that they involve placing intuitively derived rank values upon specific vegetation types or communities (e.g. Tubbs and Blackwood 1971; Yapp 1973; Ratcliffe 1977). Even the apparently quantitative formulae, such as that developed by Goldsmith (1975), are inherently intuitive in approach and do not form a firm basis for comparison. As a result, the assessment of broad-scale changes in ecology remains a matter largely of personal perception. Much more research is clearly needed to evaluate the impacts of agriculture at this scale in any objective fashion.

14.4 LANDSCAPE CHANGES

The visual landscape can be considered as a mosaic of vegetational and other features (buildings, walls, etc.) embossed upon a topographic surface. As such, the landscape is susceptible to the influence of agriculture through changes in both the nature and the pattern of the vegetation, as well as associated changes in farm buildings. Undoubtedly, over the centuries of agricultural land-use, the landscape has undergone fundamental changes, but the appreciation of these changes is an essentially subjective and personal response. As a result, much discussion about agricultural impacts on the landscape has been highly emotive.

Many attempts have been made to overcome these difficulties of subjectiveness in landscape evaluation, and it is now widely accepted that modal preferences for particular landscapes can be identified for specific groups of people. Little work has been done on the cultural and social factors influencing these preferences, though Wellman and Buhyoff (1980) have recently suggested that preferences are not significantly affected by degree of familiarity with the landscape. Since the early studies by Fines (1968) and Linton (1968), however, a number of methods have been devised for assessing the visual quality of the landscape. One of the most widely used methods is that initially proposed by Warwickshire County Council (1971) and subsequently developed by the University of Manchester (1976). This uses a set of observers to assign scores to selected sample landscapes, then applies multiple regression techniques to establish the relationships between these scores and measurable components of the landscape. The regression equation developed takes the general form:

$$Q = a + b_1X_1 + b_2X_2 + \ldots + b_nX_n \qquad [14.1]$$

where Q is the index of quality
$X_1, X_2 \ldots X_n$ are landscape components (e.g. % woodland cover, relief amplitude)
$a, b_1, b_2 \ldots b_n$ are constants.

This approach has been applied in a wide range of situations and shown to work successfully (e.g. Blacksell and Gilg 1975; Durham County Council 1976; South Yorkshire County Council 1978; Briggs and France 1980, 1981). On the basis of results obtained from such studies it is possible to identify the effects of agricultural land use on the visual quality of the landscape. Blacksell and Gilg (1975), for example, assessed an area of south-east Devon. Positive contributions to landscape quality were provided by features including farmland, woodland, heathland, hedgerows and farm buildings. Negative features included most aspects of industrial and urban use. Similarly in the study of the Coventry–Solihull–Warwickshire district (Warwickshire County Council 1971), strong positive influences were identified for hedgerow trees, parkland and woodland, and somewhat weaker positive influences for farmland and heathland. Briggs and France (1981), in a study of south Yorkshire, found that the strongest positive contributions to landscape quality were made by pasture and woodland, whilst negative effects were made by hedgerows; arable land had no detectable affect on landscape quality.

Table 14.3   Relationships between environmental factors and landscape quality

Environmental factor	Relationship with landscape quality			
	Warwickshire	Macclesfield	Durham	South Yorkshire*
Relief	+	+	+	
Unimproved pasture	−			+
Industrial land	−	−		−
Wasteland	−			−
Moorland		−	−	
Marshland		−		
Deciduous woodland		+		+
Bare rock		+	−	
Hedgerows				−
Streams			+	
Powerlines		−		
Residential land				−
Railways		−		−
Coastline		+		

* Results from Briggs and France 1981
+ positive relationship with landscape quality
− negative relationship with landscape quality

Source: Based on data from South Yorkshire County Council 1979

As the results for a range of different areas presented in Table 14.3 indicate, the effects of individual components of the landscape upon landscape quality vary from area to area. This reflects not only the preferences of different groups of observers but also the interactions between landscape features, which have not been allowed for in these studies. Nonetheless, the results do indicate that, in any area, changes in agricultural land use may have significant effects upon landscape quality. In the Macclesfield and South Yorkshire areas, for example, conversion of deciduous woodland or pasture to arable land seems likely to reduce the visual quality of the landscape. Conversely the results imply that reclamation of moorland for grazing may raise landscape quality. A cautionary word should, however, be added, for the studies quoted here relate to assessment of visual quality over space at a single point in time; it cannot be assumed that equivalent relationships will be valid for changes in landscape conditions at a single place over time. What can be concluded is that agricultural practices *can* alter the visual appearance of the landscape and thus play a role in changing the aesthetic quality of the environment. More research is needed on the ways in which these changes are perceived by the public and the extent to which public preferences subsequently adapt to the altered landscape before more specific assessments of agricultural impact can be made. There will, however, always remain the problem that individual perceptions vary and what may be acceptable to one individual may be unacceptable to another. This concept of individuality applies, of course, not only to those members of the community at large in their assessment of the landscape, but also to farmers themselves.

14.5 PERCEPTION AND THE FARMER

It would be wrong to suggest in a book on this subject that problems concerned with agriculture-environment relationships can be understood without any appreciation of the wider sociological and political context. It has already been emphasised that management decisions are largely influenced by economic considerations, and these considerations are themselves determined in part by government policies. Moreover, farming practices and attitudes to the environment (as well as to economic questions) are also strongly influenced by the individual farmer's perception of the world in which he operates. Westmacott and Worthington (1974) illustrate the attitude of farmers to wildlife and landscape conservation and draw attention to the low priority that many farmers give to this aspect. Thus, they comment that in response to questions on the reasons for removal of trees on Fenland farms, 'the reasons were so varied that it is surprising that any trees remain'. Reasons given for removal included interference with access to ditches for maintenance; damage to drains; harbouring pests; encouragement of potato blight; interference with spraying; interference with cultivations and harvesting; reduction in yields; and uneven ripening of crops. In response to the question where farmers would prefer new tree planting on their land, many commented that they did not want any new planting even if incentives were offered, and one farmer is reported to have suggested that the questioner should 'go elsewhere for amenity' (Westmacott and Worthington 1974, p. 51).

In a different context, Burton and Kates (1964) have shown that farmers' attitudes to and willingness to adopt soil conservation measures depend upon their tripartite perception of the problem, of the effectiveness of the remedial action and of their ability to pay for implementation of the remedy. More generally, it can be suggested that decisions about land management – for example the choice of crop, the cultivation method, fertiliser practice and so on – are based upon some 'vision of the future'. This vision involves the evaluation of the benefits and disbenefits to the farmer of different possible management strategies, on the assumption of certain yields and taking into account predicted future changes in market prices, government policy, input costs and personal circumstances. These potential benefits and disbenefits are likely to include not only economic aspects, but also social considerations, environmental relationships and many individual and possibly only partially formulated psychological factors. Clearly it is impossible to analyse all these influences in detail, though much research has been devoted to examining the factors that determine farmers' perceptions. It is, however, apparent that agricultural impacts upon the environment operate through the actions of the farmer, and the way in which he acts depends in turn upon these internal (psychological) and external (economic, social, etc.) influences. Without doubt, in this context, one of the most potent external influences on farming practice is government policy. In conclusion, therefore, it is useful to consider attitudes to the environment and changing agricultural practices, in the light of changing government policies.

14.6   GOVERNMENT AGRICULTURAL POLICY AND THE ENVIRONMENT IN BRITAIN

The development of government agricultural policy in Britain has been summarised by Bowler (1979). In Britain, the most recent government statements of policy – *Food from our own resources* (HMSO 1975) and *Farming and the nation* (HMSO 1979) – both advocated substantial development of agriculture. However, there is some change in emphasis with regard to the attitude toward environmental issues and it is important to draw attention to it. It could be suggested that the earlier White Paper was somewhat naive in its assessment of the environmental restraint on agricultural expansion:

> The Government are aware of concern in the agricultural industry lest investment and higher output from our farms might be prejudiced by . . . developments such as more rigorous control of effluents and agrochemicals and for the preservation of the traditional appearance of the countryside. The projected increase in the output of British agriculture should not result in any undesirable changes in the environment; the continuing improvement of grazing land and hill land can contribute to a better looking as well as to a more productive countryside. The Government's conclusions in the White Paper can be reconciled with their commitment to proper safeguards for the environment. (HMSO 1975: 16)

In the face of growing concern about the environmental effects of modern agriculture, the 1979 White Paper is rather more circumspect in its treatment of the environmental issues than was its 1975 predecessor (and it should perhaps be remarked that both were produced by a government of the same political colour). Nevertheless, there are many apparent inconsistencies in the 1979 White Paper. For example, the document refers several times to the need to strike a balance between agriculture and environmental interests; yet, without being specific, it appears to suggest that this can be achieved by government grant aid both to increase land in cultivation and somehow to subsidise the environmental interest. (In the event, detailed proposals were overtaken by a change to a government with strong views about the distribution of public funds although, in reality, there seems to have been little or no change in grant aid to agriculture.) Finally, the Ministry of Agriculture was still emphasising the changes that would have to take place. The area of ploughed cereal growth was forecast to increase by between 119,000 and 144,000 ha between 1977 and 1983 – growth of between 3 and 4 per cent (MAFF 1979a).

As was noted in section 14.2, at the same time as pressure on production has increased there has also been a considerable loss of agricultural land to other uses, principally urbanisation. Thus the pressure on the remaining land has increased from two directions.

Although the total area of farmland is becoming smaller the average size of individual farms is getting bigger and by 1975 the largest 17 per cent of farms contributed about 60 per cent of the total output and covered about 50 per cent of the agricultural land (ACAH 1978). The clear indication is that these bigger farms are more efficient crop and livestock producers. In general this increased efficiency will tend to increase the pressure on wildlife.

The complex issue of land ownership should also not be ignored. In Britain there has been a gradual and continuing transition to institutional and also to foreign land-owners. This, and other sociological issues, has been examined by Newby (1979). Newby describes the impact of changes in ownership and increasing urbanisation – not simply of the land itself but also the impact of urban ideas – on agriculture.

14.7 AGRICULTURAL POLICY AND THE ENVIRONMENT IN THE EEC

The relationship between government policy, farming practice and the environment is not, of course, limited to Britain. Within the wider context of the European Economic Community, for example, this relationship has particular significance. Community policy to farming is framed in the Common Agricultural Policy, which through its pricing system and through its support for agricultural modernisation, is motivating intensification of farming systems. As a consequence it has tended to encourage extension of cultivation onto previously uncultivated (natural or semi-natural) land, removal of local habitats, drainage of wetlands and increased use of agricultural chemicals. Interestingly, these developments have been encouraged independently of the Community's short-term needs for agricultural produce, with the consequence that large internal food surpluses have been created. It is also notable that these policies – or at least their environmental consequences – are to some extent in conflict with the Community's environmental policies which include the provision to encourage the development of 'non-damaging use and rational management of land' (Official Journal of the European Communities 1979).

There is now evidence to indicate that this conflict between agricultural and environmental objectives is being appreciated. Muntingh (1982), in a draft proposal on EEC aid for Scottish hill farming, for example, stresses the need to avoid developments which damage the environment. Moreover, methods are being sought which might allow environmental factors to be taken into consideration within the Common Agricultural Policy, and which would provide a basis for identifying environments needing protection within the Community (CEC 1983).

Whether such attempts are successful in influencing agricultural practices and environmental impacts cannot yet be predicted. In the USA, where by the mid-1970s there was far more stringent legislation on the environment than in Europe, there has nevertheless been a significant deterioration in environmental conditions (Wille 1982), largely, it seems, as a result of changes in government attitudes. Recent years have shown, however, that concern about agricultural impacts on the environment is growing. To date, the debate has, on occasion, been ill informed and prejudiced. What is needed is more careful analysis of the processes involved. The scale of modern agriculture is such that its potential for environmental impacts is considerable, but by the same token the importance of modern agriculture is fundamental, not only to the producer but also to the consumer. The necessity for the future is that the priorities and requirements of the farmer be understood more clearly, but also that the environmental consequences of farming practices be evaluated and predicted more precisely. In the long term it is vital to ensure that agricultural systems operate to the benefit of mankind as a whole.

REFERENCES

ACAH (1978) *Agriculture and the countryside* (The Strutt report). London: Advisory Council for Agriculture and Horticulture in England and Wales.

ASAE (1971) *Compaction of agricultural soils*. St Joseph, Michigan: American Society of Agricultural Engineers.

Aarstad, J. S. and Miller, D. E. (1978) 'Corn residue management to reduce erosion in irrigation furrows', *Journal of Soil and Water Conservation*, **33**, 289–91.

Abbott, I. and Parker, C. A. (1981) 'Interactions between earthworms and their soil environment', *Soil Biology and Biochemistry*, **13**, 191–7.

Abbott, I., Parker, C. A. and Sills, I. D. (1979) 'Changes in the abundance of large soil animals and physical properties of soils following cultivation', *Australian Journal of Soil Research*, **17**, 343–53.

Abel, W. (1978) *Agricultural fluctuations in Europe from the thirteenth to the twentieth centuries* (transl. O. Ordish). London: Methuen.

A'Brook, J. and Heard, A. J. (1975) 'The effects of ryegrass mosaic virus on the yield of perennial ryegrass swards', *Annals of Applied Biology*, **80**,163–8.

Adams, E., Blake, G., Martin, W. and Boelter, D. (1960) 'Influence of soil compaction on crop growth and development', in *Seventh International Congress of Soil Science*, Madison, vol. **1**, 607–15.

Adams, J. E., Richardson, C. W. and Burnett, E. (1978) 'Influence of row spacing of grain sorghum on ground cover, runoff and erosion', *Soil Science Society of America, Proceedings*, **42**, 959–62.

Adams, R. M. (1962) 'Agriculture and urban life in early southwestern Iran', *Science*, **13** (3511), 109–22.

Ahlgren, H. L. (1952) 'Comparisons of productivity of permanent and rotation pasture on plowable land', *Proceedings of the 6th International Grassland Congress,* **1**, 356–70.

Alberda, T. and Sibma, L. (1968) 'Dry matter production and light interception of crop surfaces. 3: Actual herbage production in different years as compared with potential values', *Journal of the British Grassland Society*, **23**, 206–15.

Alder, F. E. and Chambers, D. T. (1958) 'Studies in calf management. 1: Preliminary studies of post-weaning grazing', *Journal of the British Grassland Society*, **13**, 13–20.

Alder, F. E., McLeod, D. St L. and Gibbs, B. G. (1969) 'Comparative feeding value of silages made from wilted and unwilted grass and grass/clover herbage', *Journal of the British Grassland Society*, **24**, 199–206.

Alexander, M. (1969) 'Microbial degradation and biological effects of pesticides in soils', in *Soil biology*, Paris: Unesco, 209–40.

Allen, S. E. (1964) 'Chemical aspects of heather burning', *Journal of Applied Ecology*, **1**, 347–67.

Allen, S. E., Evans, C. C. and Grimshaw, H. M. (1969) 'The distribution of mineral nutrients in soil after heather burning', *Oikos*, **20**, 16–25.

Allgood, F. P. and Gray, F. (1978) 'Utilization of soil characteristics in computing productivity ratings of Oklahoma soils', *Soil Science*, **125**, 359–66.

Allison, K. J., Beresford, M. W. and Hirst, J. G. (1965) *The deserted villages of Oxfordshire*. Leicester: Leicester University Press.

Alston, A. M. (1979) 'Effects of soil water content and foliar fertilization with nitrogen and phosphorus in late season on the yield and composition of wheat', *Australian Journal of Agricultural Research*, **30**, 577–85.

Anderson, T. J., Barrett, G. W., Clark, C. S., Elia, V. J. and Majeti, V. A. (1982) 'Metal concentration in tissues of meadow voles from sewage-treated fields', *Journal of Environmental Quality*, **11**, 272–7.

Anslow, R. C. and Green, J. O. (1967) 'The seasonal growth of pasture grasses', *Journal of Agricultural Science, Cambridge*, **68**, 109–22.

Aranda, J. M. and Coutts, J. R. H. (1963) 'Micrometeorological observations in an afforested area in Aberdeenshire: rainfall characteristics', *Journal of Soil Science*, **14**, 124–33.

Armitage, E. R. and Templeman, W. G. (1964) 'Response of grassland to nitrogenous fertiliser in the west of England', *Journal of the British Grassland Society*, **19**, 291–7.

Armstrong, D. E. and Chesters, G. (1968) 'Adsorption catalyzed hydrolysis of atrazine', *Environment, Science and Technology*, **2**, 683–9.

Arnold, G. W. (1970) 'Regulation of food intake in grazing ruminants', in *Physiology of digestion in the ruminant* (A. T. Phillipson, ed.). Newcastle-upon-Tyne: Oriel Press, 264–7.

Arnold, M. H. (1980) 'Plant breeding', in *Perspectives in world agriculture*. Slough: Commonwealth Agricultural Bureaux, 67–89.

Arnon, I. (1972) *Crop production in dry regions, Vol. 1: Background and principles*. London: Leonard Hill (Plant Science Monographs).

Aronovici, V. S. and Donnan, W. W. (1946) 'Soil permeability as a criterion for drainage design', *Transactions, American Geophysical Union*, **27**, 95–101.

Aspinall, D. (1961) 'The control of tillering in barley plant, 1. The pattern of tillering and its relation to nutrient supply', *Australian Journal of Biological Science*, **14**, 493–505.

Astbury, A. K. (1957) *The black fens*. Cambridge: Golden Head Press.

Austin, R. B. (1978) 'Actual and potential yields of wheat and barley in the United Kingdom', *ADAS Quarterly Review*, **29**, 76–87.

Avery, B. W. (1980) *Soil classification in England and Wales*. Harpenden: Soil Survey of England and Wales, Technical Monograph No. 14.

Aymonin, G. G. (1974) *Etudes sur les régressions d'espèces végetales en France. Rapport No. 2 Listes préliminaires des espèces endemiques et des espèces menacées en France*. Paris: Ministère de la Qualité de la Vie.

Aymonin, G. G. (1977) *Etudes sur le régressions d'espèces végetales en France. Rapport No. 3. Liste générale des espèces justifiant des mesures de protection*. Paris: Ministère de la Qualité de la Vie.

Baeumer, K. and Bakermans, W. A. P. (1973) 'Zero tillage', *Advances in Agronomy*, **25**, 78–123.

Baier, W. (1969) 'Concepts of soil moisture availability and their effects on soil moisture estimates from a meteorological budget.' *Agricultural Meteorology*, **6**, 165–78.

Baier, W. and Robertson, G. W. (1967) 'Relationships between soil moisture, actual and potential evapotranspiration',*Proceedings, Hydrological Symposium of Canada, Ottawa*, 155–204.

Baier, W., St Pierre, J. C. and Lovering, J. H. (1980) 'Analysis of environmental factors affecting timothy yields', *Agricultural Meteorology*, **22**, 317–39.

Bailey, A. D., Dennis, C. W., Harris, G. L. and Horner, M. W. (1980) *Pipe size design and field drainage*. Cambridge: Agricultural Development and Advisory Service, Land Drainage Series, Report No. 5.

Bailey, S., Bunyon, P. J., Hamilton, O. A., Jennings, D. M. and Stanley, P. I. (1972) 'Accidental poisoning of wild geese in Perthshire, November 1971', *Wildfowl*, **23**, 88–91.

Baker, R. L., Powell, J., Morrison, J. D. and Stritzke, J. F. (1980) 'Effects of atrazine 2,4-D and fertilizer on crude protein content of Oklahoma tall-grass prairie', *Journal of Range Management*, **33**, 404–7.

Bakermans, W. A. P. and de Wit, C. T. (1970) 'Crop husbandry on naturally compacted soils', *Netherlands Journal of Agricultural Science*, **18**, 225–4.

Baldwin, J. H. (1979) 'The chemical control of wild oats and blackgrass: a review based on results from ADAS agronomy trials', *ADAS Quarterly Review*, **32**, 69–95.

Ball, D. F., Dale, J., Sheail, J., Dickson, K. E. and Williams, M. W. (1981a) *Ecology of vegetation change in upland landscapes. Part 1. General synthesis*. Bangor: Institute of Terrestrial Ecology, Occasional Paper No. 2.

Ball, D. F., Dale, J., Sheail, J. and Williams, M. W. (1981b) *Ecology of vegetation change in upland landscapes. Part 2. Study areas*. Bangor: Institute of Terrestrial Ecology, Occasional Paper No. 3.

Ball D. F., Dale, J., Sheail, J. and Heal, O. W. (1982) *Vegetation change in upland landscapes*. Cambridge: Institute of Terrestrial Ecology.

Barker, M. G. (1963) 'A drainage investigation on a clay soil', *Journal of the Royal Agricultural Society of England*, **124**, 50–9.

Barley, K. P. and Jennings, A. C. (1959) 'Earthworms and soil fertility, III', *Australian Journal of Agricultural Research*, **10**, 364–7.

Barry, R. G. (1969) 'Precipitation', in *Water, earth and man* (R. J. Chorley, ed.). London: Methuen, 169–84.

Barry, R. G. and Chambers, R. E. (1966) 'A preliminary map of summer albedo over England and Wales', *Quarterly Journal of the Royal Meteorological Society*, **92**, 543–8.

Bates, G. H. (1948) 'An investigation into the cause and prevention of deterioration of leys', *Journal of the British Grassland Society*, **3**, 177–83.

Batey, T. and Davies, B. D. (1971) 'Soil structure and the production of arable crops', *Journal of the Royal Agricultural Society, England*, **132**, 106–22.

Baumgartner, A. (1965) 'The heat, water and carbon dioxide budget of plant cover: methods and measurement', in *Methodology of plant eco-physiology*, Montpellier Symposium. *Proceedings, Unesco Arid Zone Research*, **25**, 495–512.

Baver, L. D., Gardner, W. H. and Gardner, W. R. (1976) *Soil physics*. New York: Wiley.

Bayliss-Smith, T. P. (1982) *The ecology of agricultural systems*. Cambridge: Cambridge University Press.

Beckett, P. H. T. and Webster, R. (1971) 'Soil variability: a review', *Soils and Fertilisers*, **34**, 1–15.

Beeton, A. M. (1971) 'Eutrophication of the St Lawrence Great Lakes', in *Man's impact on the environment* (T. R. Detwyler, ed). New York: McGraw Hill, 233–45.

Ben Harrath, A., Frankinet, M. and Ledieu, J. (1977) 'Incidence de la sécheress de 1976 sur les conditions d'humidité du sol et les rendements agricoles', *Pedologie*, **27**, 207–24.

Bendelow, V. C. and Hartnup, R. (1977) 'The assessment of climatic limitations in land use capability classification', *Proceedings of the North of England Soils Discussion Group*, **13**, 19–28.

Bergmans, W. (1978) 'Herpetogeografische mededelingen. 1. De verspreiding van de Vroedmeersterpad, *Alytes obstetricans)* (Laurenti 1768) in Nederland', *Lacarta*, **37**, 35—a–44.

Berryman, C. (1965) *Composition of organic manures and waste products used in agriculture*. London: HMSO, National Agricultural Advisory Service, Advisory Paper No. 2.

Best, J. A., Weber, J. B. and Monaco, T. J. (1975) 'Influence of soil pH on s-triazine availability to plants', *Weed Science*, **23** 378–82.

Best, R. H. (1981) *Land use and living space*. London: Methuen.

Beven, K. (1980) *The Grendon Underwood field drainage experiment*. Wallingford: Institute of Hydrology, Report No. 65.

Beyer, W. N. and Gish, C. D. (1980) 'Persistence in earthworms and potential hazards to birds in soil applied DDT, dieldrin and heptachlor', *Journal of Applied Ecology*, **17**, 295–307.

Bhat, K. S., Flowers, T. H. and O'Callaghan, J. R. (1980) 'A model for the simulation of the fate of nitrogen in farm wastes on land application', *Journal of Agricultural Science, Cambridge*, **94**, 183–93.

Bibby, C. J. (1978) 'Conservation of the Dartford warbler on English lowland heaths: a review', *Biological Conservation*, **13**, 299–307.

Bibby, J. S. and Mackney, D. (1969) *Land use capability classification*. Harpenden: Soil Survey of England and Wales, Technical Bulletin, No. 1.

Biederbeck, V. O., Campbell, C. A., Bowren, K. E., Schnitzer, M. and McIver, R. N. (1980) 'Effect of burning cereal straw on soil properties and grain yields in Saskatchewan', *Soil Science Society of America, Journal*, **44**, 103–11.

Bigmore, P. (1979) *The Bedfordshire and Huntingdonshire landscape*. London: Hodder and Stoughton.

Bijay Singh and Sekhon, G. S. (1977) 'Some measures of reducing leaching losses of nutrients beyond potential rooting zone. III. Proper crop rotations', *Plant and Soil*, **47**, 585–91.

Blackmore, D. K. (1963) 'The toxicity of some chlorinated hydrocarbon insecticides to British wild foxes (*Vulpes vulpes*)', *Journal of Comparative Pathology and Therapeutics*, **73**, 391–409.

Blacksell, M. and Gilg, W. (1975) 'Landscape evaluation in practice: the case of south-east Devon', *Transactions, Institute of British Geographers*, **66**, 135–40.

Blacksell, M. and Gilg, W. (1981) *The countryside: planning and change*. London: Allen and Unwin.

Blackwood, J. W. and Tubbs, C. R. (1970) 'A quantitative survey of chalk downland in England', *Biological Conservation*, **3**, 1–5.

Blad, B. L. and Rosenberg N. J. (1974) 'Evapotranspiration by subirrigated alfalfa pasture in the east Central Great Plains', *Agronomic Journal*, **66**, 248–52.

Blood, T. F. (1963) 'Effect of height of cutting on the subsequent regrowth of a sward', *National Agricultural Advisory Service, Quarterly Review*, **14**, 139–43.

Boorman, L. A. and Fuller, R. M. (1981) 'The changing status of reedswamp in the Norfolk Broads', *Journal of Applied Ecology*, **18**, 241–69.

Borman, F. H. and Likens, G. E. (1979) *Pattern and process in a forested ecosystem: disturbance, development and the steady state based on the Hubbard Brook ecosystem study*. New York: Springer-Verlag.

Bosch, J. M. and Hewlett, J. D. (1982) 'A review of catchment experiments to determine the effect of vegetation changes on water yield and evapotranspiration', *Journal of Hydrology*, **55**, 3–23.

Bouma, J., de Laat, P. J. M., Awater, R. H. C. M., van Heesen, H. C., van Holst, H. C. and van de Nes, T. J. (1980) 'Use of soil survey data in a model for simulating regional soil moisture regimes', *Soil Science Society of America, Journal*, **44**, 808–13.

Bower, C. A. and Wilcox, L. B. (1969) 'Nitrate content of the upper Rio Grande as influenced by nitrogen fertilisation of adjacent beds', *Soil Science Society of America, Proceedings*, **33**, 971–3.

Bower, M. M. (1962) 'The cause of erosion in blanket peat bogs. A review of evidence in the light of recent work in the Pennines', *Scottish Geographical Magazine*, **78**, 33–43.

Bowers, J. K. and Cheshire, P. (1983) *Agriculture, the countryside and land use: an economic critique*. London: Methuen.

Bowler, I. R. (1979) *Government and agriculture: a spatial perspective*. London: Longman.

Boyd, A. E. W. (1973) 'Potato blight control in east and south-east Scotland, 1959–68, *Annals of Applied Biology*, **74**, 41–58.

Boyle, Sir R. (1661) *The sceptical chymist*. London: J Cadwell for J. Crooke.

Bradford, J. M. (1980) 'The penetration resistance in a soil with well-defined structural units', *Soil Science Society of America, Journal*, **44**, 601–5.

Brady, N. C. (1974) *The nature and properties of soils* (8th ed). New York: Macmillan.

Bremner, J. M. and Shaw, K. (1958) 'Denitrification in soil. II. Factors affecting denitrification', *Journal of Agricultural Science, Cambridge*, **51**, 40–52.

Briggs, D. J. (1977) *Soils*. London: Butterworth.

Briggs, D. J. (1978) 'Edaphic effects of poaching by cattle', Proceedings, *North of England Soils Discussion Group*, **14**, 51–62.

Briggs, D. J. (1981) 'Environmental influences on the yield of spring barley in England and Wales', *Geoforum*, **12**, 99–106.

Briggs, D. J. and France, J. (1980) 'Landscape evaluation: a comparative study', *Journal of Environmental Management*, **10**, 263–75.

Briggs, D. J. and France, J. (1981) 'Assessing landscape attraction: a South Yorkshire study', *Landscape Research*, **6**, 2–5.

Briggs, D. J. and France, J. (1982) 'Mapping soil erosion by wind for regional environmental planning', *Journal of Environmental Management*, **15**, 159–68.

Briggs, D. J. (1983) *Biomass potential of the European Community. A report of a case study to develop a bio-climatic method of assessing the potential for biomass production*. Sheffield: University of Sheffield.

Broadbent, F. E. (1978) 'Mineralization, immobilization and nitrification', in *Management of nitrogen in irrigated agriculture* (P. F. Pratt, ed.). Washington, DC: National Science Foundation, 109–37.

Brockington, N. R. (1978) 'Simulation models in crop research', *Acta Agriculturae Scandinavica*, **28**, 33–40.

Broekhuizen, S. (1982) 'Studies on the population ecology of hares in The Netherlands', in *Research Institute for Nature Management, Annual Report 1981*. Arnhem, 94–102.

Brooks, D. H. (1970) 'Powdery mildew of barley and its control', *Outlook on Agriculture*, **6**, 122–7.

Brown, L. (1976) *British birds of prey*. London: Collins.

Browning, E. M. (1977) 'History and challenges in soil and water conservation and management', *Soil Science Society of America, Journal*, **41**, 254–9.

Brun, L. J., Kanemasu, E. T. and Powers, W. L. (1972) 'Evapotranspiration from soybean and sorghum fields', *Agronomy Journal*, **64**, 145–8.

Bullen, E. R. (1967) 'Break crops in cereal production', *Journal of the Royal Agricultural Society of England*, **128**, 77–85.

Bullen, E. R. (1974) 'Burning cereal crop residues in England', *Proceedings of the Annual Tall Timbers Fire Ecology Conference*, **13**, 223–35.

Bunting, A. H. and Drennan, D. S. H. (1965) 'Some aspects of the morphology and physiology of cereals in the vegetative phase', in *The growth of cereals and grasses* (F. L. Milthorpe and J. D. Ivins, eds). London: Butterworth, 20–38.

Burford, J. R., Greenland, D. J. and Pain, B. F. (1976) 'Effects of heavy dressings of slurry and inorganic fertilizers applied to grassland on the composition of drainage waters and the soil atmosphere', in *Agriculture and water quality*. London: HMSO, MAFF Technical Bulletin 32, 432–43.

Burke, W., Galvin, J. and Galvin, L. (1964) 'Measurement of structure stability of pasture soils', *8th International Congress of Soil Science, Bucharest, Commissions I and VI*, 581–6.

Burkhard, N. and Guth, J. A. (1981) 'Rate of volatilisation of pesticides from soil surfaces: comparison of calculated results with those determined in a laboratory model system', *Pesticide Science*, **12**, 37–44.

Burns, I. G. (1974) 'A model for predicting the redistribution of salts applied to fallow soils after excess rainfall or evaporation', *Journal of Soil Science*, **25**, 165–78.

Burns, I. G. (1977) 'Nitrate movement in soil and its agricultural significance', *Outlook on Agriculture*, **9**, 144–8.

Burns, I. G. (1980a) 'A simple model for predicting the effects of leaching of fertilizer nitrate during the growing season on the nitrogen fertilizer need of crops', *Journal of Soil Science*, **31**, 175–85.

Burns, I. G. (1980b) 'A simple model for predicting the effects of winter leaching of residual nitrate on the nitrogen fertilizer need of spring crops', *Journal of Soil Science*, **31**, 187–202.

Burns, I. G. (1980c) 'Influences of the spatial distribution of nitrate on the uptake of N by plants: a review and a model for rooting depth', *Journal of Soil Science*, **31**, 155–73.

Burton, I. and Kates, R. (1964) 'The perception of natural hazards in resource management', *Natural Resources Journal*, **3**, 412–41.

CEC (1982a) *The agricultural situation in the Community, 1981 report*. Brussels: Commission of the European Communities.

CEC (1982b) *A methodological approach to an information system on the state of the European environment* (Draft synthesis report on the results of work undertaken in the framework of the project 'Ecological Mapping' of the European Community). Brussels: Commission of the European Communities.

Cadbury, C. J. (1976) 'Botanical implications of bracken control', *Botanical Journal of the Linnaean Scoiety*, **73**, 285–94.

Cadbury, C. J. (1980) 'The status and habitats of the corncrake in Britain, 1978–79', *Bird Study*, **27**, 203–18.

Campbell, A. G. (1966) 'Effects of treading by dairy cows on pasture production and botanical structure, on a Te Kowhai soil', *New Zealand Journal of Agricultural Research*, **9**, 1009–24.

Campbell, C. A. (1978) 'Soil organic carbon, nitrogen and fertility', in *Soil organic matter* (M. Schnitzer and S. U. Khan, eds). Amsterdam: Elsevier, 173–271.

Campbell, C. A., Biederbeck, V. O. and Hinnam, W. C. (1975) 'Relationships between nitrate in summer-fallowed surface soil and some environmental variables', *Canadian Journal of Soil Science*, **55**, 213–23.

Campini, A. (1774) *Saggi d'agricoltura*. (Publisher not cited.)

Cannell, R. Q. (1969a) 'The tillering pattern in barley varieties, I. Production, survival and contribution to yield by component tillers', *Journal of Agricultural Science, Cambridge*, **72**, 405–22.

Cannell, R. Q. (1969b) 'The tillering pattern in barley varieties, II. The effect of temperature, light intensity and daylength on the frequency of occurrence of the coleoptile node and second tiller in barley', *Journal of Agricultural Science, Cambridge*, **72**, 423–35.

Cannell, R. Q. (1975) 'Current research on soil environment and root growth at ARC Letcombe Laboratory', in *Soil physical conditions and crop production*. London: HMSO, MAFF Technical Bulletin, No. 29, 439–48.

Cannell, R. Q. (1981) 'Potentials and problems of simplified cultivation and conservation tillage', *Outlook on Agriculture*, **10**, 379–84.

Cannell, R. Q. and Finney, J. R. (1975) 'Effects of direct drilling and reduced cultivation on soil conditions for root growth', *Outlook on Agriculture*, **8**, 184–9.

Cannell, R. Q., Davies, D. B. and Pidgeon, J. D. (1979) 'The suitability of soils for sequential direct drilling of combine harvested crops in Britain: a provisional classification', in *Soil survey applications* (M. G. Jarvis and D. Mackney, eds). Harpenden: Soil Survey of England and Wales, Technical Monograph No. 13, 1–23.

Carr, A. J. H. (1979) 'Causes of sward change: diseases', in *Changes in sward composition and productivity* (A. H. Charles and R. J. Haggar, eds).

Hurley: British Grassland Society, Occasional Symposium, No. 10, 161–6.

Carson, R. (1962) *Silent spring*. Boston: Houghton-Mifflin.

Carstensen, N. (1976) '*Bombina bombina* review in Funen, Denmark', *Norwegian Journal of Zoology*, **24**, 235.

Centre for Agricultural Strategy (1976) *Land for agriculture*. Reading: University of Reading, Centre for Agricultural Strategy, Report No. 1.

Centre for World Food Studies (1980) *The model of physical crop production*. Wageningen: Centre for World Food Studies, Research Report, SOW-80-5.

Chapman, S. B. (1967) 'Nutrient budgets for a dry heath ecosystem in the south of England', *Journal of Ecology*, **55**, 677–89.

Charles, A. H. (1979) 'Treading as a factor in sward deterioration', in *Changes in sward composition and productivity* (A. H. Charles and R. J. Haggar, eds). Hurley; British Grassland Society, Occasional Symposium, No. 10, 137–40.

Chepil, W. S. (1953) 'Field structure of cultivated soils with special reference to erodibility by wind', *Soil Science Society of America, Proceedings*, **17**, 185–91.

Chepil, W. S. (1954) 'Seasonal fluctuations in soil structure and erodibility of soil by wind', *Soil Science Society of America. Proceedings*, **18**, 13–16.

Chepil, W. S. (1955) 'Factors that influence clod structure and erodibility of soil by wind. V. Organic matter at various stages of decomposition', *Soil Science*, **80**, 413–21.

Chepil, W. S. and Woodruff, N. P. (1963) 'The physics of wind erosion and its control', *Advances in Agronomy*, **15**, 211–302.

Chisci, G. and Zanchi, C. (1981) 'The influence of different tillage systems and different crops on soil losses on hilly silty-clayey soil', in *Soil conservation: problems and perspectives* (R. P. C. Morgan, ed.). Chichester: Wiley, 211–17.

Chopra. U. K. and Chaudhary, T. N. (1980) 'Effect of soil temperature alteration by soil covers on seedling emergence of wheat (*Triticum aestivum* L.) sown on two dates', *Plant and Soil*, **57**, 125–9.

Clark, O. R. (1940) 'Interception of rainfall by prairie weeds, grasses and certain crop types', *Ecological Monographs*. **10** 243–77.

Clarke, A. L., Greenland, D. J. and Quirk, J. P. (1967) 'Changes in some physical properties of the surface of an impoverished red-brown earth under pasture', *Australian Journal of Soil Research*, **5**, 59–68.

Clarke, G. R. (1940) *Soil Survey of England and Wales: field handbook*. Oxford: Oxford University Press.

Clarke, G.R. (1951) 'The evaluation of soils and the definition of quality classes from studies of the physical properties of the soil profile in the field', *Journal of Soil Science*, **2**, 50–60.

Clement, C. R. and Williams, T. E. (1958) 'An examination of the method of aggregate analysis by wet sieving in relation to the influence of diverse leys on arable soils', *Journal of Soil Science*, **9**, 252–66.

Clement, C. R. and Williams, T. E. (1959) 'Crumb stability within the profile of arable soils under leys', *Mededelingen van de Landbouwhogeschool Gent*, **24**,166–75.

Clement, C. R. and Williams, T. E. (1964) 'Leys and soil organic matter. I. The accumulation of organic carbon in soil under different leys', *Journal of Agricultural Science, Cambridge*, **63**, 377–83.

Clements, R. O. (1980) 'Pests: the unseen enemy', *Outlook on Agriculture*, **10**, 219–23.

Collingbourne, R. H. (1976) 'Radiation and sunshine', in *Climate of the British Isles* (T.C. Chandler and S. Gregory, eds). London: Longman, 74–95.

Colquhoun, M. K. and Morley, A. (1941) 'The density of downland birds', *Journal of Animal Ecology*, **10**, 35–46.

Columella, L. I. M. (*c.* AD 100) *De re rustica* (1941 edition, H. B. Ash (ed) London: Loeb).

Conacher, A. J., Combes, P. L., Smith, P. A. and McLellan, R. C. (1983)

'Evaluation of throughflow interceptors for controlling secondary soil and water salinity in dryland agricultural areas of southwestern Australia. I. Questionnaire surveys', *Applied Geography*, **3**, 29–44.

Conway, V. M. (1954) 'The stratigraphy and pollen analysis of southern Pennine blanket peats', *Journal of Ecology*, **42**, 117–47.

Cooke, A. S. (1973) 'Shell thinning in avian eggs by environmental pollutants', *Environmental Pollution*, **4**, 85–152.

Cooke, A. S. (1975) 'Pesticides and eggshell formation', *Symposium of the Zoological Society of London*, **35**, 334–61.

Cooke, A. S. (1979) 'Population decline of the magpie, *Pica pica* in Huntingdonshire and other parts of eastern England', *Biological Conservation*, **15**, 317–24.

Cooke, G. W. (1974) *The control of soil fertility*. London: English Language Book Society and Crosby Lockwood Staples.

Cooke, G. W. (1975) *Fertilising for maximum yields*. London: English Language Book Society and Crosby Lockwood Staples.

Cooke, G. W. (1976) 'A review of the effect of agriculture on the chemical composition and quality of surface and underground waters', in *Agriculture and Water Quality*. London: HMSO, MAFF Technical Bulletin, **32**, 5–57.

Cooke, G. W. (1977) 'Waste of fertilisers', *Philosophical Transactions of the Royal Society of London*, **B 281**, 231–41.

Cooke, G. W. (1979) 'Some priorities for British soil science', *Journal of Soil Science*. **30**, 187–214.

Cooke, G. W. (1983) *Fertilising for maximum yields* (3rd edit.) London: Granada.

Cooke, G. W. and Williams, R. J. B. (1971) 'Problems with cultivation and soil structure at Saxmundham', *Report of the Rothamsted Experimental Station for 1970*, Part 2, 121–42.

Cooke, R. U. and Reeves, R. W. (1976) *Arroyos and environmental change*. Oxford: Oxford University Press.

Cooper, M. M. (1961) 'Grazing systems', *Journal of the Kings College Agricultural Society*, **15**, 5–10.

Copeman, G. J. F. (1979) 'Causes of sward change: climate', in *Changes in sward composition and productivity* (A. H. Charles and R. J. Haggar, eds). Hurley: British Grassland Society, Occasional Symposium, No. 10, 151–5.

Copland, S. (1866) *Agriculture ancient and modern*, London: Virtue, 2 vols.

Coppock, J. T. (1976) *An agricultural atlas of England and Wales* (rev. edn). London: Faber.

Corbett, W. M. (1973) *Breckland forest soils*. Harpenden: Soil Survey of England and Wales, Special Survey.

Costin, A. B. (1979) 'Runoff and soil and nutrient losses from an improved pasture at Ginninderra, Southern Tablelands, New South Wales', *Australian Journal of Agricultural Research*, **31**, 533–46.

Countryside commission (1978) *Upland land use in England and Wales*. Cheltenham: Countryside Commission.

Courtney, F. M. (1981) 'Developments in forest hydrology', *Progress in Physical Geography*, **5**, 217–41.

Cramp, S. and Simmons, K. E. L. (eds) (1980) *The birds of the western Palearctic*, vol. 2. Oxford: Oxford University Press.

Crampton, C. B. (1972) *Soils of the Vale of Glamorgan* (sheets 212 and 263). Harpenden Soil Survey (Memoirs of the Soil Survey of Great Britain).

Crescentius (*c.* 1300) *Opus ruralium Commodorum sive de Agricultura*. (1471 edn) Libri xii, Augsburg.

Crisp, D. T. (1966) 'Input and output of minerals for an area of Pennine moorland', *Journal of Applied Ecology*, **3**, 327–48.

Cromack, H. T. H., Mudd, C. H. and Strickland, M. J. (1970) 'A comparison of types of lime and their frequency of application to grassland', *Experimental Husbandry*, **19**, 40–8.

Crowdy, S. H. (1967) 'Control of pests and diseases', in *Potential crop production* (Wareing, D. F. and Cooper. J. P., eds). London: Heinemann, 319–30.

Cumber, R. H. (1964) 'The egg-parasite complex (Scelionidae: Hymnoptera) of shield bugs (Pentatomidae, Acanthosomidae: Hepteroptera) in New Zealand', *New Zealand Journal of Soil Science*, **7**, 536–54.

Currie, J. A. (1975) 'Soil respiration', in *Soil physical conditions and crop production*, London: HMSO, MAFF Technical Bulletin No. 29, 461–8.

Curtis, J. T. (1956) 'The modification of mid-latitude grasslands and forests by man', in *Man's role in changing the face of the earth* (W. L. Thomas, ed). Chicago: University of Chicago Press, 721–36.

Curtis, L. F. and Walker, A. J. (1982) *Moorland conservation an Exmoor. The Porchester maps: their construction and policies*. Dulverton: Exmoor National Park Committee.

Curtis, L. F., Courtney, F. M. and Trudgill, S. T. (1976) *Soils in the British Isles*. London: Longman.

Daniel, J. (1977) 'Effect of herbicides and fertilization on culinary potato quality' (in Czech), *Rostlinná Výoba*, **23**, 317–22.

Danz, W. and Henz, H. R. (1981) *Integrated development of mountain areas: the alpine region*. Brussels: Commission of the European Communities, Regional Policy Series, No. 20.

Darby, H, C, (1969) *The draining of the fens* (2nd edn). Cambridge: Cambridge University Press.

Dasmann, R. F. (1972) *Environmental conservation*. New York: Wiley.

Daubeny, C. G. B. (1845) 'Memoir on the rotation of crops, and on the quantity of inorganic manures abstracted from the soil by various plants under different circumstances', *Philosophical Transactions of the Royal Society of London*, 179–252.

Davies, D. A. (1979) 'Effect of species and varieties on yields of grass', in *Changes in sward composition and productivity* (H. Charles and R. J. Haggar, eds). Hurley: British Grassland Society, Occasional Symposium, No. 10, 77–83.

Davies, D. B. (1975) 'Field behaviour of medium textured and "silty" soils', in *Soil physical conditions and crop production*. London: HMSO, MAFF Technical Bulletin, No. 29, 52–75.

Davies, D. B. and Cannell, R. Q. (1975) 'Review of experiments on reduced cultivation and direct drilling in the United Kingdom (1957–74)', *Outlook on Agriculture*, **8**, 216–20.

Davies, E. B. and Hogg, D. E. (1960) 'Potassium losses in pasture', *New Zealand Journal of Agriculture*, **100**, 491–2.

Davies, E. B., Hogg, D. E. and Hopewell, H. G. (1962) 'Extent of return of nutrient elements by dairy cattle: possible leaching losses', in *Joint Meeting of the International Soil Science Society, Commissions IV and V*, New Zealand, 715–20.

Davies, E. T. and Dunford, W. J. (1962) *Some physical and economic considerations of field enlargement*. Exeter: University of Exeter, Department of Agricultural Economics. Research Report, No. 133.

Davies, G. E., Newbould, P. and Baillie, G. J. (1979) 'The effect of controlling bracken (*Pteridium aquilinum* (L) Kuhn) on pasture production', *Grass and Forage Science*, **34**, 163–71.

Davies, G. R. (1978) 'Results of subsoiling experiments on the Cottam (Salop) series in North Wales', *Proceedings of the North of England Soils Discussion Group*, **14**, 63–71.

Davies, P. W. and Davis, P. E. (1973) 'The ecology and conservation of the Red Kite in Wales', *British Birds*, **66** (5), 183–224, and **66** (6), 241–70.

Davy, Sir Humphry (1813) *Elements of agricultural chemistry, in a course of lectures for the Board of Agriculture*. London: Longman, Hurst, Rees, Orme and Browne; Edinburgh: Constable.

Day, W., Legg, B. J., French, B. K., Johnston, A. E., Lawler, D. W. and de Jeffers, W. (1978) 'A drought experiment using mobile shelters: the effect of drought on barley yield, water use and nutrient uptake', *Journal of Agricultural Science, Cambridge*, **91**, 599–623.

DeBach, P. (1974) *Biological control by natural enemies*. London: Cambridge University Press.

de Boer, P. B., van Dijk, H., and Oostendorp, D. (1979) *The statistics of cattle and sheep farming in the Netherlands*. Leylstad: Research and Advisory Institute for Cattle Husbandry.

Dempster, J. P. (1968) 'The control of *Pieris vapae* with DDT. II. Survival of the young stages of *Pieris* after spraying', *Journal of Applied Ecology*, **5**, 451–62.

Denmead, O. T. and Shaw, R. H. (1959) 'Evapotranspiration in relation to the development of the corn crop', *Agronomy Journal*, **51**, 725–6.

Department of Agriculture for Scotland (1952) *Types of farming in Scotland*. Edinburgh: HMSO.

de Ploey, J. (1981) 'Crusting and time-dependent rainwash mechanisms on loamy soils', in *Soil conservation: Problems and perspectives* (R. P. C. Morgan, ed.). Chichester: Wiley, 139–52.

Dettman, M. G. and Emerson, W. W. (1959) 'A modified permeability test for measuring the cohesion of soil crumbs', *Journal of Soil Science*, **10**, 215–26.

de Wit, C. T. and Goudriaan, J. (1974) *Simulation of ecological processes*. Wageningen: Pudoc, Simulation Monographs.

Disney, R. H. L. (1978) 'A new species of scuttle fly (Diptera: Phoridae) from Gloucestershire', *Entomologists' Gazette*, **29**, 153–5.

Djeniyi, S. O. and Dexter, A. R. (1979) 'Soil factors affecting the macro structure produced by tillage', *Transactions, American Society of Agricultural Engineers*, **22**, 339–43.

Dormaar, J. F. and Pittman, U. J. (1980) 'Decomposition of organic residues as affected by various dryland spring wheat-fallow rotations', *Canadian Journal of Soil Science*. **10**, 97–106.

Dormaar, J. F., Pittman, U. J. and Spratt, E. D. (1979) 'Burning crop residues: effect on selected characteristics and long-term wheat yields', *Canadian Journal of Soil Science*, **59**, 79–86.

Dowdell, R. J. and Cannell, R. Q. (1975) 'Effect of ploughing and direct drilling on soil nitrate content', *Journal of Soil Science*. **26**, 53–61

Dowdell, R. J., Smith, K. A., Crees, R. and Restall, S. W. F. (1972) 'Field studies of ethylene in the soil atmosphere: equipment and preliminary results' *Soil Biology and Biochemistry*. **4**, 325–31.

Down, K. M. and Lazenby, A. (1981) 'Appendix IV. Fertiliser usage on grassland in England and Wales', in *Grasslands in the British Economy* (J. L. Jollans, ed.). Reading: Centre for Agricultural Strategy, Paper No. 10, 564–6.

Drabble, P. (1981) 'Is the badger doomed?' *Observer Colour Magazine* (*21. 2. 81*), 10–12.

Drabløs, D. and Tollan, A. (*eds*) (1980) *Ecological impact of acid precipitation*. Proceedings of an International Conference, Norway, March 11–14, 1980. Oslo, SNSF.

Draycott, A. P. and Messon, A. B. (1977) 'Response to sugar beet by irrigation, 1965–75', *Journal of Agricultural Science, Cambridge*, **89**, 481–93.

Draycott, A. P., Durrant, M. J. and Webb, D. J. (1978) 'Long term effects of fertilisers at Broom's Barn, 1971–76', *Report of the Rothamsted Experimental Station for 1977*, Part 2, 15–30.

Draycott, A. P., Hull, R., Messien, A. B. and Webb, D. J. (1970) 'Effects of soil compaction on yield and fertiliser requirement of sugar beet', *Journal of Agricultural Science, Cambridge*, **75**, 533–7.

Drummond, J. M. and Soper, D. (1978) 'The long-term benefits of bracken (*Pteridium aquilinum* (L) Kuhn) removal with asulam on hill farms in northern Britain, *Proceedings 1978 British Crop Protection Conference-weeds, Brighton*. 317–24.

Duckham, A. N. and Masefield, G. B. (1970) *Farming systems of the world*. London: Chatto and Windus.

Duffey, E., Morris, M. G., Sheail, J., Ward, L. K., Wells, D. A. and Wells, T. C. E. (1974) *Grassland ecology and wildlife management*. London: Chapman and Hall.

Duffus, J. H. (1980) *Environmental toxicology*. London: Edward Arnold.

du Hamel du Manceau, H. L. (1759) *A practical treatise of husbandry* (transl. J. Mills). London: (publisher not cited).

Dupuit, J. (1863) *Etudes theoretiques et practiques sur le mouvement des eaux* (2nd edn). Paris: Dunod.

Durham County Council (1976) *County structure plan: report of survey*. Durham: County Planning Office.

Dyke, G. V. (1964) 'Broadbalk', *Report, Rothamsted Experimental Station for 1963*, 183–6.

Dyke, G. V. (1968) 'Field experiments and increases in yield of crops', *Journal of the National Institute of Agricultural Botany*, **11**, 329–42.

Eadie, J. (1967) 'The nutrition of grazing hill sheep: utilisation of hill pastures', *Report of the Hill Farming Research Organisation*, **4** 38–45.

Eagle. D. J. (1975) 'ADAS ley fertility experiments', in *Soil physical conditions and crop production*. London: HMSO, MAFF Technical Bulletin, No. 29, 344–59.

Eddowes, M. (1976) *Crop production in Europe*. Oxford: Oxford University Press.

Edmond, D. B. (1958) 'The influence of treading on pasture: a preliminary study', *New Zealand Journal of Agricultural Research*, **1**, 319–28.

Edmond, D. B. (1963) 'Effects of treading perennial ryegrass (*Lolium perenne* L.) and white clover (*Trifolium repens* L.) pastures in winter and summer at two moisture levels', *New Zealand Journal of Agricultural Research*, **6**, 265–76.

Edwards, A. J. (1970) 'Field size and machine efficiency' in *Hedges and hedgerow trees* (M. D. Hooper and M. W. Holdgate, eds). London: Nature Conservancy Council, Monks Wood Experimental Station Symposium No. 4, 28–32.

Edwards, A. M. C. (1974) 'Long term changes in the water quality of agricultural catchments'. Unpublished paper presented to the Institute of British Geographers, Symposium on Applied Physical Geography, 5–6 January 1974, Norwich.

Edwards, C. A. (1969a) 'Problems of insecticidal residues in agricultural soils', *NAAS Quarterly Review*. **86**, 47–54.

Edwards, C. A. (1969b) 'Soil pollutants and soil animals', *Scientific American*, **220** (4) 88–99.

Edwards, C. A., Dennis, E. B. and Empson, D. W. (1967) 'Pesticides and the soil fauna. I. Effects of aldrin and DDT in an arable field', *Annals of Applied Biology*, **60**, 11–22.

Edwards, R. W., Hughes, B. D. and Read, M. W. (1975) 'Biological survey in the detection and assessment of pollution', in *The ecology of resource degradation and renewal* (M. J. Chadwick and G. T. Goodman, eds). Oxford: Blackwell, Symposium of the British Ecological Society, No 15, 139–56.

Ehlers, W. (1975) 'Observations on earthworm channels and infiltration on tilled and untilled loess soils', *Soil Science*, **119**, 242–9.

Ehlers, W. (1976) 'Evapotranspiration and drainage in tilled and untilled loess soil with winter wheat and sugar beet', *Zeitschrift für Acker-und Pflanzenbau*, **142**, 285–303.

Ehlers, W., Khosla, B. K., Kopke, U., Stulpnagel, R., Bohm, W. and Baeumer, K. (1980–81) 'Tillage effects on root development, water uptake and growth of oats', *Soil and Tillage Research*, **1**, 19–34.

Ekern, D. C., Robins, J. S. and Staple, W. T. (1967) 'Soil and cultural factors affecting evapotranspiration', in *Irrigation of agricultural lands* (R. M. Hagan, H. R. Haise and T. W. Edminster, eds). Madison: American Society of Agronomy, 11, 522–33.

Elliott, J. G. (1972) 'Wild oats, where next?', *Proceedings, 11th British Weed*

Control Conference, Brighton, **3**, 965–76.

Elliott, J. G., Oswald, A. K., Allen, G. P. and Haggar, R. J. (1974) 'The effect of fertiliser and grazing on the botanical composition and output of an *Agrostis/Festuca* sward', *Journal of the British Grassland Society*, **29**, 29–35.

Elliott, R. J. (1953) *The effects of burning on heather moors of the southern Pennines*. Unpublished PhD thesis, University of Sheffield.

Ellis, E. A. (1965) *The Broads*. London: Collins.

Ellis, F. B., Elliott, J. G., Pollard, F., Cannell, R.Q. and Barnes, B. T. (1979) 'Comparison of direct drilling, reduced cultivation and ploughing on the growth of cereals. 3. Winter wheat and spring barley on a calcareous clay', *Journal of Agricultural Science, Cambridge*. **93**, 391–401.

Elwell, H. McMurphy, W. E. and Santelmann, P. W. (1970) *Burning and 2,4,5-T on post and blackjack oak rangeland in Oklahoma*. Bulletin of the Oklahoma Agriculture Experiment Station. B–675.

Emerson, W. W., Bond, R. D. and Dexter, A. R. (eds) (1978) *Modification of soil structure*. Chichester: Wiley.

Emery, F. (1974) *The Oxfordshire landscape*. London: Hodder and Stoughton.

Eno, C. F. (1957) *Field accumulation of insecticide residues in soils. Effect of soil applications of carbamate fungicides on the soil microflora*. Miami: Florida Agricultural Experiment Station, Report 142.

Ernle, Lord (1961) *English farming past and present* (6th edn). London: Heinemann.

Ernst, P, Le Du, Y. L. P. and Carlier, L. (1980) 'Animal and sward production under rotational and continuous grazing management: a critical appraisal', in *Proceedings of the European Grassland Federation Symposium on the Role of Nitrogen in Intensive Grassland Production* (W. H. Prins and G. H. Arnold, eds). Wageningen: Pudoc, 119–26.

Espinoza, W. (1979) 'Effect of plant population on maize evapotranspiration when irrigated in the dry season on the Cerrado of the Federal District', (in Portuguese). *Pesguisa Agroecuana Brasileira*, **14**, 343–50

Evans, A. C. and Guild, W. J. M. (1947) 'Studies on the relationships between earthworms and soil ferility. 1. Biological studies in the field', *Annals of Applied Biology*. **34**, 307–30.

Evans, C. C. and Allen, S. E. (1971) 'Nutrient losses in smoke produced during heather burning', *Oikos*, **22**, 149–54.

Evans, I. A. (1976) 'Relationship between bracken and clover', *Botanical Journal of the Linnaean Society*, **73**, 105–12.

Evans, J. G. (1975) '*The environment of early man in the British Isles*. London: Elek.

Evans, K. (1979) 'Nematode problems in the Woburn ley-arable experiment and changes in *Longidorus leptocephalus* population density associated with time, depth, cropping and soil type', *Report of the Rothamsted Experimental Station for 1978*, Part 2, 27–45.

Evans, R. (1977) 'Overgrazing and soil erosion on hill pastures with particular reference to the Peak District', *Journal of the British Grassland Society*, **32**, 65–76.

Evans, S. A. and Neild, J. R. A. (1981) 'The achievement of very high yields of potatoes in the UK', *Journal of Agricultural Science*, **97**, 391–6.

Evelyn, J. (1676) *A philosophical discourse of Earth relating to the culture and improvement of it for vegetation and the propagation of plants*. London.

Everett, M. (1971) 'The golden eagle survey in Scotland, 1964–68', *British Birds*, **64**, 49–56.

Exmoor National Park Committee (1977a) *Exmoor National Park plan*. Dulverton: Exmoor National Park Committee.

Exmoor National Park Committee (1977b) *Exmoor National Park plan: supplement*. Dulverton: Exmoor National Park Committee.

FAO (1966) *Statistics of crop responses to fertilisers*. Rome: United Nations Food and Agricultural Organisation.

FAO (1970) *Fertilizers. An annual review of world production, consumption trade and prices, 1969*. Rome: United Nations Food and Agricultural Organisation.

Farrimond, M. S. (1980) 'Impact of man in catchments. III. Domestic and industrial wastes', in *Water quality in catchment ecosystems* (A. M. Gower, ed.). Chichester: Wiley, 113–44.

Fawcett, R. G. (1978) 'Effect of cultivation, stubble retention and environment on the accumulation of fallow water', in *Modification of soil structure* (W. W. Emerson, R. D. Bond and A. R. Dexter, eds). Chichester: Wiley, 403–10.

Feare, C, J. (1980) 'The economics of starling damage', in *Bird problems in agriculture* (E. W. Wright, ed.). Croydon: British Crop Protection Council Publications.

Feddes, R. A. and van Wijk, A. J. M. (1976) 'An integrated model approach to the effect of water management on crop yields', *Agricultural Water Management*, **1**, 3–30.

Ferguson-Lees, I. J. (1951) 'The Peregrine population of Great Britain'. *Bird Notes* **24**, 200–5; 309–14.

Ferguson-Lees, I. J. (1963) 'Changes in the status of birds of prey in Europe', *British Birds*, **56**, 1400–48.

Ferns, P. N., Green, G. H. and Round, P. R. (1979) 'Significance of the Somerset and Gwent levels in Britain as feeding areas for migrant whimbrels (*Numenius phaeopus*)', *Biological Conservation*, **16**, 7–22.

Fick, G. W. (1980) 'A pasture production model for use in whole farm simulations', *Agricultural Systems*, **5**, 137–61.

Fines, K. D. (1968) 'Landscape evaluation: a research project in east Sussex', *Regional Studies*, **2**, 41–55.

Fitzherbert, (1523) *The boke of husbandrye* (1882 edn, W. W. Skeats, ed.) London: English Dialect Society.

Fletcher, W. W. (1974) *The pest war*. Oxford: Blackwell.

Floate, M. J. S. (1970) 'Mineralisation of nitrogen and phosphorus from organic materials of plant and animal origin and its significance in the nutrient cycle in grazed upland and hill soils', *Journal of the British Grassland Society*, **25**, 295–302.

Forbes, T. J., Dibb, C., Green, J. O. and Fenlow, K. A. (1977) *The permanent pasture project: objectives and methods*. Hurley: Grassland Research Institute/Agricultural Development and Advisory Service, Joint Permanent Pasture Group, Permanent Grassland Studies, 1.

Forbes, T. J., Dibb, C., Green, J. O., Hopkins, A. and Peel, S. (1980) *Factors affecting the productivity of permanent grassland: a national farm survey*. Hurley: Grassland Research Institute/Agricultural Development and Advisory Service, Joint Permanent Pasture Group.

Fornstrom, K. J. and Boennke, R. D. (1976) 'A grazing mulch and tillage system to reduce wind erosion losses of sugar beets', *Journal, American Society of Sugar Beet Technicians*, **19**, 64–73.

Foster, S. S. D., Cripps, A. C. and Smith-Carington, A. (1982) 'Nitrate leaching to groundwater', *Philosophical Transactions of the Royal Society of London*, **B 296**, 477–82.

France, J., Brockington, N. R. and Newton, J. E. (1981) 'Modelling grazed grassland systems: wether sheep grazing perennial ryegrass', *Applied Geography*, **1**, 133–50.

Francois, J. and Renard, C. (1979) 'Etude en milieu contrôlé du comportement d'un tapis de *Festuca arundinacea* Schreb en régime d'assèchement', *Oecologia Plantarum*, **14**, 417–33.

Frank, A. B. and Willis, W. O. (1978) 'Effect of winter and summer windbreaks on soil water gain and spring wheat yield', *Soil Science Society of America, Journal*, **42**, 950–3.

Freebairn, D. M. and Boughton, W. C. (1981) 'Surface runoff experiments on the eastern Darling Downs', *Australian Journal of Soil Research*, **19**, 133–46.

French, N. (ed) (1979) *Ecological studies 26: perspectives in grassland ecology.* New York: Springer.

French, R. J. (1978) 'The effect of fallowing on the yield of wheat. I. The effect on soil water storage and nitrate supply', *Australian Journal of Agricultural Research*, **29**, 653–68.

Fretey, J. (1975) *Guide des reptiles et batriciens de France.* Paris: Hatier.

Frissel, M. J. (ed.) (1978) *Cycling of mineral nutrients in agricultural ecosystems.* Amsterdam: Elsevier, Proceedings of the First International Environmental Symposium of the Royal Netherlands Land Development Society, Amsterdam, 31 May–4 June 1976. Reprinted from *Agroecosystems* **4** (1/2), 1–354.

Fryer, J. D. (1977) 'Recent developments in the agricultural use of herbicides in relation to agricultural effects', in *Ecological effects of pesticides* (F. H. Perring and K. Mellanby, eds). London: Academic Press, Linnaean Society Symposium, **5**, 27–45.

Fryer, J. D. and Chancellor, R. J. (1974) 'Herbicides and our changing weeds', in *The flora of a changing Britain* (F. Perring, ed.), London : Botanical Society of the British Isles No. 11, 105–18.

Fryer, J. D., Ludwig, J. W., Smith, P. D. and Hance, R. J. (1980) 'Tests of soil fertility following repeated applications of MCPA, tri-allate, simazine and linuron', *Weed Research, UK*, **20**, 111–16.

Fuller, R. J. (1982) *Bird habitats in Britain.* Calton: T. and A.D. Poyser.

Fussell, G. E. (1972) *The classical tradition in West European farming.* Newton Abbot: David and Charles.

Fussell, G. E. (1973) *Jethro Tull: his influence on mechanized agriculture.* Reading: Osprey.

Gair, R., Jenkins, J. E. E. and Lester, E. (1972) *Cereal pests and diseases.* Ipswich: Farming Press.

Garwood, E. A. (1979) 'The effect of irrigation on grassland productivity', in *Water control and grassland productivity*, Winter Meeting of the British Grassland Society, Hurley, 1979, **2**, 1–10.

Garwood, E. A. and Tyson, K. C. (1979) 'Productivity and botanical composition of a grazed ryegrass/white clover sward over 24 years as affected by soil conditions and weather', in *Changes in sward composition and productivity* (A. H. Charles and R. J. Haggar, eds). Hurley: British Grassland Society, Occasional Symposium No. 10, 41–6.

Garwood, E. A. and Williams, T. E. (1967a) 'Soil water use and the growth of a grass sward', *Journal of Agricultural Science, Cambridge*, **68**, 281–92.

Garwood, E. A. and Williams, T. E. (1967b) 'Growth, water use and nutrient uptake from the subsoil by grass swards', *Journal of Agricultural Science, Cambridge*, **69**, 125–30.

Garwood, E. A., Tyson, K. C. and Clement, C. R. (1977) *A comparison of yield and soil conditions during 20 years of grazed grass and arable cropping.* Hurley: Grassland Research Institute, Technical Report, No. 21.

Gasser, J. K. R. (1980) 'Impact of man in catchments. I. Agriculture', in *Water quality and catchment ecosystems* (A. M. Gower, ed.). Chichester: Wiley, 49–71.

Gasser, J. K. R. (1982) 'Agricultural productivity and the nitrogen cycle', *Philosophical Transactions of the Royal Society of London*, **B 296**, 303–14.

Gerard, B. M. and Hay, R. K. M. (1979) 'The effect on earthworms of ploughing, tined cultivation, direct drilling and nitrogen in a barley monoculture system', *Journal of Agricultural Science, Cambridge*, **93**, 147–55.

Getzin, L. W. (1968) 'Persistence of diazinon and zinophos in soil: effects of autoclaving, temperature, moisture and acidity', *Journal of Economic Entomology*, **61**, 1560–5.

Geurink, J. H., Malestein, A., Kemo, A. and van 't Klooster, A Th. (1979) 'Nitrate poisoning in cattle. 3. The relationship between nitrate intake with hay or fresh roughage and the speed of intake on the formation of methemoglobin', *Netherlands Journal of Agricultural Science*, **27**, 268–76.

Gifford, G. F. and Hawkins, R. H. (1978) 'Hydraulic impact of grazing on infiltration: a critical review', *Water Resources Research*, **14**, 305–13.

Gifford, G. F. and Hawkins, R. H. (1979) 'Deterministic hydrologic modelling of grazing system impacts on infiltration rates', *Water Resources Bulletin*, **15**, 924–34.

Gillingham, A. G. and During, C. (1973) 'Pasture production and transfer of fertility within a long established hill pasture', *New Zealand Journal of Experimental Agriculture*, **1**, 227–32.

Gimingham, C. H. (1972) *Ecology of heathlands*. London: Chapman and Hall.

Gliessman, S. R. (1976) 'Alleopathy in a broad spectrum of environments as illustrated by bracken', *Botanical Journal of the Linnaean Society*, **73**, 95–104.

Glopper, J. and Smits, H. (1974) 'Reclamation of land from the sea and lakes in the Netherlands', *Outlook on Agriculture*, **44**, 148–55.

Glymph, L. M. and Holtan, H. N. (1969) 'Land treatment in agricultural watershed hydrology research', in *Effects of watershed changes on streamflow* (W. L. Moore and C. W. Morgan, eds). Austin: University of Texas Press, 44–68.

Glynne, M. D. (1969) 'Fungus diseases of wheat on Broadbalk, 1843–1967', *Report of the Rothamsted Experimental Station for 1968*, Part 2, 116–40.

Goldsmith, F. B. (1975) 'The evaluation of ecological resources in the countryside for conservation purposes', *Biological Conservation*, **8**, 89–96.

Golisch, G. (1977) 'Winter barley: agronomy and fertiliser application'(in German), *DLG – Mitteilungen*, **92**, 186–8.

Gonner, E. C. K. (1912) *Common land and inclosure*. London: Macmillan.

Gordon, C. H., Kane, E. A., Derbyshire, J. C., Jacobson, W. C., Melin, C. J. and McCalmont, J. R. (1959) 'Nutrient losses, quality and feeding values of wilted and direct-cut orchard grass stored in bunker and tower silos', *Journal of Dairy Science*, **42**, 1703–11.

Gosling, L. M. and Baker, S. J. (1980) 'Acidity fluctuations at a Broadland site in Norfolk', *Journal of Applied Ecology*, **17**, 479–90.

Goss, M. J., Howse, K. C. and Harris, W. (1978) 'Effects of cultivation on soil water retention and water use by cereals in clay soils', *Journal of Soil Science*, **29**, 457–88.

Gradwell, M. W. (1968) 'Compaction of pasture topsoils under winter grazing', *Transactions of the 9th International Congress of Soil Science, Adelaide*, **3**, 429–36.

Graham–Bryce, I. J. (1977) 'Recent developments in the chemical control of agricultural pests and diseases in relation to ecological effects', in *Ecological effects of pesticides* (F. H. Perring and K. Mellanby, eds). London: Academic Press, Linnaean Society Symposium Series, No. 5, 47–60.

Grant, S. A., Lamb, W. I. C., Kerr, C. D. and Bolton, G. R. (1976) 'The utilisation of blanket bog vegetation by grazing sheep', *Journal of Applied Ecology*, **13**, 857–69.

Gray, D. M. (1973) *Handbook on the principles of hydrology*. New York: Water Information Center Inc.

Gray, E. M. and Morgan-Jones, M. (1980) 'A comparative study of nitrate levels at three adjacent ground-water sources in a Chalk catchment area west of London', *Ground Water*, **18**, 159–67.

Green, B. H. (1981) *Countryside conservation*. London: George Allen and Unwin.

Green, F. H. W. (1973) 'Aspects of the changing environment: some factors affecting the aquatic environment in recent years', *Journal of Environmental Mangement*, **1**, 377–91.

Green, F. H. W. (1979) 'Field drainage and the hydrological cycle', in *Man's impact on the hydrological cycle in the United Kingdom* (G. E. Hollis, ed.). Norwich: Geo Abstracts, 9–17.

Green, F. H. W. (1980) 'Current field drainage in northern and western Europe', *Journal of Environmental Management*, **10**, 149–53.

Green, J. O. and Baker, R. D. (1981) 'Classification, distribution and productivity of UK grassland', in *Grassland in the British economy* (L. Jollans, ed.). Reading: Centre for Agricultural Strategy, Paper 10, 237–47.

Greenland, D. J. (1977) 'Soil damage by intensive arable cultivation: temporary or permanent?', *Philosophical Transactions of the Royal Society of London*, **B 281**, 193–208.

Greenland, D. J. (1981) 'Soil management and soil degradation', *Journal of Soil Science*, **32**, 301–22.

Greenwood, D. J. and Goodman, D. (1971) 'Studies on the supply of oxygen to the roots of mustard seedlings (*Sinapis alba* L.)', *New Phytologist*, **70**, 85–96.

Grieve, I. C. (1980) 'The magnitude and significance of soil structural stability declines under cereal cropping', *Catena*, **7**, 79–85.

Griffiths, W. (1972) 'Weed control in cereals. A manufacturer's view of farmer benefits', *Proceedings of the 11th British Weed Control Conference*, Brighton, **3**, 900–7.

Grigg, D. B. (1974) *The agricultural systems of the world: an evolutionary approach*. Cambridge: Cambridge University Press.

Guild, W. J.McL. (1951) 'The distribution and population density of earthworms (Lumbricadae) in Scottish pasture fields', *Journal of Animal Ecology*, **20**, 88–97.

Guild, W. J. McL. (1952) 'Variations in earthworm numbers within field populations', *Journal of Animal Ecology*, **21**, 169–81.

Gupta, S. C., Onstad, C. A. and Larsen, W. E. (1979) 'Predicting the effects of tillage and crop residue management on soil erosion', *Journal of Soil and Water Conservation*, **34**, 77–9.

HMSO (1975) *Food from our own resources*. London: HMSO, Cmnd 6020.

HMSO (1978) *A study of Exmoor*. London: HMSO (The 'Porchester Report').

HMSO (1979) *Farming and the nation*. London: HMSO, Cmnd 7458.

Haghiri, F., Miller, R. A. and Logan, T. J. (1978) 'Crop response and quality of soil leachate as affected by land application of beef cattle wasters', *Journal of Environmental Quality*, **7**, 406–12.

Hamblin, A. P. and Davies, D. B. (1977) 'Influence of organic matter on the physical properties of some East Anglian soils of high silt content', *Journal of Soil Science*, **28**, 11–22.

Hammad, H. Y. (1962) 'Depth and spacing of tile drain systems', *Journal of Irrigation Drainage Division, American Society of Civil Engineers*, **89**, 15–21.

Hance, R. J. (1979) 'Effect of pH on the degradation of atrazine, dichlorprop, linuron and propyzamide in soil', *Pesticide Science*, **10**, 83–6.

Hance, R. J., Smith, P. D., Byast, T. H. and Cotterill, E. G. (1978a) 'Effects of cultivations on the persistence and phytotoxicity of atrazine and propyzamide', *Proceedings of the 14th British Weed Control Conference*, Brighton, **2**, 541–7.

Hance, R. J., Smith, P. D., Cotterill, E. G. and Reid, D. C. (1978b) 'Herbicide persistence: effects of plant cover, previous history of the soil and cultivation', *Mededelingen van de Faculteit Landbouwwetenschappen Rijksuniversiteit Gent*, **43**, 1127–34.

Hanks, R. J. and Hill, R. W. (1980) *Modeling crop responses to irrigation*. Oxford: Pergamon Press, International Irrigation Information Center.

Hanks, R. J. amd Puckeridge, D. W. (1980) 'Prediction of the influence of water, sowing date and planting density on dry matter production of wheat', *Australian Journal of Agricultural Research*, **31**, 1–11.

Harmeson, R. H. and Larsen, T. E. (1970) 'Existing levels of nitrates in waters: the Illinois situation', *Proceedings of the 12th Sanitary Engineering Conference, University of Illinois*, 27–38.

Harris, C. R. (1964) 'Influence of soil moisture on the toxicity of insecticides in a mineral soil to insects', *Journal of Economic Entomology*, **57**, 946–50.

Harris, C. R. (1966) 'Influence of soil type on the activity of insecticides in the soil', *Journal of Economic Entomology*, **59**, 1221–5.

Harris, C. R. (1970) 'Laboratory evaluation of candidate materials as potential soil insecticides, III', *Journal of Economic Entomology*, **63**, 782–7.

Harris, C. R. (1973) Laboratory evaluation of candidate materials as potential soil insecticides, IV', *Journal of Economic Entomology*, **66**, 216–21.

Harris, C. R. and Mazurek, J. H. (1964) 'Comparison of the toxicity to insects of certain insecticides applied by contact and in the soil', *Journal of Economic Entomology*, **57**, 698–702.

Harris, C. R. and Mazurek, J. H. (1966) 'Laboratory evaluation of candidate materials as potential soil insecticides', *Journal of Economic Entomology*, **59**, 1215–21.

Harrod, M. F. (1975) 'Field behaviour of light soils', in *Soil physical conditions and crop production*. London: HMSO, MAFF Technical Bulletin, No. 29, 22–51.

Hartmann, R., Verplancke, H. and de Boodt, M. (1981) 'Influence of soil surface structure on infiltration and subsequent evaporation under simulated laboratory conditions', *Soil and Tillage Research*, **1**, 351–9.

Harvey, P. N. (1959) 'The disposal of cereal straw', *Journal of the Royal Agricultural Society of England*, **120**, 55–63.

Harvey, P. N. and Wellings, L. W. (1970) 'Irrigation of sugar beet on light sand soil', *Experimental Husbandry*, **19**, 1–12.

Haun, J. R. (1975) 'Potato growth: environmental relationships', *Agricultural Meteorology*, **15**, 325–32.

Hauser, V. L. and Hiler, E. A. (1975) 'Rainfall-induced runoff computed for fallow fields', *American Society of Agricultural Engineers, Transactions*, **18**, 122–5.

Hawkes, H. A. and Davies, L. J. (1971) 'Some effects of organic enrichment on benthic invertebrate communities in stream riffles', in *The scientific management of animal and plant communities for conservation* (E. Duffey and A. S. Watt, eds). 11th Symposium of the British Ecological Society, University of East Anglia, Norwich, 7–9 July 1970, 271–93.

Hawkesworth, D. L., Rose, F. and Coppins, B. J. (1973) 'Changes in the lichen flora of England and Wales attributable to pollution of the air by sulphur dioxide', in *Air pollution and lichens* (B. W Ferry, M. S. Baddeley, and D. L. Hawkesworth, eds). London: Athlone Press, 330–67.

Hawkins, J. C. (1980) 'Agricultural engineering', in *Perspectives in world agriculture*. Slough; Commonwealth Agricultural Bureaux, 345–66.

Hay, R. K. M. (1977) 'Effects of tillage and direct drilling on soil temperature in winter', *Journal of Soil Science*, **28**, 403–9.

Hay, R. K. M., Holmes, J. C. and Hunter, F. A. (1978) 'The effects of tillage, direct drilling and nitrogen fertiliser on soil temperature under a barley crop', *Journal of Soil Science*, **29**, 174–83.

Heapy, L. A., Robertson, J. A., McBeath, D. K., Maydell, U. M. von, Love, H. C. and Webster, G. R. (1976a) 'Development of a barley yield equation for central Alberta. 1. Effects of soil and fertilizer N and P', *Canadian Journal of Soil Science*, **56**, 233–47.

Heapy, L. A., Webster, G. R., Love, H. C., McBeath, D. K., Maydell, U. M. von and Robertson, J. A. (1976b) 'Development of a barley yield equation for central Alberta. 2. Effects of soil moisture stress', *Canadian Journal of Soil Science*, **56**, 249–56.

Heath, G. W. (1962) 'The influence of ley management on earthworm populations', *Journal of the British Grassland Society*, **17**, 237–44.

Heikurainen, L. (1972) 'Peatland classification for forestry in Finland', *Proceedings of the 4th International Peat Congress, Helsinki*, **3**, 435–50.

Helling, C. S., Kearney, R. C. and Alexander, M. (1971) 'Behaviour of pesticides in soils', *Advances in Agronomy*, **23**, 147–240.

Helliwell, J. M. (1977) 'Change in natural and managed ecosystems: detection, measurement and assessment', *Proceedings of the Royal Society of London*, **B 197**, 31–57.

Henderson, I. F. and Clements, R. O. (1977) 'Grass growth in different parts of Britain in relation to invertebrate numbers and pesticide treatment', *Journal of the British Grassland Society*, **32**, 89–98.

Henderson, I. F. and Clements, R. O. (1979) 'Differential susceptibility to pest damage in agricultural grasses', *Journal of Agricultural Science, Cambridge*, **73**, 465–72.

Herring, K. (ed) (1977) *Fenland farming, wild life and landscape conservation.* March, Cambs: Ministry of Agriculture, Fisheries and Food.

Herriott, J. B. D., Wells, D. A. and Crooks, P. (1965) 'Gülle as a grassland fertilizer, Part III', *Journal of the British Grassland Society*, **20**, 129–38.

Hewitt, J. S. and Dexter, A. R. (1979) 'An improved model of root growth in structured soil', *Plant and Soil*, **52**, 325–43.

Hewitt, J. S. and Dexter, A. R. (1980) 'Effects of tillage and stubble management on the structure of a swelling soil', *Journal of Soil Science*, **31**, 203–21.

Hickey, J. J., Keith, J. A. and Coon, F. B. (1966) 'An exploration of pesticides in a Lake Michigan ecosystem', *Journal of Applied Ecology*, **3** (suppl.), 141–54.

Hilder, E. J. (1966) 'Distribution of excreta by sheep at pasture', *Proceedings of the 10th International Grassland Congress, Helsinki*, Paper No. 39, Section 4, 977–81.

Hill, A. R. (1976) 'The environmental impacts of agricultural land drainage', *Journal of Environmental Management*, **4**, 251–74.

Hill, A. R. (1978) 'Factors affecting the export of nitrate-nitrogen from drainage basins in southern Ontario', *Water Research*, **12**, 1045–57.

Hill, T. A. (1977) *The biology of weeds.* London: Edward Arnold.

Hodgson, D. R., Holliday, R. and Cope, F. (1963) 'The reclamation of land covered with pulverised fuel ash: the influence of soil depth on crop performance', *Journal of Agricultural Science, Cambridge*, **61**, 299–308.

Hodgson, D. R., Proud, J. R. and Browne, S. (1977) 'Cultivation systems for spring barley with special reference to direct drilling (1971–1974)', *Journal of Agricultural Science, Cambridge*, **88**, 631–44.

Hodgson, J. (1974) 'Grazing management', in *Silver Jubilee Report 1949–1974* (C. R. W. Spedding and R. D. Williams, eds). Hurley: Grassland Research Institute, 116–28.

Hodgson, J. and Ollerenshaw, J. A. (1969) 'The frequency and severity of defoliation of individual tillers in set-stocked swards', *Journal of the British Grassland Society*, **24**, 226–34.

Holmes, W. (1980) 'Grazing management', in *Grass: its production and utilization* (W. Holmes, ed.). Oxford: Blackwell, 125–73.

Honeggar, R. (1978) *Threatened amphibians and reptiles in Europe.* Strasbourg: Council of Europe, Nature and Environment Series, No. 15.

Hood, A. E. M. (1982) 'Fertilizer trends in relation to biological productivity within the UK', *Philosophical Transactions of the Royal Society of London*, **B 296**, 315–28.

Hood, A. E. M. and Proctor, J. (1961) 'An intensive cereal growing experiment', *Journal of Agricultural Science*, **57**, 241–7.

Hoogerkamp, M. (1973) *Accumulation of organic matter under grassland and its effect on grassland and on arable crops.* Wageningen: Pudoc, Agricultural Research Report, No. 80.

Hoogerkamp, M. (1974) *Ley, periodically reseeded grassland or permanent pasture.* Wageningen: Pudoc, Agricultural Research Report, No. 812.

Hoogerkamp, M. (1979) 'Avoiding the lean years', in *Changes in sward composition and productivity* (A. H. Charles and R. J. Haggar, eds). Hurley: British Grassland Society, Occasional Symposium No. 10, 199–205.

Hooghoudt, S. B. (1940) 'Review of the problem of detail drainage and subirrigation by means of parallel drains, trenches, ditches and canals', *Landbouwk, Onderzoek*, **46**, 515–707.

Hooper, M. D. (1970) 'Rates of hedgerow removal', in *Hedges and hedgerow trees* (M. D. Hooper and M. W. Holdgate, eds). London: Nature Conservancy Council, Monks Wood Experimental Station, Symposium No. 4.

Hooper, M. D. (1979) 'Hedgerows and small woodlands', in *Conservation and agriculture* (J. Davidson and R. Lloyd, eds). Chichester: Wiley, 45–57.

Hopkins, A. and Green, J. O. (1979) 'The effect of soil fertility and drainage on sward changes', in *Changes in sward composition and productivity* (H. Charles and R. J. Haggar, eds). Hurley: British Grassland Scoiety, Occasional Symposium No. 10 115–29.

Horton, R. E. (1933) 'The role of infiltration in the hydrologic cycle', *Transactions of the American Geophysical Union*, **14**, 446–60.

Hoskins, W. G. (1976) *The making of the English landscape*. London: Hodder and Stoughton.

Howe, G. M., Slaymaker, H. O. and Harding, D. M. (1966) 'Flood hazard in mid-Wales', *Nature*, **212**, 584–5.

Howe, G. M., Slaymaker, H. O. and Harding, D. M. (1967) 'Some aspects of the flood hydrology of the upper catchments of the Severn and Wye', *Transactions, Institute of British Geographers*, **41**, 33–58.

Hughes, G. P. and Redford, R. A. (1952) 'Controlled grazing for beef production', *Agriculture*, **58**, 505–9.

Hunt, E. G. and Bischoff, A. I. (1960) 'Inimical effects on wildlife of periodic DDD application to Clear Lake', *California Fish and Game*, **4**, 91–106.

Hunter, R. F. (1964) 'Home range behaviour in hill sheep', in *Grazing in terrestrial and marine environments* (D. J. Crisp, ed.). Bangor: British Ecological Society, 155–72.

Hutchinson, J. N. (1980) 'The record of peat wastage in the East Anglian fenlands at Holme Post 1848–1978 AD', *Journal of Ecology*, **68**, 229–49.

IAHS (1980) *The influence of man on the hydrological regime with special reference to representative and experimental basins. Proceedings of the Helsinki Symposium, International Association of Hydrological Sciences, 23–26 June 1980*. Dorking: IAHS Publication No. 130.

ICBP (1981) *Important bird areas in the European Community*. Cambridge: International Council for Bird Preservation, Report prepared for the Environment and Consumer Protection Service of the Commission of the European Communities.

Ibn-Al Awan (n.d.) *Le livre de l'agriculture* (Kitab al Felahah) (1864 edn, transl. J. J. Clement-Mullet).

Imeson, A. C. (1971) 'Heather burning and soil erosion in the North York Moors', *Journal of Applied Ecology*, **8**, 537–42.

Jablonska-Ceglarek, R. (1976) 'Effectiveness of irrigation on cabbage and celeriac given different rates of fertilisers and farmyard manure' (in Polish). *Biuletyn Warzywriczy*, **19**, 143–55.

Jarvis, M. G. (1973) *Soils of the Wantage and Abingdon district*. Harpenden: Soil Survey (Memoirs of the Soil Survey of Great Britain).

Jefferies, D. J. and Pendlebury, J. B. (1968) 'Population fluctuations of stoats, weasels and hedgehogs'. *Journal of Zoology, London*, **156**, 513–49.

Jefferies, D. J., Stainsby, B. and French, M. C. (1973) 'The ecology of small mammals in arable fields drilled with winter wheat and the increase in their dieldrin and mercury levels', *Journal of Zoology*, **171**, 513–39.

Jeffreys, H. (1917) 'On the vegetation of four Durham Coal Measure fells. III. Water supply as an ecological factor', *Journal of Ecology*, **5**, 129–55.

Jenkins, D. (1980) 'Ecology of otters in northern Scotland. I. Otter (*Lutra lutra*) breeding and dispersion in mid-Deeside, Aberdeenshire in 1974–79', *Journal of Animal Ecology*, **49**,713–35.

Jenkins, D. and Watson, A. (1967) 'Population control in Red Grouse and Rock Ptarmigan in Scotland', *Finnish Game Research*, **30**, 121–41.

Jenkins, J. E. E. (1973) 'An evaluation of the control of cereal leaf diseases by fungicides in England and Wales', in *Proceedings of the 7th British Insecticide and Fungicide Conference, Brighton*, **3**, 781–90.

Jenkinson, D. S. (1965) 'Studies on the decomposition of plant material in soil, I', *Journal of Soil Science*, **16**, 104–15.

Jenkinson, D. S. (1968) 'Studies on the decomposition of plant material in soil.

III. The distribution of labelled and unlabelled carbon in soil incubated with ^{14}C-labelled ryegrass', *Journal of Soil Science*, **19**, 25–39.

Jenkinson, D. S. (1971) 'The accumulation of organic matter in soil left uncultivated', *Report, Rothamsted Experimental Station for 1970*, Part **2**, 113–47.

Jenkinson, D. S. and Johnston, A. E. (1977) 'Soil organic matter in the Hoosfield continuous barley experiment', *Report of the Rothamsted Experimental Station for 1976*, Part **2**, 87–102.

Jenkinson, D. S. and Rayner, J. H. (1977) 'The turnover of soil organic matter in some of the Rothamsted classical experiments', *Soil Science*, **123**, 298–305.

Jensen, C. R., Letey, J. and Stolzy, L. H. (1964) 'Labelled oxygen: transport through growing corn roots', *Science*, **144**, 550–2.

Jensen, N. F. (1978) 'Limits to growth in world food production', *Science*, **201**, 317–20.

Jewiss, O. R. (1972) 'Vegetative growth' in *Grasses and legumes in the British economy* (C. R. W. Spedding and E. C. Diekmahns, eds), Farnham Royal:Commonwealth Agricultural Bureaux, Commonwealth Bureau of Pastures and Field Crops, Bulletin 49.

Johnson, C. B., Mannering, J. V. and Moldenhauer, W. C. (1979) 'Influence of surface roughness and clod stability on soil and water loss', *Soil Science Society of America, Journal*, **43**, 772–7.

Johnston, A. E. (1972) 'The effects of leys and arable cropping systems on the amounts of soil organic matter in the Rothamsted and Woburn ley arable experiments', *Report of the Rothamsted Experimental Station for 1971*, Part **2**, 131–59.

Johnston, A. E. (1973) 'The effects of ley and arable cropping systems on the amount of soil organic matter in the Rothamsted and Woburn ley arable experiments', *Report of the Rothamsted Experimental Station for 1972*, Part **2**, 131–67.

Johnston, A. E. and Poulton, P. R. (1977) 'Yields on the exhaustion land and changes in the NPK contents of the soils due to cropping and manuring, 1852–1975', *Report of the Rothamsted Experimental Station for 1976*, Part **2**, 53–85.

Johnstone, J. (1797) *An account of the mode of draining land according to the system practised by Mr J. Elkington*. Edinburgh.

Jones, G. E. (1965) 'Diffusion of technological change', in *Readings in resource management and conservation* (I. Burton and R. W. Kates, eds). Chicago: University of Chicago Press, 493–501.

Jones, H. G. and Kirby, E. J. H. (1977) 'Effects of manipulation of number of tillers and water supply on grain yield in barley', *Journal of Agricultural Science*, Cambridge, **88**, 391–7.

Jones, J. A. A. (1981) *The nature of soil piping: a review of research*. Norwich: Geo Books, BGRG Research Monograph, 3.

Jones, M. E. (1976) 'Topographic climates: soils, slopes and vegetation', in *Climate of the British Isles* (T. J. Chandler and S. Gregory, eds). London: Longman, 288–306.

Jones, R. J. A. (1979) 'Soils of the western Midlands grouped according to ease of cultivation', in *Soil survey applications* (M. G. Jarvis and D. Mackney, eds). Harpenden: Soil Survey of England and Wales, Technical Monograph 13, 24–42.

Kaiser, W. G. (1930) 'Recent progress in agricultural engineering', *Agricultural Engineering* **11**, 231–4.

Keeney, D. R. and Bremner, J. M. (1964) 'Effect of cultivation on the nitrogen distribution in soils', *Soil Science Society of America, Proceedings*, **28**, 653–6.

Keith, J. O., Wood, L. A. and Hunt, E. G. (1970) 'Reproductive failure in Brown Pelican on the Pacific Coast', *Transactions, North American Wildlife and Natural Resources Conference*, **35**, 56–64.

Keller, W. D. and Smith, G. E. (1967) *Ground-water contamination by dissolved nitrate*. Columbia, Missouri: University of Missouri, Geological Society of America, Special Paper No. 90.

Kellett, A. J. (1978) *Poaching of grassland and the role of drainage*. London: MAFF, Field Drainage Experimental Unit, Technical Report 78/1.

Kells, J. J., Rieck, C. E., Blevins, R. L. and Muir, W. M. (1980) 'Atrazine dissipation as affected by surface pH and tillage', *Weed Science*, **28**, 101–4.

Kemp, A., Guerink, J. H., Malestein, A. and van't Klooster, A. Th. (1978) 'Grassland production and nitrate poisoning in cattle', *7th General Meeting of the European Grassland Federation, Gent*, **9**, 1–14.

Kenworthy, J. B. (1964) *A study of the changes in plant and soil nutrients associated with moor burning and grazing*. Unpublished PhD thesis, University of St Andrews.

Kerridge, E. (1967) *The agricultural revolution*. London: George Allen and Unwin.

Kerry County Committee of Agriculture (1972) *County Kerry agricultural resource survey*. Tralee: County Kerry Committee of Agriculture.

Ketcheson, J. W. and Webber, L. R. (1978) 'Effects of soil erosion on yield of corn', *Canadian Journal of Soil Science*, **58**, 459–63.

Keuren, R. W. van, McGuinness, J. L. and Chichester, F. W. (1979) 'Hydrology and chemical quality of flow from small pastured watersheds', *Journal of Environmental Quality*, **8**, 162–6.

Kinako, P. D. S. and Gimingham, C. H. (1980) 'Heather burning and soil erosion on upland heaths in Scotland', *Journal of Environmental Management* **10**, 277–84.

King, J. E. (1973) 'Cereal foliar disease surveys', *Proceedings of the 7th British Insecticide and Fungicide Conference, Brighton*, **3**, 771–80.

King, K. L. and Hutchinson, K. J. (1980) 'Effects of superphosphate and stocking intensity on grassland microarthropods', *Journal of Applied Ecology*, **17**, 581–91.

Kirby, E. J. M. (1967) 'The effect of plant density upon the growth and yield of barley', *Journal of Agricultural Science, Cambridge*, **68**, 318–24.

Kirby, E. J. M. (1969) 'The effect of sowing date and plant density on barley', *Annals of Applied Biology*, **63**, 513–21.

Kirby, E. J. M. (1973) 'The control of leaf and ear size in barley', *Journal of Experimental Botany*, **24**, 567–78.

Kirby, E. J. M. and Faris, D. E. (1972) 'The effect of plant density on tiller growth and morphology in barley', *Journal of Agricultural Science, Cambridge*, **78**, 281–8.

Kirby, E. J. M. and Jones, H. E. (1977) 'The relations between main shoots and tillers in barley plants', *Journal of Agricultural Science, Cambridge*, **88**, 381–9.

Kirkham, D. (1950) 'Seepage into ditches in the case of a plant water table and an impervious substratum', *Transactions, American Geophysical Union*, **31**, 425–30.

Kittredge, J. (1973) *Forest influences*. New York: Dover Publications.

Klapp, E. (1971) *Wiesen und Weiden*. Berlin: Paul Paray.

Knight, C. G. and Richard, T. J. (1971) 'Perception and ethnography in SW Kansas', *Proceedings, Association of American Geographers*, **3**, 96–100.

Koelliker, J. K., Miner, J. R., Beer, C. E. and Hazen, T. E. (1971) 'Treatment of livestock-lagoon effluent by soil filtration', in *Livestock water management and pollution abatement*. St Joseph, Michigan: American Society of Agricultural Engineers, 329–33.

Koike, H, (1961) 'The effects of fumigants on nitrate production in soil', *Soil Science Society of America, Proceedings*, **25**, 204–6.

Kolb, H. H. and Hewson, R. (1980) 'A study of fox populations in Scotland from 1971 to 1976', *Journal of Applied Ecology*, **17**, 7–19.

Kolenbrander, G. J. (1972) 'Eutrophication from agriculture with special reference to fertilisers and animal waste', *FAO Soils Bulletin*, **1**, 305–27.

Kononova, M. M. (1966) *Soil organic matter*. New York: Pergamon.

Kontorshchikov, A. S. and Eremina, K. A. (1963) 'Interception of precipitation by spring wheat during the growing season', *Soviet Hydrology*, **2**, 400–9.

Kruuk, H., Parish, T., Brown, C. A. J. and Carrera, J. (1979) 'The use of pasture by the European badger (*Meles meles*)', *Journal of Applied Ecology*, **16**, 453–9.

Kulkarni, B. K. and Savant, N. K. (1977) 'Effect of soil compaction on root-cation exchange capacity of crop plants', *Plant and Soil*, **48**, 269–78.

Kuwatsuka, S. and Igarishi, M. (1975) 'Degradation of PCP in soils. II. The relationship between the degradation of PCP and the properties of soils, and the identification of the degradation products of PCP', *Soil Science and Plant Nutrition*, **21**, 405–14.

Lack, D. (1935) 'The breeding bird population of British heaths and moorland', *Journal of Animal Ecology*, **4**, 43–51.

Laflen, J. M. and Moldenhauer, W. C. (1979) 'Soil and water losses from corn-soybean rotations', *Soil Science Society of America, Journal*, **43**, 1213–15.

Laflen, J. M., Bauer, J. L., Hartwig, R. O., Buchele, W. F. and Johnson, H. P. (1978) 'Soil and water loss for conservation tillage systems', *American Society of Agricultural Engineers, Transactions*, **21**, 881–5.

Lal, R. (1976) 'Soil erosion on Alfisols in Western Nigeria. IV. Nutrient element losses in runoff and eroded sediments', *Geoderma*, **16**, 403–17.

Lauenroth, W. K. (1979) 'Grassland primary production: North American grasslands in perspective', in *Perspectives in grassland ecology* (N. R. French, ed.). New York: Springer-Verlag, 32, 3–24.

Lavake, D. E. and Wiese, A. F. (1979) 'Influence of weed growth and tillage interval during fallow on water storage, soil nitrates and yield', *Soil Science Society of America, Journal*, **43**, 565–9.

Lazenby, A. and Down, K. M. (1982) 'Realizing the potential of British grasslands: some problems and perspectives', *Applied Geography*, **2**, 171–88.

Leach, G. and Slesser, M. (1973) *Energy equivalents of network inputs to agriculture*. Glasgow: Strathclyde University.

Leafe, E. L. (1978) 'Physiological, environmental and management factors of importance to maximum yield of the grass crop', in *Maximising yields of crops* (J. K. R. Gasser and B. Wilkinson, eds). London: HMSO, 37–49.

Lee, G. F., Rast, W. and Jones, R. A. (1978) 'Eutrophication of water bodies: insights for an age-old problem', *Environmental Science and Technology*, **12**, 910–18.

Legg, B. J., Day, W., Lawlor, D. W. and Parkinson, K. J. (1979) 'The effects of drought upon barley growth: models and measurements showing the relative importance of leaf area and photosynthetic rate', *Journal of Agricultural Science, Cambridge*, **92**, 703–16.

Leith, H. and Box, E. (1972) 'Evapotranspiration and primary productivity: C. W. Thornthwaite memorial model', *Publications in Climatology*, **25** (3), 36–44.

Leland, John (1710–12) *The itinerary of John Leland the Antiquary*, published by T. Hearne. To which is prefixed Mr. Leland's New-Year's Gift. Oxford: The Theater, 9 vols.

Leonard, P. L. and Stokes, C. (1979) 'Landscape and agricultural change', in *Conservation and agriculture* (J. Davidson and R. J. Lloyd, eds). Chichester: Wiley, 121–35.

Lewin, J. and Lomas, J. (1974) 'A comparison of statistical and soil moisture techniques in long-term studies of wheat yield performance under semi-arid conditions', *Journal of Applied Ecology*, **11**, 1081–90.

Lichtenstein, E. P. (1958) 'Movement of insecticides in soils under leaching and non-leaching conditions', *Journal of Economic Entomology*, **51**, 380–3.

Liebig, J. F. von (1842) *Animal chemistry, or organic chemistry in its application to physiology and pathology* (1964 edn, W. Gregory, ed.). New York: Johnson.

Likens, G. E. (1970) 'Effects of forest cutting and herbicide treatment on nutrient budgets in the Hubbard Brook watershed-ecosystem', *Ecological Monographs*, **40**, 23–47.

Likens, G. E., Bormann, E. H., Johnson, N.M. and Pierce, R. S. (1967) 'The Ca, Mg, K and Na budgets for a small forested catchment', *Ecology*, **48**, 772–85.

Likens, G. E., Bormann, F. H., Pierce, R. S., Eaton, J. S. and Johnson, N. M. (1977) *Biochemistry of a forested ecosystem*. New York: Springer-Verlag.

Linton, D. L. (1968) 'The assessment of scenery as a natural resource', *Scottish Geographical Magazine*, **84**, 219–38.

Lloyd, R. J. and Wibberley, G. P. (1979) 'Agricultural change', in *Conservation and agriculture* (J. Davidson and R. J. Lloyd, eds). Chichester: Wiley, 3–22.

Lockhart, J. A. R. and Wiseman, A, J. L. (1978) *Introduction to crop husbandry*. Oxford: Pergamon.

Lockie, J. D. and Ratcliffe, D. A. (1964) 'Insecticides and Scottish Golden Eagles', *British Birds*, **57**, 89–101.

Lockie, J. D., Ratcliffe, D. A. and Balharry, R. (1969) 'Breeding success and organochlorine residues in Golden Eagles in west Scotland', *Journal of Applied Ecology*, **6**, 381–9.

Loehr, R. (1977) *Pollution control for agriculture*. New York: Academic Press.

Lomas, J., Schlesinger, E. and Lewin, J. (1974) 'Effects of environmental and crop factors on the evapotranspiration rate and water-use efficiency of maize', *Agricultural Meteorology*, **13**, 239–51.

Long, F. L. and Huck, M. G. (1980) 'Nitrate movement under corn and fallow conditions', *Soil Science Society of America, Journal*, **44**, 787–92.

Loos, M. A. (1975) 'Indicator media for micro-organisms degrading chlorinated pesticides', *Canadian Journal of Microbiology*, **21**, 104–7.

Low, A. J. (1954) 'The study of soil structure in the field and the laboratory', *Journal of Soil Science*, **5**, 57–74.

Low, A. J. (1955) 'Improvements in the structural state of soils under leys', *Journal of Soil Science*, **6**, 179–99.

Low, A. J. (1972) 'The effect of cultivation on the structure and other physical characteristics of grassland and arable soils (1945–70)', *Journal of Soil Science*, **23**, 363–80.

Low, A. J. (1973) 'Soil structure and crop yield', *Journal of Soil Science*, **24**, 249–59.

Low, A. J. (1975) 'Ley fertility experiments at Jealott's Hill', in *Soil physical conditions and crop production*. London: HMSO, MAFF Technical Bulletin No. 29, 360–87.

Low, A. J. and Armitage, E. R. (1970) 'The composition of the leachate through cropped and uncropped soils in lysimeters compared with that of rain', *Plant and Soil*, **33**, 393–411.

Low, A. J., Piper, F. J. and Roberts, P. (1963) 'Soil changes in ley-arable experiments', *Journal of Agricultural Science, Cambridge*, **60**, 229–38.

Lowday, J. E. and Wells, T. C. E. (1977) *The management of grassland and heathland in country parks*. Cheltenham: Countryside Commission.

Lull, H. W. (1964) 'Ecological and silvicultural aspects', in *Handbook of applied hydrology* (V. T. Chow, ed.). New York: McGraw Hill, Section 6.

Lund, J. W. G (1971) 'Eutrophication', in *The scientific management of animal and plant communities for conservation* (E. Duffy and A. S. Watt, eds). 11th Symposium of the British Ecological Society, University of East Anglia, Norwich, 7–9 July 1970, 225–40.

Luthin, J. N. (1959) 'The falling water table in tile drainage. II. Proposed criteria for spacing tile drains', *American Society of Agricultural Engineers, Transactions*, **2**, 44–5.

Lvovitch, M. I. (1958) 'Streamflow formation factors', *International Association of Scientific Hydrology, Publications*, **45**, 122–32.

MAFF (1964a) *The farmer's weather*. London: HMSO, Ministry of Agriculture,

Fisheries and Food, Bulletin No. 165.

MAFF (1964b) 'Soil organic matter: farming at High Mowthorpe', *4th Annual Report of the High Mowthorpe Experimental Husbandry Farm*, Malton. 9–12.

MAFF (1966) *Agricultural land classification*. London: HMSO, Agricultural Land Service, Technical Report No. 11.

MAFF (1967) *Potential transpiration*. London: HMSO, MAFF Technical Bulletin No. 16.

MAFF (1968) *A century of agricultural statistics Great Britain 1866–1966*. London: HMSO.

MAFF (1970a) *Modern farming and the soil*. London: HMSO, Agricultural Advisory Council.

MAFF (1970b) *Beneficial insects and mites*. London: HMSO, MAFF Bulletin No. 20.

MAFF (1974) *Bracken and its control*. London: HMSO, MAFF Advisory Leaflet No. 190.

MAFF (1975) *Fenland notebook*. March, Cambs.: Ministry of Agriculture, Fisheries and Food.

MAFF (1976) *Agriculture and water quality*, London: HMSO, MAFF Technical Bulletin No. 32.

MAFF (1977) *Gleadthorpe Experimental Husbandry Farm. Sandland irrigation*. Mansfield: Ministry of Agriculture, Fisheries and Food.

MAFF (1978) *Grazing management for lowland sheep*. London: Agricultural Development and Advisory Service, Grassland Practice No. 12, Short Term Leaflet No. 205.

MAFF (1979a) *Possible patterns of agricultural production in the United Kingdom by 1983*. London: HMSO.

MAFF (1979b) *Bovine tuberculosis in badgers, 3rd report*. London: HMSO.

MAFF (1980a) *Output and utilisation of farm produce in the UK*. London: HMSO.

MAFF (1980b) *Minimum requirements for the structural design of bunker silos for forage*. Leeds: Agricultural Development and Advisory Service, Farm Buildings Group.

MAFF (1981) *List of approved products and their uses for farmers and growers, 1981*. London: HMSO.

Mabey, R. (1980) *The common ground: a place for nature in Britain's future?* London: Hutchinson in association with the Nature Conservancy Council.

McCauley, G. N., Stone, J. F. and Chin Choy, E. W. (1978) 'Evapotranspiration reduction by field geometry effects in peanuts and grain sorghum' *Agricultural Meteorology*, **19**, 295–304.

McDonald, A. T. (1973) *Some views on the effect of peat drainage*. Leeds: University of Leeds, Department of Geography, Occasional Paper No. 40.

MacDonald, S. M., Mason, C. F. and Coghill, I. S. (1978) 'The otter and its conservation on the River Teme catchments', *Journal of Applied Ecology*, **15**, 373–84.

McEwan, F. L. and Stephenson, G. R. (1979) *The use and significance of pesticides in the environment*. Chichester: Wiley.

McGowan, M. and Williams, J. B. (1980a) 'The water balance of an agricultural catchment. I. Estimation of evaporation from soil water records', *Journal of Soil Science*, **31**, 217–30.

McGowan, M. and Williams, J. B. (1980b) 'The water balance of an agricultural catchment. II. Crop evaporation: seasonal and soil factors', *Journal of Soil Science*, **31**, 231–44.

McGowan, M., Williams, J. B. and Monteith, J. L. (1980) 'The water balance of an agricultural catchment. III. The water balance', *Journal of Soil Science*, **31**, 245–62.

MacIntyre, D. S. (1958) 'Soil splash and the formation of surface crusts by raindrop impact', *Soil Science*, **85**, 261–6.

Mackney, D. (1969) 'The agronomic significance of soil mapping units', in *The soil ecosystem* (J. G. Sheals, ed.). London: Systematics Association, **8**, 55–62.

McLaren, J. S. (1981) 'Field studies on the growth and development of winter wheat', *Journal of Agricultural Science, Cambridge*, **97**, 85–97.

MacLusky, D. S. (1960) 'Some estimates of the areas of pasture fouled by the excreta of dairy cows', *Journal of the British Grassland Society*, **15**, 181–8.

McMeekan, C. P. and Walshe, M. J. (1963) 'The inter-relationships of grazing method and stocking rate on the efficiency of pasture utilization by dairy cattle', *Journal of Agricultural Science, Cambridge*, **61**, 147–66.

McMurphy, W. E. and Anderson, K. K. L. (1965) 'Burning Flint Hills range', *Journal of Range Management*, **18**, 265–9.

Maizlish, N. A., Fritton, D. D. and Kendall, W. (1980) 'Root morphology and early development of maize at varying levels of nitrogen', *Agronomy Journal*, **72**, 25–31.

Maltby, E. (1975) 'Numbers of soil micro-organisms as ecological indicators of changes resulting from moorland reclamation on Exmoor, UK', *Journal of Biogeography*, **2**, 117–36.

Margules, C. and Usher, M. B. (1981) 'Criteria used in assessing wildlife conservation potential: a review', *Biological Conservation*, **21**, 79–109.

Marquiss, M., Newton, I. and Ratcliffe, D. A. (1978) 'The decline of the Raven, *Corvus corax*, in relation to afforestation in southern Scotland and northern England'. *Journal of Applied Ecology* **15**, 129–44.

Marshall, W. (1787) *The rural economy of Norfolk, comprising the management of landed estates and the present practice of husbandry in that county*. London, 2 vols.

Marshall, W. (1790) *The rural economy of the Midland Counties, including the management of the livestock in Leicestershire and its environs, with minutes on agriculture and planting in the district of the Midland Station*. London, 2 vols.

Marston, D. and Hird, C. (1978) 'Effect of stubble management on the structure of black cracking clays', in *Modification of soil structure* (W. W. Emerson, R. D. Bond and A. R. Dexter, eds). Wiley: Chichester, 411–17.

Martel, Y. A. and MacKenzie, A. F. (1980) 'Long-term effects of cultivation and land use on soil quality in Quebec', *Canadian Journal of Soil Science*, **60**, 411–20.

Martin, D. J. (1976) 'Control of bracken', *Botanical Journal of the Linnaean Society*, **73**, 241–6.

Martin, H. (1964) *The scientific principles of crop protection*. London: Edward Arnold.

Martin, J. H. and Leonard, W. H. (1967) *Principles of field crop production*. New York: Macmillan.

Mason, H. J. (1975) 'Farming in 1970', in *Soils of the Ely district* (R. S. Seale, ed). Harpenden: Soil Survey (Memoirs of the Soil Survey of England and Wales) 57–68.

Massee, T. W. and Cary, J. W. (1978) 'Potential for reducing evaporation during summer fallow', *Journal of Soil and Water Conservation*, **33**, 126–9.

Massey, H. F. and Jackson, M. L. (1952) 'Selective erosion of soil fertility constituents', *Soil Science Society of America, Proceedings*, **16**, 353–6.

Mattingly, G. E. G. (1965) 'The influence of intensity and capacity factors on the availability of soil phosphorus', in *Soil phosphorus*. London: HMSO, MAFF Technical Bulletin No. 13, 1–9.

Mattingly, G. E. G., Chater, M. and Johnston, A. E. (1975) 'Experiments made on Stackyard Field, Woburn, 1867–1974. III. Effects of NPK fertilisers and farmyard manure on soil carbon, nitrogen and organic phosphorus', *Report, Rothamsted Experimental Station for 1974*, Part 2, 61–77.

Mattingly, G. E. G., Russell, R. D. and Jephcott, B. M. (1963) 'Experiments on cumulative dressings of fertilisers on calcareous soils in south-west England', *Journal of Science, Food and Agriculture*, **14**, 629–37.

Mech, S. J. (1949) 'Effect of slope and length of run on erosion under irrigation', *Agricultural Engineering*, **30**, 379–83.

Mech, S. J. and Smith, D. M. (1967) 'Water erosion under irrigation', in *Irrigation of agricultural lands* (R. M. Hagan, H. R. Haise, and T. W. Edminster, eds). Madison: American Society of Agronomy, II, 950–63.

Meikle, R. W., Youngson, C. R., Hedlund, R. T., Hamauer, J. W. and Addington, W. W. (1973) 'Measurement and prediction of picloram disappearance rates from soil', *Weed Science*, **21**, 549–55.

Mellanby, K. (1981) *Farming and wildlife*. London: Collins.

Miller, G. R. (1964) 'The management of heather moorlands', in *Land use in the Scottish Highlands. Advancement of Science, London*, **21** (9), 163–9.

Miller, G. R. and Miles, J. (1970) 'Regeneration of heather (*Calluna vulgaris* (L) Hull) at different ages and seasons in north-east Scotland', *Journal of Applied Ecology*, **7**, 51–60.

Minson, D. J., Harris, C. E., Raymond, W. F. and Milford, R. (1964) 'The digestibility and voluntary intake of S. 22 and H. I. ryegrass, S. 170 tall fescue, S. 28 timothy, S. 215 meadow grass and germinal cocksfoot', *Journal of the British Grassland Society*, **19**, 298–305.

Monteith, J. L. (1973) *Principles of environmental physics*. London: Edward Arnold.

Monteith, J. L. (1977) 'Climate and the efficiency of crop production in Britain', *Philosophical Transactions of the Royal Society of London*, **B 281**, 277–94.

Monteith, J. L. and Szeicz, G. (1961) 'The radiation balance of bare soil and vegetation', *Quarterly Journal of the Royal Meteorological Society*, **87**, 159–70.

Monteith, J. L., Szeicz, G. and Yabuki, K. (1964) 'Crop photosynthesis and the flux of carbon dioxide below the canopy', *Journal of Applied Ecology*, **1**, 321–37.

Moore, N. W. (1957) 'The buzzard in Britain', *British Birds*, **50**, 173–97.

Moore, N. W. (1962) 'The heaths of Dorset and their conservation', *Journal of Ecology*, **50**, 369–91.

Moore, N. W. (1977a) 'Arable land', in *Conservation and agriculture* (J. Davidson and R. J. Lloyd, eds). Chichester: Wiley, 23–43.

Moore, N. W. (1977b) 'The future prospect for wildlife', in *Ecological effects of pesticides* (F. H. Perring and K. Mellanby, eds). London: Academic Press, Linnaean Society Symposium Series, No. 5, 175–80.

Moore, N. W. (1980) '*Lestes dryas* Kirby – a declining species of dragonfly (Odonata) in need of conservation: notes on its status and habitat in England and Ireland', *Biological Conservation*, **17**, 143–8.

Moore, P. D. and Bellamy, D. J. (1974) *Peatlands*. London: Elek Science.

Morgan, R. P. C. (1977) *Soil erosion in the United Kingdom: field studies in the Silsoe area, 1973–75*. Silsoe: National College of Agricultural Engineering, Occasional Paper No. 5.

Moriarty, F. (1977) 'Prediction of ecological effects by pesticides', in *Ecological effects of pesticides* (F. H. Perring and K. Mellanby, eds). London: Academic Press, Linnaean Society Symposium Series, No. 5, 165–73.

Morris, M. G. (1979) 'Responses of grassland invertebrates to management by cutting. II. Heteroptera', *Journal of Applied Ecology*, **16**, 417–32.

Morrison, J., Jackson, M. and Sparrow, P. E. (1980) *The response of perennial ryegrass to fertiliser nitrogen in relation to climate and soil*, Hurley: Grassland Research Institute, Technical Report, No. 27.

Moss, M. R. and Strath Davis, L. (1982) 'The potential and actual primary productivity of southern Ontario's agro-ecosystem', *Applied Geography*, **2**, 17–38.

Muirhead, R. H., Gallagher, J. and Burns, K. J. (1974) 'Tuberculosis in wild badgers in Gloucestershire: epidemiology', *Veterinary Record*, **95**, 552–5.

Müller, P. (1976) 'Araealveränderingen von Amphibien und Reptilien in der Bundesrepublik Deutschland', *Schriftenreihe für Vegetationskunde*, **10**, 269–93.

Munro, J. M., Davies, D. A. and Thomas, T. A. (1973) 'Potential pasture production in the uplands of Wales. 3. Soil nutrient resources and limitations', *Journal of the British Grassland Society*, **28**, 247–55.

Munro, P. (1958) 'Irrigation of grassland: the influence of irrigation and nitrogen treatments on the yield and utilization of a riverside meadow', *Journal of the British Grassland Society*, **13**, 213–21.

Muntingh, H. (1982) *Draft opinion on Resolution 1–1021/81 of 15 February 1981 on the crisis in agriculture in the Highlands and Islands of Scotland*. Strasbourg: European Parliament, PE 80. 477.

Murdoch, J. C. (1980) 'The conservation of grass', in *Grass: its production and utilization* (W. Holmes, ed.). Oxford: Blackwell, 174–215.

Murton, R. K. (1971) *Man and birds*. London: Collins.

Murton, R. K. (1977) 'Pesticides and wildlife: current ITE research', in *Some aspects of research on pesticides* (L. Iyengar and P. J. W. Saunders, eds). Swindon: NERC, IRCCOPR Seminar Report No. 3 (Report of a seminar held at the Agricultural Research Council, 25 Oct 1977).

Musgrave, G. W. (1938) 'Field research offers significant new findings', *Soil Conservation*, **3**, 210–14.

Nageli, W. (1946) 'Weitere untersuchungen über die Windverhaltnisse im Bereich von Windschutzst reifen', *Mitteilungen Schweizerischen Anstalt für das Forstliche Versuchswesen* **24**, 659–737.

Nature Conservancy Council (1977) *Nature conservation and agriculture*. London: Nature Conservancy Council.

Nature Conservancy Council (1979) *The changing face of Devon*. Exeter: Devon County Planning Department.

Nature Conservancy Council (1982) *A draft Community list of threatened species of wild flora and vertebrate fauna*, 2 vols. London: NCC (unpublished report prepared for the Commission of the European Communities).

Newbould, C. (1979) 'Wetlands and agriculture, in *Conservation and agriculture* (J. Davidson and R. J. Lloyd, eds). Chichester: Wiley, 59–77.

Newby, H. (1979) *Green and pleasant land? Social change in rural England*. London: Hutchinson.

Newman, J. F. (1976) 'Assessment of the environmental impact of pesticides', *Outlook on Agriculture*, **9**, 9–15.

Newton, I. (1974) 'Changes attributed to pesticides in the nesting success of the sparrowhawk in Britain', *Journal of Applied Ecology*, **11**, 95–101.

Newton, I. (1976) 'Breeding of sparrowhawks (*Accipter nisus*) in different environments'. *Journal of Animal Ecology* **45**, 831–49.

Newton, I. (1979) *Population ecology of raptors*. Berkhampsted: T. and A. D. Poyser.

Newton, I. and Bogan, J. (1978) 'The role of different organo-chlorine compounds in the breeding of British sparrowhawks', *Journal of Applied Ecology*, **15**, 105–16.

Newton, I., Davis, P. E. and Moss, D. (1981) 'Distribution and breeding of Red Kites in relation to land use in Wales', *Journal of Applied Ecology*, **18**, 173–86.

Newton, I., Marquiss, M. and Moss, D. (1979) 'Habitat, female age, organo-chlorine compounds and breeding of European sparrowhawks', *Journal of Applied Ecology*, **16**, 777–93.

Neyra, C. A. and Dobereiner, J. (1977) 'Nitrogen fixation in grasses', *Advances in Agronomy*, **29**, 1–38.

Nichols, M. L. (1929) *Methods of research in soil dynamics*, Birmingham: Alabama Agricultural Experimental Station Bulletin, 229.

Norman, M. J. T. and Green, J. O. (1958) 'The local influence of cattle dung and urine upon the yield and botanical composition of permanent pasture', *Journal of the British Grassland Society*, **13**, 39–45.

Nutman, P. S. and Hearne, R. (1980) 'Persistence of nodule bacteria in soil under long term cereal cultivation', *Report of the Rothamsted Experimental Station for 1979*, Part 2, 71–90.

Nwa, E. U. and Twocock, J. G. (1969) 'Drainage design theory and practice', *Journal of Hydrology*, **9**, 259–76.

Odum, P. E. (1973) *Fundamentals of ecology*. Philadelphia: W. B. Saunders.

Official Journal of the European Communities (1979) 'Council Directive of 2 April 1979 on the conservation of wild birds (79/409/EEC: Article 1–4 and related considerations)', *Official Journal of the European Communities*, **L103**, 25 April 1979.

Open University (1972) *Environment. Block 4. Changing environments; Block 5. Conservation*. Bletchley: Open University Press.

Orr, D. M. (1980) 'Effects of sheep grazing *Astrebla* grassland in central western Queensland. 1. Effects of grazing pressure and livestock distribution', *Australian Journal of Agricultural Research*, **31**, 797–80.

Orwin, C. S. and Sellick, R. J. (1970) *The reclamation of Exmoor Ferest*. Newton Abbot: David and Charles.

Owen, M. and Thomas, G. J. (1979) 'The feeding ecology and conservation of wigeon wintering at the Ouse Washes, England', *Journal of Applied Ecology*, **1**, 795–809.

Owens, M. (1970) 'Nutrient balances in rivers', *Water Treatment Examiner*, **19**, 239–47.

Owens, M., Garland, J. H. N., Hart, I. C. and Wood, G. (1972) 'Nutrient budgets in rivers', *Symposium of the Zoological Society of London*, **29**, 21–40.

Page, C. N. (1976) 'The taxonomy and phytogeography of bracken: a review', *Botanical Journal of the Linnaean Society*, **73**, 1–34.

Page, J. B. (1972) 'Arable crop rotations', *Journal of The Royal Agricultural Society of England*, **133**, 98–105.

Parfenov, V. I. (1979) 'Contemporary anthropogenic kinetics of the flora and vegetation of the Pripyat forest zone', *Botanicheskii Zhurnal*, **64**, 1377–89.

Parks, C. C. (1980) *Ecology and environmental management*. Folkestone: Dawson/Westview Press.

Parry, M. L., Bruce A. and Harkness, C. E. (1982a) 'Changes in the extent of moorland and roughland in the North York Moors', University of Birmingham; *Surveys of Moorland and Roughland Change No. 5*.

Parry, M. L., Bruce, A. and Harkness, C. E. (1982b) 'Changes in the extent of moorland and roughland in the Brecon Beacans National Park', University of Birmingham, *Surveys of Moorland and Roughland Change No. 7*.

Parry, M. L., Bruce, A. and Harkness, C. E. (1982c) 'Changes in the extent of moorland and roughland on Dartmoor. A supplementary survey to 1979', University of Birmingham, *Surveys of Moorland and Roughland Change No. 11*.

Parton, W. J., Lauenroth, W. K. and Smith, F. M. (1981) 'Water loss from a shortgrass steppe', *Agricultural Meteorology*, **24**, 97–109.

Pashby, P. S. (1976) 'Pink-footed Goose fatalities at the Humber Wildfowl refuge', *Naturalist*, **1976**, 15–18.

Patterson, H. D. (1960) 'An experiment on the effects of straw ploughed in or composted on a three-course rotation of crops', *Journal of Agricultural Science, Cambridge*, **54**, 222–30.

Patto, P. M., Clement, C. R. and Forbes, T. J. (1978) *Grassland poaching in England and Wales*. Hurley: Grassland Research Institute/ADAS, Joint Permanent Pasture Group, Permanent Pasture Studies, 2.

Patton, D. L. H. and Frame, J. (1981) 'The effect of grazing in winter by wild geese on improved grassland in west Scotland', *Journal of Applied Ecology*, **18**(1), 311–25.

Peacock, J. M. (1975a) 'Temperature and leaf growth in *Lolium perenne*. I. The thermal microclimate: its measurement and relation to crop growth', *Journal of Applied Ecology*, **12**, 99–114.

Peacock, J. M. (1975b) 'Temperature and leaf growth in *Lolium perenne*. II. The site of temperature perception', *Journal of Applied Ecology*, **12**, 115–24.

Penman, H. L. (1962) 'Woburn irrigation. III. Results for rotation crops', *Journal of Agricultural Science, Cambridge*, **58**, 365–79.

Penman, H. L. (1963) *Vegetation and hydrology*. Harpenden: Commonwealth Agricultural Bureau, Commonwealth Bureau of Soils, Technical Communication No. 53.

Penny, A. and Jenkyn, J. K. (1975) 'Results from two experiments with winter wheat, comparing top dressings of a liquid N-fertiliser either alone, or with added herbicide or mildew fungicide or both, and of 'Nitro-Chalk' without or with herbicide or fungicide or both', *Journal of Agricultural Science, Cambridge*, **85,** 533–9.

Pereira, H. C. (1975) 'Agricultural science and the traditions of tillage', *Outlook on Agriculture*, **6**, 211–12.

Perring, F. H. (1970) 'The last seventy years', in *The flora of a changing Britain* (F. H. Perring, ed.). Report of the Botanical Society of the British Isles, No. 11.

Perring, F. H. and Farrell, L. (1977) *Red data book*. Morges, Switzerland: International Union for Conservation of Nature and Natural Resources.

Perring, F. H. and Mellanby, K. (eds) (1977) *Ecological effects of pesticides*. London: Academic Press, Linnaean Society Symposium Series, No. 5.

Peterken, G. F. and Harding, P. T. (1975) 'Woodland conservation in eastern England: comparing the effects of changes in three study areas since 1946', *Biological Conservation*, **8**, 279–98.

Petersen, R. G., Lucas, H. L. and Woodhouse, W. W. (1956) 'The distribution of excreta by freely grazing cattle and its effect on pasture fertility. 2. Effect of returned excreta on the residual concentration of some fertiliser elements', *Agronomic Journal*, **48**, 444–9.

Petit, G. and Knoepffler, P. H. (1959) 'Sur la disparition des amphibiens et des reptiles méditerranéens', *Proceedings, 7th Technical Meeting of IUCN, Athens*. Gland: IUCN, 50–3.

Philips, G. L. (1977) 'The mineral nutrient levels in three Norfolk Broads differing in trophic status, and an annual mineral content budget for one of them', *Journal of Ecology*, **65**, 447–74.

Phillips, M. C. (1978) 'Survey in the use of minimum cultivation in Warwickshire', *ADAS Quarterly Review*, **28**, 13–19.

Pidgeon, J. and Soane, D. B. (1978) 'Soil structure and strength relations following tillage, zero tillage and wheel traffic in Scotland', in *Modification of soil structure* (W. W. Emerson, R. D. Bond and A. R. Dexter, eds). Chichester; Wiley.

Piskin, R. (1973) 'Evaluation of nitrite content of groundwater in Hall County, Nebraska', *Groundwater*, **11**, 4–13.

Pollard, E., Hooper, M. D. and Moore, N. W. (1974) *Hedges*. London: Collins.

Polous, G. P. (1977) 'Accumulation and translocation of nitrogen in winter wheat plants under the influence of a foliar spray of nitrogen', *Nauchnye Trudy, Stavropol'skii Sel'skokhozyaistvennyi Institut*, **40**, 30–3.

Porchester, Lord (1977) *A study of Exmoor*. London: HMSO (Department of the Environment/Ministry of Agriculture, Fisheries and Food).

Potts, G. R. (1971) 'Agriculture and the survival of the partridge', *Outlook on Agriculture*, **6**, 267–71.

Powell, J., Zawi, H. T., Crockett, J. J., Croy, L. I. and Morrison, R. D. (1979) *Central Oklahoma rangeland response to fire, fertilisation and grazing by sheep*. Stillwater, Oklahoma: Oklahoma State University, Bulletin of the Agricultural Experiment Station, B-744.

Power, J. F. (1981) 'Nitrogen in the cultivated ecosystem', in *Terrestrial nitrogen cycles* (F. E. Clark and T. Rosswall, eds). Stockholm: Swedish Natural Science Research Council, Ecological Bulletins, No. 33.

Prater, A. J. (1981) *Estuary birds of Britain and Ireland*. Calton, Staffs: T. and A. D. Poyser.

Pringle, J. and Coutts, J. R. H. (1956) 'The effect of grasses on aggregation in light soils', *Journal of the British Grassland Society*, **11**, 185–9.

RSPB (1981) 'Annual report and accounts, 1980–81', *Birds*, **8**(7) iv.

Rackham, O. (1976) *Trees and woodland in the British landscape*. London: Dent.

Radley, J. and Simms, C. (1967) 'Wind erosion in east Yorkshire', *Nature*, **216**, 20–2.

Raghavan, G. S. V., McKyes, E., Baxter, R. and Gendron, G. (1979) 'Traffic-soil-plant (maize) relations', *Journal of Terramechanics*, **16**, 181–9.

Rahman, A. (1978) 'Effect of climatic and edaphic factors on the persistence and movement of ethofumesate', *Proceedings of the 14th British Weed Control Conference, Brighton*, **2**, 557–63.

Ralph, N. (1982) *Assessment of ancient land use in abandoned settlements and fields: a study of prehistoric and medieval land use and its influence upon soil properties on Holne Moor, Dartmoor, England*. Unpublished PhD thesis, University of Sheffield.

Rasmussen, P. E., Allmara, R. R., Rohde, C. R. and Roager, N. C. (1980) 'Crop residue influences on soil carbon and nitrogen in a wheat-fallow system', *Soil Science Society of America, Journal*, **44**, 596–600.

Ratcliffe, D. A. (1962) 'Breeding densities in the Peregrine *Falco peregrinius* and Raven *Corvus corax*', *Ibis*, **104**, 13–39.

Ratcliffe, D. A. (1963) 'The status of the Peregrine in Great Britain', *Bird Study*, **10**, 56–90.

Ratcliffe, D. A. (1965) 'The Peregrine situation in Great Britain', 1963–64', *Bird Study*, **12**, 66–82.

Ratcliffe, D. A. (1967a) 'Decrease in eggshell weight in certain birds of prey', *Nature*, **215**, 208–10.

Ratcliffe, D. A. (1967b) 'The Peregrine situation in Great Britain, 1965–66', *Bird Study*, **14**, 238–46.

Ratcliffe, D. A. (1970) 'Changes attributable to pesticides in egg breakage frequency and eggshell thickness in some British birds', *Journal of Applied Ecology*, **7**, 67–115.

Ratcliffe, D. A. (1977) *A nature conservation review*. Cambridge: Cambridge University Press (2 vols.).

Ratcliffe, F. N. (1965) 'Biological control', *Australian Journal of Science*, **28**, 237–40.

Rawes, M. (1981) 'Further results of excluding sheep from high-level grasslands in the north Pennines', *Journal of Ecology*, **69**, 651–69.

Rawes, M. (1983) 'Changes in two high altitude blanket bogs after the cessation of sheep grazing', *Journal of Ecology* **71**, 219–35.

Rawes, M. and Hobbs, R. (1979) 'Management of semi-natural blanket bog in the northern Pennines', *Journal of Ecology*, **67**, 789–807.

Rehm, G. V., Sorensen, R. C. and Wiese, R. A. (1981) 'Application of phosphorus, potassium and zinc to corn grown for grass or silage: early growth and yield', *Soil Science Society of America, Journal*, **45**, 523–8.

Reid, D. (1966) 'Studies on the cutting management of grass clover swards. 4. The effects of close and lax cutting on the yield of herbage from swards cut at different frequencies', *Journal of Agricultural Science, Cambridge*, **66**, 101–6.

Reid, I. (1979) 'Seasonal changes in microtopography and surface depression storage in soils', in *Man's impact on the hydrological cycle in the United Kingdom* (G. E. Hollis, ed.). Norwich: Geo Abstracts, 19–30.

Reid, R. L., June, G. A. and Murray, S. J. (1966) 'Nitrogen fertilization in relation to the palatability and nutritive value of orchard grass (*Dactylis glomerata*)', *Journal of Animal Science*, **25**, 636–45.

Richards, I. R. (1977) 'Influence of soil and sward characteristics on the response to nitrogen', *Proceedings of the International Meeting on Annual Production from Temperate Grassland, Dublin*, 45–9.

Richardson, E. B. and Bovey, O. R. W. (1979) 'Hydrological effects of brush control on Texas rangelands', *American Society of Agricultural Engineers, Transactions*, **22**, 315–19.

Richardson, S. J. and Smith, J. (1977) 'Peat wastage in the East Anglian fens', *Journal of Soil Science*, **28**, 485–9.

Rickard, D. S. (1973) 'Plant and soil water: information required in agronomic trials', *Proceedings of the 1973 soil plant water symposium, Wellington, New Zealand, DSIR Information Series*, No. 96, 107–12.

Ridley, A. O. and Hedlin, R. A. (1980) 'Crop yields and soil management on the Canadian Prairies, past and present', *Canadian Journal of Soil Science*, **60**, 393–402.

Riggio, S. (1976) 'Il discoglosso in Sicilia', in *SOS Fauna – Animali in pericolo in Italia* (F. Pedrotti, ed). Camerino: WWF, 417–64.

Riggs, T. J., Hanson, P. R., Start, N. D., Miles, D. M., Morgan, C. L. and Ford, M. A. (1981) 'Comparison of spring barley varieties grown in England and Wales between 1880 and 1980', *Journal of Agricultural Science, Cambridge*, **97**, 599–610.

Room Singh and Ghildyal, B. P. (1977) 'Influence of soil edaphic factors and their critical limits on seedling emergence of corn (*Zea mays* L.)', *Plant and Soil*, **47**, 125–36.

Rorison, I. H. (1971) 'The use of nutrients in the control of the floristic composition of grassland', in *The scientific management of animal and plant communities for conservation* (E. Duffey and A. S. Watt, eds). 11th Symposium of the British Ecological Scoiety, University of East Anglia, Norwich, 7–9 July 1970, 65–77.

Rose, C. W., Begg, J. E., Byrne, G. F., Torsell, B. W. R. and Goncz, J. H. (1972) 'A simulation model of growth-field environment relationships for Townsville stylo (*Stylosanthes humilis* HBK) pasture', *Agricultural Meteorology*, **10**, 161–83.

Rose, D. A. (1968) 'Water movement in dry soils. 1. Physical factors affecting sorption of water by dry soil', *Journal of Soil Science*, **19**, 81–93.

Rose, F. (1974) 'The epiphytes of oak', in *The British oak* (M. G. Morris and F. H. Perring, eds). Published for the BSBI by Edward Classey.

Rosenberg, N. J. (1974) *Microclimate: the biological environment*. New York: Wiley.

Rothschild, M. (1963) 'A rise in the flea index on the hare (*Lepus europaeus*) with relevant notes on the fox (*Vulpes vulpes*) and the woodpigeon (*Colomba palumbus*) at Ashton, Peterborough', *Journal of Zoology, London*, **140**, 341–6.

Rowley, T. (1972) *The Shropshire landscape*. London: Hodder and Stoughton.

Russell, E. J. (1966) *A history of agricultural science in Great Britain*. London: George Allen and Unwin.

Russell, E. W. (1971a) 'Soil structure: its maintenance and improvement', *Journal of Soil Science*, **22**, 137–51.

Russell, E. W. (1971b) *The world of the soil*. London: Fontana.

Russell, E. W. (1973) *Soil conditions and plant growth*. London: Longman.

Russell, E. W. (1977) 'The role of organic matter in soil fertility', *Philosophical Transactions of the Royal Society of London*, **B 281**, 209–19.

Ryan B. and Gross, N. C. (1943) 'The diffusion of hybrid seed corn in two Iowa communities', *Journal of Rural Sociology*, **8**, 15–24.

Ryecroft, D. W. (1975) *The evidence in literature for the effect of field drainage on river flow*. London: MAFF, Field Drainage Experimental Unit, Technical Bulletin 76/1.

Ryecroft, D. W. and Massey, W. (1975) *The effect of field drainage on river flow*. London: ADAS, Field Drainage Experimental Unit, Technical Bulletin 75/9.

Rymer, L. (1976) 'The ecological status of bracken', *Botanical Journal of the Linnaean Society*, **73**, 151–76.

Saarinen, T. F. (1966) *Perception of the drought hazard on the Great Plains*. Chicago: University of Chicago, Department of Geography Research Report No. 106.

Saini, G. R. and Grant, W. J. (1980) 'Long term effects of intensive cultivation on soil quality in the potato-growing areas of New Brunswick (Canada) and Maine (USA)', *Canadian Journal of Soil Science*, **60**, 421–8.

Salmon, A. J. (1980) *An investigation into cattle trampling and feeding on a grassland system in Portland, Dorset*. Unpublished BSc Dissertation, University of Sheffield.

Salter, P. J. and Williams, J. B. (1963) 'The effect of FYM on the moisture characteristics of a sandy loam soil', *Journal of Soil Science*, **14**, 73–81.

Salter, R. M. and Green, T. C. (1933) 'Factors affecting the accumulation and loss of nitrogen and organic carbon in cropped soils', *Journal of the American Society of Agronomy*, **25**, 622–30.

Sandford, H. (1979) 'Some effects of fertiliser nitrogen on the botanical composition and yield of hill and upland swards', in *Changes in sward composition and productivity* (A. H. Charles and R. J. Haggar, eds). Hurley: British Grassland Society, 61–4.

Sawhney, B. L. (1979) 'Leaching of phosphate from agricultural soils to groundwater', *Water, Air and Soil Pollution*, **9**(4), 499–505.

Schilfgaarde, J. van (1963) 'Design of tile drainage for falling water tables', *Journal of the Irrigation Division, American Society of Civil Engineers*, **89**(IR 2), 1–11.

Schonrok-Fischer, R. and Sachse, B. (1980) 'Investigations into the occurrence and injurious effect of cereal cyst nematodes (*Heterodera avenae* Woll.) in relation to the proportion of cereals and intensification measures in the crop rotation. I. The influence of specialised cropping on the contamination of the soil with *Heterodera avenae* and on crop yield', *Archiv für Acher- und Pflanzenbau und Bodenkunde*, **24**, 359–66.

Schwerdtle, F. (1969) 'Investigations on the population density of earthworms in relation to conventional tillage and direct drilling' (in German), *Zeitchrift für Pflanzenkrankheiten, Pflanzenpathologie und Pflanzshutz*, **76**, 635–41.

Scorgie, H. R. A. (1980) 'Ecological effects of the aquatic herbicide cyanatryn on a drainage channel', *Journal of Applied Ecology* **17**, 207–25.

Scott, D. A. (1980) *A preliminary inventory of wetlands of international importance for waterfowl in West Europe and Northwest Africa*. Slimbridge: IRWB Special Publication 2.

Seale, R. S. (1975) *Soils of the Ely district (Sheet 173)*, Harpenden: Soil Survey (Memoirs of the Soil Survey of England and Wales).

Sears, P. D. and Evans, L. T. (1953) 'The influence of red and white clovers, superphosphate, lime, and dung and urine on soil composition, and on earthworm and grass-grub populations', *New Zealand Journal of Science and Technology, No. 1*, **35** (suppl. 1), 42–52.

Seebohm, M. E. (1952) *The evolution of the English farm*, London: George Allen and Unwin.

Selby, M. J. (1972) 'Relationships between land-use and erosion in central North Island, New Zealand', *Journal of Hydrology, New Zealand*, **11**, 73–87.

Selim, H. M. and Voth, R. D. (1980) 'Soil water behavior under different tillage practices', *Louisiana Agriculture*, **23**, 14–15.

Semple, A. T. (1970) *Grassland improvement*. London: Leonard Hill Books (Plant Science Monographs).

Sharrock, J. T. R. (1976) *The atlas of breeding birds of Britain and Ireland*. Berkhamsted: T. and A. D. Poyser.

Sheail, J. (1971) *Rabbits and their history*. Newton Abbot: David and Charles.

Shepherd, F. W. (1970) 'Physical effects of shelter', in *Hedges and hedgerow trees* (M. D. Hooper and M. W. Holdgate, eds), Monks Wood Experimental Station, Symposium No. 4. London: Nature Conservancy Council.

Shoard, M. (1980) *The theft of the countryside*. London: Temple Smith.

Sibma, L. (1977) 'Maximization of arable crop yield in the Netherlands', *Netherlands Journal of Agricultural Science*, **25**, 278–87.

Simmons, I. G. (1979) *Biogeography: natural and cultural*. London: Edward Arnold.

Simmons, I. G. (1980) 'Ecological-functional approaches to agriculture in geographical contexts', *Geography*, **65**(4), 305–16.

Simonson, R. W. (1968) 'Concept of soil', *Advances in Agronomy*, **20**, 1–47.

Singh, D. P. (1978) 'Evapotranspiration as affected by soil moisture and fertiliser use', *Indian Journal of Plant Physiology*, **21**, 253–60.

Sinha, M. K., Sinha, D. P. and Sinha, H. (1977) 'Organic matter transformations in soils. V. Kinetics of carbon and nitrogen mineralization in soils amended with different organic materials', *Plant and Soil*, **46**, 579–90.

Sinsch, U. (1978) 'Die Amphibien des Hülser Bruchs (Kreis: Krefeld)', *Decheniana*, **131**, 147–54.

Skidmore, E. L., Kumar, M. and Larsen, W. E. (1979) 'Crop residue management for wind erosion control in the Great Plains', *Journal of Soil and Water Conservation*, **34**, 90–4.

Slesser, M. (1975) 'Energy requirements of agriculture', in *Food, agriculture and the environment*. (J. Lenihan and W. W. Fletcher, eds). Glasgow: Blackie, 1–20.

Sly, J. M. A. (1977) 'Changes in the use of pesticides since 1945', in *Ecological effects of pesticides* (F. H. Perring and K. Mellanby, eds). London: Academic Press, Linnaean Society Symposium Series, No. 5, 1–6.

Smith, A. E. (1980) 'Persistence studies with (^{14}C) 2,4-D in soils previously treated with herbicides and pesticides', *Weed Research*, **20**, 355–9.

Smith, C. J. (1980) *Ecology of the English Chalk*. London: Academic Press.

Smith, D. W. and Bowes, G. C. (1974) 'Loss of some elements in fly-ash during old-field burns in southern Ontario', *Canadian Journal of Soil Science*, **54** (2), 215–24.

Smith, D. W. and Wischmeier, W. H. (1962) 'Rainfall erosion', *Advances in Agronomy*, **14**, 109–48.

Smith, K. A. and Robertson, P. D. (1971) 'Effects of ethylene on root extension of cereals', *Nature*, **234**, 148–9.

Smith, K. A. and Scott Russell, R. (1969) 'Occurrence of ethylene and its significance in anaerobic soil', *Nature*, **222**, 769–71.

Smith, L. P. (1967) 'Meteorology and the pattern of British grassland farming', *Agricultural Meteorology*, **4**, 321–38.

Smith, L. P. (1972) 'The effect of weather, drainage efficiency and duration of spring cultivations on barley yields in England', *Outlook on Agriculture*, **7**, 79–83.

Smith, L. P. (1976) *The agricultural climate of England and Wales: areal averages 1941–70*. London: HMSO, MAFF Technical Bulletin 35.

Smith, O. L. (1979a) 'An analytical model of the decomposition of soil organic matter', *Soil Biology and Biochemistry*, **11**, 585–606.

Smith, O. L. (1979b) 'Application of a model of the decomposition of soil organic matter', *Soil Biology and Biochemistry*, **11**, 607–18.

Smith, P. M. (1972) 'Serology and species relationships in annual bromes (*Bromus* L. sect. *Bromus*)', *Annals of Botany*, **36**, 1–30.

Smith, P. M. (1973) 'Observations on some critical Brome grasses', *Watsonia*, **9**, 319–32.

Smith, R. V. (1976) 'Nutrient budget in the River Moine, Co. Antrim', in *Agriculture and water quality*. London: HMSO, MAFF Technical Bulletin 32, 315–39.

Sneesby, N. J. (1966) 'Erosion control in the Black Fens', *Agriculture*, **73**, 391–4.

Soane, B. D. (1973) 'Techniques for measuring changes in the packing state and cone resistance of soil after the passage of wheels and tracks', *Journal of Soil Science*, **24**, 311–23.

Soane, B. D. (1975) 'Studies on some soil physical properties in relation to cultivations and traffic', in *Soil physical conditions and crop production*. London: HMSO, MAFF Technical Bulletin 29, 160–82.

Soane, B. D., Butson, M. J. and Pidgeon, J. D. (1975) 'Soil/machine interactions in zero-tillage for cereals and raspberries in Scotland', *Outlook on Agriculture*, **7**, 221–6.

Söderlund, R. and Svensson, B. H. (1976) 'The global nitrogen cycle', in *Nitrogen, phosphorus and sulphur global cycles* (B. H. Svensson and R. Söderlund, eds). Stockholm: Swedish National Science Research Council Ecological Bulletin No. **22**, 23–73.

South Yorkshire County Council (1979) *Environmental mapping of the European Community. South Yorkshire case study*. Barnsley: South Yorkshire County Council Environment Department.

Spedding, C. R. W. (1971) *Grassland ecology*. Oxford: Clarendon Press.

Spedding, C. R. W. and Diekmahns, E. C. (eds) (1972) *Grasses and legumes in British agriculture*. Farnham Royal: Commonwealth Agricultural Bureau. Commonwealth Bureau of Pastures and Field Crops, Bulletin 49.

Spence, M. T. (1955) 'Wind erosion in the Fens', *Meteorological Magazine* **84**, 304–7.

Spencer, W. F. and Cliath, M. M. (1974) 'Factors affecting vapour loss of trifluralin from soil', *Journal of Agricultural and Food Chemistry*, **22**, 987–91.

Spoor, G. (1975) 'Fundamental aspects of cultivations', in *Soil physical conditions and crop production*. London: HMSO, MAFF Technical Bulletin 29, 128–44.

Spoor, G. (1979) 'Soil type and workability', in *Soil survey applications* (M. G. Jarvis and D. Mackney, eds). Harpenden: Soil Survey of England and Wales, Technical Monograph No. 13.

Stahlecker, B. (1977) 'Auswirkung unterschiedlicher organischer Düngung massnahmen auf den Aufgang von Zuckerrüben', *Zuckerrübe*, **26**, 18–19.

Stanford, G. and Epstein, E. (1974) 'Nitrogen mineralization-water relations in soil', *Soil Science Society of America, Proceedings*, **38**, 103–7.

Stanford, G., Frere, M. H. and Schwaninger, D. H. (1973) 'Temperature coefficient of soil nitrogen mineralization', *Soil Science*, **115**, 321–3.

Stanhill, G. (1965) 'The concept of potential evapotranspiration in arid zone agriculture', in *Methodology of plant eco-physiology*, Montpellier Symposium. *Proceedings, Unesco Arid Zone Research*, **25**, 109–17.

Stevenson, F. J. (1967) 'Organic acids in soil', in *Soil biochemistry* (A. D. McLaren and G. H. Peterson, eds). London: Edward Arnold, 119–46.

Stewart, T. A. and Haycock, R. E. (1980) The economics of beef from grass/clover swards, *Agriculture in Northern Ireland*, **54(7)**, 186–91.

Stewart, A. J. A. and Lance, A. N. (1983) 'Moor draining. A review of impacts on land use', *Journal of Environmental Management*, **17**, 81–99.

Stewart, B. A., Viets, F. G. Jr. and Hutchinson, G. L. (1968) 'Agriculture's effect on nitrate pollution of groundwater', *Journal of Soil and Water Conservation*, **23**, 13–15.

Stewart, J. B. (1971) 'The albedo of a pine forest', *Quarterly Journal of the Royal Meteorological Society*, **97**, 561–4.

Stiles, W. and Williams, T. E. (1965) 'The response of a ryegrass/white clover sward to various irrigation regimes', *Journal of Agricultural Science, Cambridge*, **65**, 351–64.

Street, A. G. (1936) *Farmers glory*. London: Faber.

Stryk, F. G. von and Bolton, E. F. (1977) 'Atrazine residues in tile-drain-water from corn plots as affected by cropping practices and fertility levels', *Canadian Journal of Soil Science*, **57**, 249–53.

Sturrock, F. and Cathie, J. (1980) *Farm modernisation and the countryside: the impact of increasing field size and hedge removal on arable farms*. Cambridge: University of Cambridge, Department of Land Economy, Occasional Paper No. 12.

Summers, D. (1976) *The Great Level: a history of drainage and land reclamation in the Fens*. Newton Abbot: David and Charles.

Swain, R. W. (1975) 'Subsoiling', in *Soil physical conditions and crop production*. London: HMSO, MAFF Technical Bulletin 29, 189–204.

Sykes, G. B. (1979) 'Yield losses in barley, wheat and potatoes associated with field populations of 'large form' *Longidorus leptocephalus*', *Annals of*

Applied Biology, **91**, 237–41.

Tallis, J. H. (1964) 'Studies on southern Pennine peats. II. The pattern of erosion', *Journal of Ecology*, **52**, 333–44.

Tallis, J. H. (1973) 'Studies on southern Pennine peats. V. Direct observations on peat erosion and peat hydrology at Featherbed Moss, Derbyshire', *Journal of Ecology*, **61**, 1–22.

Tanchandrphongs, S. and Davidson, J. M. (1970) 'Bulk density, aggregate stability and organic matter content as influenced by two wheatland soil management practices', *Soil Science Society of America, Proceedings*, **34**, 302–5.

Tansley, A. E. (1935) 'The use and abuse of certain vegetational concepts and terms', *Ecology*, **16**, 284–307.

Tansley, A. E. (1953) *The British Isles and their vegetation* (2 vols). Cambridge: Cambridge University Press.

Tapper, S. (1979) 'The effect of fluctuating vole numbers (*Microtus agrestis*) on a population of weasels (*Mustela nivalis*) on farmland', *Journal of Animal Ecology*, **48**, 603–17.

Taylor, C. (1975) *Fields in the English landscape*. London: Dent.

Taylor, J. A. (1980) 'Bracken – an increasing problem and a threat to health', *Outlook on Agriculture*, **10**, 298–304.

Terry, R. V., Powers, W. L., Olson, R. V., Murphy, L. S. and Rubison, R. M. (1981) 'The effect of beef feedlot runoff on the nitrate-nitrogen content of a shallow aquifer', *Journal of Environmental Quality*, **10**, 22–6.

Thomasson, A. J. (1975) 'Soil properties affecting drainage design', in *Soils and field drainage* (A. J. Thomasson, ed). Harpenden: Soil Survey of England and Wales. Technical Monograph 7, 18–29.

Thornthwaite, C. W. (1948) 'An approach towards a rational classification of climate', *Geographical Review*, **38**, 55–94.

Thow, R. F. (1963) 'The effect of tilth on the emergence of spring oats', *Journal of Agricultural Science, Cambridge*, **60**, 291–5.

Tinbergen, J. M. and Drent, R. H. (1980) 'The starling as a successful forager', in *Bird problems in agriculture* (E. N. Wright, ed.). Croydon: British Crop Protection Society.

Tisdale, S. L. and Nelson, W. L. (1975) *Soil fertility and fertilizers* (3rd ed). New York: Macmillan.

Tivy, J. (1973) *The organic resources of Scotland: their nature and evaluation*. Edinburgh: Oliver and Boyd.

Tomlinson, T. E. (1970) 'Trends in nitrate concentrations in English rivers in relation to fertiliser use', *Water Treatment Examiner*, **19**, 277–89.

Toutain, J. C. (1961) *Le produit de l'agriculture française de 1700 à 1958, Histoire quantitative de l'économique française*. Paris: Cahiers de l'Institut de Science Economique Apliquée.

Trafford, B. D. (1970) 'Field drainage', *Journal of the Royal Agricultural Society of England*, **131**, 129–52.

Trafford, B. D. (1972) *Field drainage experiments in England and Wales*. London: MAFF, Field Drainage Experimental Unit, Cambridge, Technical Bulletin 72/12.

Trafford, B. D. (1975) 'Drainage experiments and drainage design', in *Soil physical conditions and crop production*. London: HMSO, MAFF Technical Bulletin 29, 417–33.

Trafford, B. D. (1977) 'Recent progress in field drainage. Part 1', *Journal of the Royal Agricultural Society of England*, **138**, 27–42.

Trafford, B. D. (1978) 'Recent progress in field drainage. Part II', *Journal of the Royal Agricultural Society of England*, **139**, 30–42.

Trafford, B. D. and Dennis, C. W. (1974) *The role of the pipe in field drainage*. London: MAFF, Field Drainage Experimental Unit, Cambridge, Technical Bulletin 74/15.

Trafford, B. D. and Walpole, R. A. (1975) 'Drainage design in relation to soil series', in *Soils and field drainage* (A. J. Thomasson, ed.). Harpenden: Soil

Survey of England and Wales, Technical Monograph 7, 49–61.

Treleaven, R. B. (1961) 'Notes on the Peregrine in Cornwall'. *British Birds*, **54(4)**, 136–42.

Trist, P. J. O. and Boyd, D. A. (1966) 'The Saxmundham rotation experiment: rotation 1', *Journal of Agricultural Science, Cambridge*, **66**, 327–36.

Troughton, A. (1961) 'Studies on the roots of leys and the organic matter and structure of the soil', *Empirical Journal of Experimental Agriculture*, **29**, 165–74.

Troughton, A. (1963) 'The root weight under swards of equal age in successive years', *Empirical Journal of Experimental Agriculture*, **31**, 274–81.

Tubbs, C. R. (1974) *The buzzard*. Newton Abbot: David and Charles.

Tubbs, C. R. and Blackwood, J. W. (1971) 'Ecological evaluation of land for planning purposes', *Biological Conservation*, **3**, 169–72.

Tull, Jethro (1731) *The New Horse-Houghing Husbandry: or an Essay on the Principle of Tillage and Vegetation, wherein is shown a Method of introducing a sort of Vineyard culture into the Corn-fields*. London.

Ulmanis, G. A. (1983) Sources and distribution of calcium, magnesium, potassium and sodium in grazed and ungrazed grasslands, Moor House N.N.R., Cumbria, England. Unpublished PhD Thesis, University of Sheffield.

Unger, P. W. (1978) 'Straw-mulch rate effect on soil water storage and sorghum yield', *Soil Science Society of America, Journal*, **42**, 486–91.

University of Manchester (1976) *Landscape evaluation: the landscape research project, 1970–75*. Manchester: University of Manchester.

Ursic, S. J. and Dendy, F. E. (1965) 'Sediment yields from small watersheds under various land use and forest covers', *Proceedings of the Federal Inter-Agency Sedimentation Conference, USDA Miscellaneous Publications*, **970**, 47–52.

Utomo, W. H. and Dexter, A. R. (1981a) 'Tilth mellowing', *Journal of Soil Science*, **32**, 187–201.

Utomo, W. H. and Dexter, A. R. (1981b) 'Soil friability', *Journal of Soil Science*, **32**, 203–13.

Van den Bergh, J. P. (1979) 'Changes in the composition of mixed populations of grassland species', in *The study of vegetation* (M. J. A. Werger, ed.). The Hague: Dr W. Junk Publishers, 59–80.

Varro, M. T. (35 BC) *Res rusticae*, (1960 edn W. D. Hooper and H. B. Ash, eds. London: Loeb).

Verma, H. N., Singh, R., Prihar, S. S. and Chaudhary, T. N. (1979) 'Runoff as affected by rainfall characteristics and management practices on gently sloping sandy loam', *Journal of the Indian Society of Soil Science*, **27**, 18–22.

Viets, F. G. (1977) 'A perspective on two centuries of progress in soil fertility and plant nutrition', *Soil Science Society of America, Journal*, **41**, 242–9.

Virgil (n. d.) *Georgics, Book 1*. (1979 edn, H. H. Huxley, ed. Bristol: Bristol Classical Press, University of Bristol).

Vorob'ev, S. A. and Saforov, A. F. (1977) 'The aggregate composition and water permeability of soil under field crops in a rotation and in a monoculture' (in Russian), *Izvestiya Timiryazevskoö Sel'skokhozaistvennoi Akademii*, **5**, 56–62.

WHO (1970) *European standards for drinking water*. Geneva: World Health Organisation.

Wahlenburg, W. G., Greene, S. W. and Reed, H. R. (1939) *Effects of fire and cattle grazing on longleaf pine lands as studied at McNeil, Mississippi*. Washington: USDA Technical Bulletin, 683.

Wali, A. M. O. El, Le Grand, F. and Gascho, G. J. (1980) 'Nitrogen leaching from soil and uptake by sugarcane from various urea-based fertilisers', *Soil Science Society of America, Journal*, **44**, 119–22.

Walker, A. (1974) 'A simulation model for prediction of herbicide persistence', *Journal of Environmental Quality*, **3**, 396–401.

Walker, A. (1976a) 'Simulation of herbicide persistence in soil, II. Simazine and linuron in long-term experiments', *Pesticide Science*, **7**, 50–8.

Walker, A. (1976b) 'Simulation of herbicide persistence in soil, III. Propyzamide in different soil types', *Pesticide Science*, **7**, 59–64.

Walker, A. (1977) 'Herbicide persistence – the weather and the soil', *ADAS Quarterly Review*, **27**, 168–79.

Walker, A. and Bond, W. (1978) 'Simulation of the persistence of metamitron activity in soil', *Proceedings of the 14th British Weed Control Conference, Brighton*, **2**, 565–72.

Walker, C. H. (1976) 'The significance of pesticide residues in the environment', *Outlook on Agriculture*, **9**, 16–20.

Walling, D. E. (1980) 'Water in the catchment ecosystem', in *Water quality in catchment ecosystems* (A. M. Gower, ed.). Chichester: Wiley, 1–47.

Walling, D. E. and Foster, I. D. L. (1978) 'The 1976 drought and nitrate levels in the River Exe basin', *Journal of the Institution of Water Engineers and Scientists*, **32**, 341–52.

Ward, R. C. (1975) *Principles of hydrology* (2nd edn). London: McGraw-Hill.

Wareing, P. F. and Allen, E. J. (1977) 'Physiological aspects of crop choice', *Philosophical Transactions of the Royal Society of London*, **B 281**, 107–19.

Warren, R. G. and Johnston, A. E. (1962) 'Barnfield', *Report of the Rothamsted Experimental Station for 1961*, 227–47.

Warwickshire County Council (1971) *Coventry-Solihull-Warwickshire: a strategy for the sub-region. Supplementary Report No. 5. Countryside*. Warwick: Warwickshire County Council Planning Department.

Watson, D. J., Thorne, G. N. and French, S. A. W. (1958) 'Physiological causes of differences in grain yield between varieties of barley', *Annals of Botany*, **22**, 321–51.

Watson, E. R. and Lapins, P. (1969) 'Losses of nitrogen from urine on soils from south-western Australia', *Australian Journal of Experimental Agriculture and Animal Husbandry*, **9**, 85–91.

Watson, M. and Cant, G. (1972) 'Variations in productivity of Waikato dairy farms: an empirical study of multiple causation', *Proceedings of the Seventh New Zealand Geography Conference, Hamilton*, 165–75.

Watson, S. J. and Nash, M. J. (1960) *Conservation of grass and forage crops*. Edinburgh: Oliver and Boyd.

Watt, A. S. (1936) 'Studies in the ecology of Breckland. I. Climate, soil and and vegetation', *Journal of Ecology*, **24**, 117–38.

Watt, A. S. (1937) 'Studies in the ecology of Breckland. II. On the origin and development of blow-outs', *Journal of Ecology*, **25**, 91–112.

Watt, A. S. (1938) 'Studies in the ecology of Breckland. III. The origin and development of the *Festuca-Agrostidetum* on eroded sand', *Journal of Ecology*, **26**, 1–37.

Watt, A. S. (1940) 'Studies in the ecology of Breckland. IV. The grass-heath', *Journal of Ecology*, **28**, 42–70.

Watt, A. S. (1981a) 'A comparison of grazed and ungrazed grassland A in East Anglian Breckland', *Journal of Ecology*, **69**, 499–508.

Watt, A. S. (1981b) 'Further observations on the effects of excluding rabbits from grassland A in East Anglian Breckland: the pattern of change and factors affecting it (1939–73)', *Journal of Ecology*, **69**, 509–36.

Watts, D. (1971) *Principles of biogeography*. London: McGrow-Hill.

Way, J. M. (1970) 'Roads and the conservation of wildlife', *Journal of the Institute of Highway Engineers*, **17**, 5–11.

Way, J. M. (1978) 'Roadside verges and conservation in Britain: a review', *Biological Conservation*, **12**, 65–74.

Way, M. J. and Cammell, M. E. (1979) 'Optimising cereal yields – the role of pest control', *Proceedings of the 1979 British Crop Protection Conference – Pests and Diseases, Brighton*, 663–72.

Weatherly, A. B. and Dane, J. H. (1979) 'Effect of tillage on soil water movement during corn growth', *Soil Science Society of America, Journal* **43**, 1222–5.

Webb, N. R. and Haskins, L. E. (1980) 'An ecological survey of heathlands in the Poole Basin, Dorset, England in 1978', *Biological Conservation*, **17**, 281–96.

Webster, R., Hodge, C. H., Draycott, A. P. C. and Durrant, M. J. (1977) 'The effect of soil type and related factors on sugar beet yield', *Journal of Agricultural Science, Cambridge*, **88**, 455–70.

Wederkinch, E. (1976) 'The Fire-bellied toad (*Bombina bombina*) in Denmark', *Norwegian Journal of Zoology*, **24**, 235.

Wegner, J. F. and Merriam, G. (1979) 'Movements by birds and small mammals between a wood and adjoining farmland habitats', *Journal of Applied Ecology*, **16**, 349–57.

Welbank, P. J. (1975) 'Research on root growth at Rothamsted', in *Soil physical conditions and crop production*. London: HMSO, MAFF Technical Bulletin 29, 449–68.

Welbank, P. J., Gibbs, M. J., Taylor, P. J. and Williams, E. D. (1974) 'Root growth of cereal crops', *Report of Rothamsted Experimental Station for 1973*, Part 2, 26–44.

Wellings, L. W. (1973) 'Effect of irrigation on the yield and quality of maincrop potatoes', *Experimental Husbandry*, **24**, 54–69.

Wellings, L. W. and Lapwood, D. H. (1971) 'Control of common scab by the use of irrigation', *NAAS Quarterly Review*, **91**, 128–37.

Wellings, S. R. and Bell, P. (1980) 'Movement of water and nitrate in the unsaturated zone of the upper Chalk near Winchester, Hants', *Journal of Hydrology*, **48**, 119–36.

Wellington, P. S. (1965) 'Germination and seedling emergence', in *The growth of cereals and grasses* (F. L. Milthorpe and J. D. Ivins, eds). London: Butterworths, 3–19.

Wellman, J. D. and Buhyoff, G. J. (1980) 'Effects of regional familiarity on landscape preferences', *Journal of Environmental Management*, **11**, 105–10.

Wells, D. (1981) 'Distribution of wildlife in British grasslands and its importance for nature conservation', in *Grassland in the British economy* (J. Jollans, ed). Reading: Centre for Agricultural Strategy, Payer No. 10, 577–9.

Wells, T. C. E. (1967) 'Changes in the composition of a sown pasture on the Chalk in Kent', *Journal of the British Grassland Society*, **22**, 277–81.

Wesseling, J. (1964) *A comparison for the steady state drain spacing formulae of Hooghoudt and Kirkham in connection with drainage design practice*. Wageningen: Institute of Land and Water Management, Technical Bulletin 34.

Westing, A. H. (1977) 'Ecological effects of the military use of pesticides', in *Ecological effects of pesticides* (F. H. Perring and K. Mellanby, eds). London: Academic Press, Linnaean Society Symposium Series, No. 5, 89–94.

Westmacott, R. and Worthington, T. (1974) *New agricultural landscapes*. Cheltenham: Countryside Commission.

Whitelaw, K. and Rees, J. F. (1980) 'Nitrate-reducing and ammonium-oxidizing bacteria in the vadose zone of the Chalk aquifer of England', *Geomicrobiology Journal*, **2**, 179–87.

Wickens, R. (1963) *Methods of straw disposal in an arable rotation*. London: HMSO, Progress Report of the MAFF Experimental Farms and Experimental Horticultural Stations for 1963, 29–30.

Widdowson, F. V. and Penny, A. (1965) 'Residual effects of NPK fertilizers given to potatoes and wheat', *Report of the Rothamsted Experimental Station for 1964*, Part 2, 60–1.

Widdowson, F. V. and Penny, A. (1979) 'Results from the Woburn Reference Experiment: II. Yields of the crops and recoveries of N, P, K and Mg from manures and soil, 1970–74', *Report of the Rothamsted Experimental Station for 1978*, Part 2, 67–82.

Widdowson, F. V., Penny, A. and Bird, E. (1980) 'Results from the Rothamsted Reference Experiment. II. Yields of the crops and recoveries of N, P and K from manures and soil, 1971–75', *Report of the Rothamsted*

Experimental Station for 1979, Part 2, 63–75.

Wild, A. (1972) 'Nitrate leaching under bare fallow at a site in northern Nigeria', *Journal of Soil Science,* **23**, 315–24.

Wild, A. (1977) 'Nitrate in drinking water: health hazard unlikely', *Nature*, **268**, 197–8.

Wilkins, R. J. and Bather, M. (1981) 'Potential for changes in support energy use in annual output from grassland', in *Grassland in the British economy* (J. Jollans, ed.). Reading: Centre for Agricultural Strategy, Paper No. 10, 511–20.

Wilkins, R. J. and Wilson, R. F. (1974) 'Silage', in *Silver Jubilee Report 1949–74* (C. R. W. Spedding and R. D. Williams, eds). Hurley: Grassland Research Institute, 96–106.

Wilkins, R. J., Morrison, J. and Chapman, P. F. (1981) 'Potential production from grasses and legumes', in *Grassland in the British economy* (J. Jollans, ed.). Reading: Centre for Agricultural Strategy, Paper No. 10, 390–413.

Wilkinson, B. (1974) 'Quantitative basis for land capability interpretation', in *Land capability classification*. London: HMSO, MAFF Technical Bulletin 21, 23–34.

Wilkinson, B. (1975) 'Field experience on heavy soils', in *Soil physical conditions and crop production*. London: HMSO, MAFF Technical Bulletin 29.

Wilkinson, W. (1967) *Research note*. London: ICI, TMJ 131, Agricultural Division.

Wille, C. (1982) 'America – on the downward swing?' *Birds*, **9**, 48–9.

Willey, C. R. and Tanner, C. B. (1963) 'Membrane covered electrode for measurement of oxygen concentration in soil', *Soil Science Society of America, Proceedings*, **27**, 511–15.

Williams, C. H. and Lipsett, J. (1961) 'Fertility changes in soils cultivated for wheat in southern New South Wales', *Australian Journal of Agricultural Research*, **12**, 612–29.

Williams, E. D. (1974) 'Changes in yield and botanical composition caused by the new liming scheme on Park Grass', *Report of the Rothamsted Experimental Station for 1973*, Part 2, 67–73.

Williams, G. (1980) 'Swifter flows the river', *Birds*, **8**, 19–22.

Williams, G. D. V., Joynt, M. I. and McCormick, P. A. (1975) 'Regression analysis of Canadian Prairie crop-district cereal yields, 1961–1972, in relation to weather, soil and trend', *Canadian Journal of Soil Science*, **55**, 43–53.

Williams, R. J. B. (1970) 'The chemical composition of water from land drains at Saxmundham and Woburn and the influence of rainfall upon nutrient losses', *Report of the Rothamsted Experimental Station for 1969*, Part 2, 36–67.

Williams, R. J. B. (1976) 'The chemical composition of water from land drains at Saxmundham and Woburn (1970–75)', *Report of the Rothamsted Experimental Station for 1975*, Part 2, 37–62.

Williams, R. J. B. (1978) 'Effects of management and manuring on physical properties of some Rothamsted and Woburn soils', *Report of the Rothamsted Experimental Station for 1977*, Part 2, 37–52.

Williams, T. D., Beane, J., Berry, M. M. and Webb, R. M. (1979) 'The rotation-fumigation experiment, Woburn Experimental Farm, 1969–77', *Report of the Rothamsted Experimental Station for 1978*, Part 2, 47–66.

Williams, T. E. (1960) 'Leys and subsequent arable productivity', *Journal of the British Grassland Society*, **15**, 189–94.

Williams, T. E. (1980) 'Herbage production: grasses and leguminous forage crops', in *Grass: its production and utilization* (W. Holmes, ed.). Oxford: Blackwell, 6–69.

Williamson, R. E. (1964) 'The effect of root aeration on plant growth', *Soil Science Society of America, Proceedings*, **28**, 86–90.

Williamson, R. E. and Willey, C. R. (1964) 'Effect of depth of water table on yield of tall fescue', *Agronomy Journal*, **56**, 585–8.

Wilson, H. A., Gish, R. and Browning, G. M. (1947) 'Cropping systems and season as factors affecting aggregate stability', *Soil Science Society of America, Proceedings*, **12**, 36–43.

Winton, E. F., Tardiff, R. G. and McGabe, L. J. (1971) 'Nitrate in drinking water', *Journal of the American Water Works Association*, **63**, 95–8.

Wischmeier, W. H. and Mannering, J. V. (1969) 'Relation of soil properties to its erodibility', *Soil Science Society of America, Proceedings*, **33**, 131–7.

Wischmeier, W. H. and Smith, D. D. (1965) *Predicting rainfall-erosion losses from cropland east of the Rocky Mountains*. Washington: USDA, Agricultural Research Service, Agricultural Handbook 282.

Wischmeier, W. H., Johnson, C. B. and Cross, B. V. (1971) 'A soil erodibility nomograph for farmland and construction sites', *Journal of Soil and Water Conservation*, **26**, 189–93.

Wittwer, S. H. (1978) 'The next generation of agricultural research', *Science*, **199**, 375.

Woodruff, N. P. and Siddoway, F. H. (1965) 'A wind erosion equation', *Soil Science Society of America, Proceedings*, **29**, 602–8.

Woodward, J. (1699) 'Some thoughts and experiments concerning vegetation', *Philosophical Transactions of the Royal Society of London*, **21**, 193–227.

Worthing, C. R. (ed.) (1980) *The pesticide manual – a world compendium*. Croydon: British Crop Protection Council.

Wright, C. E. (1978) 'Maximising herbage production', *British Grassland Society/British Veterinary Association Conference, Burchetts Green, 1978, Proceedings*, 11–23.

Yapp, W. B. (1973) 'Ecological evaluation of a linear landscape', *Biological Conservation*, **5**, 45–7.

Yates, M. E. (1971) 'Effects of cultural changes in Makara experimental basin: hydrological and agricultural production effects of two levels of grazing on unimproved and improved small catchments', *Journal of Hydrology, New Zealand*, **10**, 59–84.

Yiakoumettis, I. and Holmes, W. (1972) 'The effect of nitrogen and stocking rate on the output of pasture grazed by beef cattle', *Journal of the British Grassland Society*, **27**, 183–91.

Young, Arthur (1804) *General view of the agriculture of the county of Norfolk*. London: Board of Agriculture.

Young, C. P. and Gray, E. M. (1978) *Nitrate in groundwater*. Medmenham: Water Research Centre, Technical Report TR69.

Young, C. P., Hall, E. J. and Oakes, D. B. (1976) *Nitrate in groundwater. Studies in the Chalk near Winchester, Hants*. Medmenham: Water Research Centre, Technical Bulletin 31.

Zabek, S. (1979) 'Dynamics of water infiltration into sandy and loamy soil under bare fallow, lucerne or annual crops based on data from four-year field experiments' (in Polish), *Roczniki Gleboznawcze*, **30**, 85–97.

Zuckerman, Lord S. (1980) *Badgers, cattle and tuberculosis*. London: HMSO, MAFF Report.

SUBJECT INDEX

TAXONOMIC INDEX

SOCIAL SCIENCE LIBRARY

Oxford University Library Services

~~Manor Road~~

Oxford OX1 3UQ

Tel: (2)71093 (enquiries and renewals)

http://www.ssl.ox.ac.uk

WITHDRAWN

This is a NORMAL LOAN item.

We will email you a reminder before this item is due.

Please see http://www.ssl.ox.ac.uk/lending.html
for details on:

- loan policies; these are also displayed on the notice boards and in our library guide.

- how to check when your books are due back.

- how to renew your books, including information on the maximum number of renewals. Items may be renewed if not reserved by another reader. Items must be renewed before the library closes on the due date.

- level of fines; fines are charged on overdue books.

Please note that this item may be recalled during Term.

WITHDRAWN